Molecular Plant Pathology

Annual Plant Reviews

A series for researchers and postgraduates in the plant sciences. Each volume in this annual series will focus on a theme of topical importance and emphasis will be placed on rapid publication.

Titles in the series:

1. Arabidopsis
Edited by M. Anderson and J. Roberts.

2. Biochemistry of Plant Secondary Metabolism
Edited by M. Wink.

3. Functions of Plant Secondary Metabolites and their Exploitation in Biotechnology
Edited by M. Wink.

4. Molecular Plant Pathology
Edited by M. Dickinson and J. Beynon.

5. Vacuolar Compartments
Edited by D.G. Robinson and J.C. Rogers.

Molecular Plant Pathology

Edited by

MATTHEW DICKINSON
Plant Science Division
School of Biological Sciences
University of Nottingham, UK

and

JAMES BEYNON
Horticulture Research International
Wellesbourne
Warwick, UK

CRC Press

First published 2000

Published by
Sheffield Academic Press Ltd
Mansion House, 19 Kingfield Road
Sheffield S11 9AS, England

ISBN 1-84127-108-X

Published in the U.S.A. and Canada (only) by
CRC Press LLC
2000 Corporate Blvd., N.W.
Boca Raton, FL 33431, U.S.A.
Orders from the U.S.A. and Canada (only) to CRC Press LLC

U.S.A. and Canada only:
ISBN 0-8493-0510-1

Printed on acid-free paper in Great Britain by
Bookcraft Ltd, Midsomer Norton, Bath

British Library Cataloguing-in-Publication Data:
A catalogue record for this book is available from the British Library

Library of Congress Cataloging-in-Publication Data:
Molecular plant pathology / edited by Matthew Dickinson, Jim Beynon.
p. cm.
Includes bibliographical references (p.).
ISBN 0-8493-0510-1 (alk. paper)
1. Plant diseases--Molecular aspects. 2. Phytopathogenic microorganisms--Molecular aspects. I. Dickinson, Matthew. II. Beynon, Jim.
SB732.65 .M648 2000
632'.96--dc21

00-021384

Preface

Since the domestication of plants for agriculture, mankind has faced a major challenge from plant diseases. This challenge is to maintain levels and quality of production in the face of the consistent adaptation of pathogens to overcome measures taken for the control of disease. In many countries, the combined use of genetic and chemical control measures has played a major role in providing a reliable and healthy food supply. Human populations and the demand for food have continued to grow, accompanied by an increasing concern about the extensive use of chemical pathogen control. Hence the more effective utilisation of natural genetic disease resistance mechanisms, together with the continuing development of novel, more environmentally friendly active ingredients for chemical control, has become a major focus of research. As a result of these pressures, the study of plant diseases and underlying resistance mechanisms has become one of the most exciting areas of modern plant biology.

For the past century, it has been known that plants possess genetically inherited resistance mechanisms to combat phytopathogenic fungi, bacteria and viruses, and that the relationship between pathogens and host plants is often highly specialised and complex. As techniques of molecular biology have developed over the past 25 years, our understanding of the molecular basis of these relationships has advanced significantly. The use of natural variation and mutants in both pathogens and plants has enabled the identification of key genes involved in pathogenicity and resistance. Many such genes have been isolated and the structure of their protein products determined. The major challenges today are not only to continue to identify key genetic components of plant–pathogen interactions, but also to understand the biochemical role of the gene products and to develop strategies by which this information can be used to produce disease-free crops.

This volume covers the latest developments in three main areas of molecular plant pathology. The first chapters describe how pathogens cause disease, including analysis of the molecular signalling that takes place between plant and pathogen and the genes that are important in the pathogen for causing disease. Five chapters then cover the genetics and molecular basis of resistance in plants, with the main emphasis on recognition genes and signalling mechanisms. The final chapters look into future developments, examining the potential use of biotechnology to produce disease-resistant plants and the emerging technologies that will enhance our ability to understand the mechanisms underlying plant-microbe interactions.

As the genes determining pathogenicity and resistance are being identified, one of the key goals is to understand the relationship between components of signalling pathways, both within and between those pathways. The intention of this volume has therefore been to highlight the role of molecular signalling in pathogenicity and resistance mechanisms, and to focus on current and future perspectives. We believe that this will be of value to researchers and professionals with an interest in plant pathology and signalling, and will fuel their desire to make further advances in these exciting areas of plant science.

Matthew Dickinson
Jim Beynon

Contributors

Dr Anna O. Avrova Unit of Mycology, Bacteriology and Nematology, Scottish Crop Research Institute, Invergowrie, Dundee DD2 5DA, UK

Dr Pascale V. Balhadère School of Biological Sciences, University of Exeter, Washington Singer Laboratories, Perry Rd, Exeter EX4 4QG, UK

Ms Claire Barker Plant Cell Biology, Research School of Biological Sciences, Australian National University, PO Box 475, Canberra ACT 2601, Australia

Dr Paul R.J. Birch Unit of Mycology, Bacteriology and Nematology, Scottish Crop Research Institute, Invergowrie, Dundee DD2 5DA, UK

Dr Alia Dellagi Unit of Mycology, Bacteriology and Nematology, Scottish Crop Research Institute, Invergowrie, Dundee DD2 5DA, UK

Dr Robert A. Dietrich Novartis Agribusiness Biotechnology Research, Inc., 3054 Cornwallis Road, PO Box 12257, Research Triangle Park, NC 27709, USA

Dr Peter N. Dodds Division of Plant Industry, CSIRO, GPO Box 1600, Canberra ACT 2617, Australia

Dr Jeffrey G. Ellis Division of Plant Industry, CSIRO, GPO Box 1600, Canberra ACT 2617, Australia

Dr Gary D. Foster School of Biological Sciences, University of Bristol, Woodland Road, Clifton, Bristol, BS8 1UG, UK

Dr Elisabeth Huguet Institut de Recherche sur la Biologie de l'Insecte, Université F. Rabelais, Faculté des Sciences, Parc de Grandmont, 37200 Tours, France

Dr David A. Jones Plant Cell Biology, Research School of Biological Sciences, Australian National University, PO Box 475, Canberra ACT 2601, Australia

Dr Christophe Lacomme Unit of Cell Biology, Scottish Crop Research Institute, Invergowrie, Dundee DD2 5DA, UK

Dr Gregory J. Lawrence Division of Plant Industry, CSIRO, GPO Box 1600, Canberra ACT 2617, Australia

Dr Gary D. Lyon Unit of Mycology, Bacteriology and Nematology, Scottish Crop Research Institute, Invergowrie, Dundee DD2 5DA, UK

Ms Sarah B. Nettleship School of Biological Sciences, University of Bristol, Woodland Road, Clifton, Bristol BS8 1UG, UK

Dr Jane E. Parker The Sainsbury Laboratory, John Innes Centre, Norwich Research Park, Colney Lane, Norwich NR4 7UH, UK

Dr A. Pryor Division of Plant Industry, CSIRO, GPO Box 1600, Canberra ACT 2617, Australia

Dr Simon Santa Cruz Unit of Cell Biology, Scottish Crop Research Institute, Invergowrie, Dundee DD2 5DA, UK

Professor Nicholas J. Talbot School of Biological Sciences, University of Exeter, Washington Singer Laboratories, Perry Road, Exeter EX4 4QG, UK

Dr John A. Walsh Plant Pathology and Microbiology Dept, Horticulture Research International, Wellesbourne, Warwick CV35 9EF, UK

Contents

1 Fungal pathogenicity—establishing infection

Pascale V. Balhadère and Nicholas J. Talbot

1.1 Introduction

The first obstacle encountered by any plant pathogen is the outer surface of a plant, whether in the form of a tough leaf cuticle, a fibrous stem, or the root surface submerged within soil. In each case, pathogenic fungi have evolved mechanisms to breach these surfaces, by using natural openings, such as stomata, or direct rupture of the surface layers. This chapter describes the early events in plant infection and the specific morphogenetic pathways that have evolved in fungi for this purpose. It concentrates mainly on the developmental biology of the rice blast fungus, *Magnaporthe grisea*, an appressorium-forming cereal pathogen. In recent years, rapid progress has occurred in identifying genes from this fungus that are involved in the plant infection process and in determining their function. A view of the specific signal transduction pathways and gene expression patterns required for plant colonisation is, therefore, emerging and can act as a framework for understanding the biology of fungal pathogens. Based on this information, experiments can be designed to determine the conservation of pathogenic processes among diverse species. This chapter follows the chronology of plant infection by *M. grisea* and describes the key genetic components so far identified and their known, or predicted, functions. Comparisons to other fungal pathogens are made throughout the text.

1.2 The rice blast fungus, *Magnaporthe grisea*

The ascomycete, *Magnaporthe grisea* (Hebert) Barr (anamorph *Pyricularia oryzae* Sacc.), causes one of the major diseases of cultivated rice, called rice blast disease (Ou, 1985). The biology of *M. grisea* shares many features with the life cycles of other major fungal pathogens, including, most significantly:

- The differentiation of infection-specialised cells called appressoria (Figure 1.1) (Mendgen *et al.*, 1996; Howard, 1997).
- The synthesis and mobilisation of materials, such as mucilage, hydrophobins, melanin, storage carbohydrates and glycerol, during this process (Money, 1995; Yoder and Turgeon, 1996; Kershaw and Talbot, 1998).
- The mechanics of cuticle penetration (Howard, 1997).

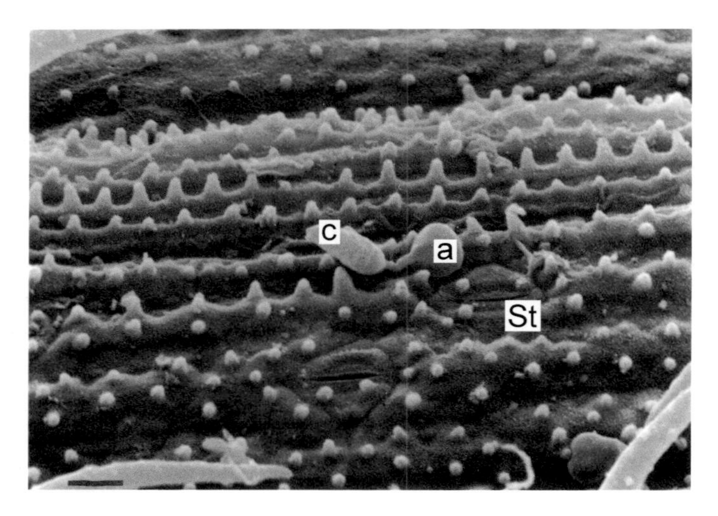

Figure 1.1 Low temperature scanning electron micrograph showing appressorium differentiation by *Magnaporthe grisea* on the surface of a rice leaf. A conidium (**c**) has landed and attached near a stoma (**St**). It has germinated through elongation of a short germ tube, whose tip has further differentiated into a dome-shaped appressorium (**a**). This melanized inflated cell is responsible for turgor-driven penetration of the leaf cuticle. (Internal scale bar = 10 μm).

- Specialised signal transduction pathways that regulate the morphogenetic changes associated with pathogenesis (Dean, 1997; Hamer and Holden, 1997; Kronstad *et al.*, 1998).
- The existence of a gene-for-gene relationship between *M. grisea* and its hosts, involving genetically dominant fungal avirulence genes and plant resistance genes (Laugé and De Wit, 1998).

The biology of rice blast has been studied using a combination of genetic and physiological approaches. The pre-penetration stages in the development of *M. grisea* can, for example, be reproducibly induced and adjusted *in vitro* on artificial surfaces, such as hydrophobic plastic membranes (Bourett and Howard, 1990) or onion epidermis (Chida and Sisler, 1987). The fungus is amenable to classical genetic analysis through sexual crosses (Valent *et al.*, 1986), by isolation of rare fertile rice pathogenic isolates of the fungus and by introgression of fertility into stable laboratory strains of *M. grisea* (Valent *et al.*, 1991). The development of a reliable DNA-mediated transformation system (Parsons *et al.*, 1987; Leung *et al.*, 1990) has also proved invaluable in allowing targeted gene replacement, gene disruption and heterologous expression of genes in *M. grisea* (Sweigard *et al.*, 1992; Mitchell and Dean, 1995; Kershaw *et al.*, 1998). The *M. grisea* genome has been extensively mapped, both genetically and physically (Sweigard *et al.*, 1993; Hayashi and Naito, 1994; Smith and Leong, 1994; Kang *et al.*, 1995; Diaz-Perez *et al.*, 1996;

Dioh *et al.*, 1996; Zhu *et al.*, 1997), and a publicly-funded international genome sequencing project has recently been initiated at several locations worldwide, coordinated from Clemson University Genome Center (Clemson, SC, USA) and in the private sector by the DuPont Company (Wilmington, DE, USA), amongst others.

Collectively, the application of these procedures has allowed a wealth of experimental data to be collected on the *M. grisea*–rice pathosystem, and a model can thus be made linking cytological, biochemical, physiological and molecular aspects of infection-related events. An understanding of how infection generally proceeds in *M. grisea* is thus beginning to emerge and should provide further insight into the general mechanisms of fungal pathogenicity, whilst also highlighting some of the inherent peculiarities of the *M. grisea*–rice interaction.

1.3 The onset of infection: infection court preparation and appressorium differentiation

M. grisea is a heterothallic pyrenomycete that can infect a wide range of grass species. The host spectrum of any particular strain of *M. grisea* is narrow, however, suggesting an ancient phylogenetic divergence between rice-pathogenic and non-pathogenic forms, as revealed by the conservation (or propagation) of MGR586 repeated sequences solely in rice pathogenic isolates of the fungus (Hamer *et al.*, 1989a). In the field, asexual reproduction of *M. grisea* predominates and the fungus reproduces by means of asexual conidia produced from disease lesions. Three-celled pyriform conidia are formed in a sympodial fashion at the tip of slender conidiophores that emerge into the air from lesions, either through stomata or by direct rupture of the cuticle (Howard and Valent, 1996). Despite the existence of sexual reproduction and reports of sexual recombination in the field (Zeigler, 1998), sexual reproductive structures seem to play little, if any, role in the infection cycle, the subsequent evolution of the disease, or the genetic variability observed among pathogen populations (Valent, 1997).

Rice blast disease is primarily propagated through conidia, which are dispersed onto healthy leaves from neighbouring sporulating sites by wind and water-splash, thus initiating multiple rounds of infection (Ou, 1985). Contact with water rapidly triggers the release and hydration of a glyco-conjugated mucilage stored in the periplasm at the spore apex. This serves as an adhesive to anchor the conidium to the highly hydrophobic rice leaf, in a water environment favourable for subsequent infection events, whilst resisting the flow of water (Hamer *et al.*, 1988). In the course of infection, mucilage is further synthesised to attach newly formed structures (germ tubes and appressoria) to the substratum. Spore tip mucilage (STM)-mediated attachment occurs regardless of whether the host is resistant or susceptible to *M. grisea*. Consistent with this, appressorium formation is induced on non-host synthetic substrates to the same extent as on host leaves, and can thus be considered a non-selective process (Gilbert *et al.*, 1996). STM-mediated attachment appears to represent a prerequisite for completion of the appressorium differentiation process and subsequent infection (Jelitto *et al.*, 1994; Xiao *et al.*, 1994a; Gilbert *et al.*, 1996).

There is now a growing body of evidence implicating particular physical and chemical characteristics of the rice leaf in the induction of appressorium development. These include: hydrophobicity (Lee and Dean, 1994); surface hardness (Xiao *et al.*, 1994b); cutin monomers and wax polar lipids (Uchiyama and Okuyama, 1990; Gilbert *et al.*, 1996). Taken singly, these properties are by no means specific to rice and are found in other plant species. However, collectively, these diverse cues may act to relay the presence of a conducive environment and later to interact with fungal-derived molecules present at the fungus/plant interface. These molecules include the class I hydrophobin, MPG1p (Talbot *et al.*, 1993; Beckerman and Ebbole, 1996; Talbot *et al.*, 1996), and several, as yet unidentified, mucilage glycoproteins with α-D mannose and α-D glucose residues (Xiao *et al.*, 1994a). The composition of the mucilage that surrounds the developing germ tube and appressorium appears to be somewhat peculiar to *M. grisea*, containing an abundance of neutral lipids and glycolipids (Ebata *et al.*, 1998). The significance of mannose-substituted glycoproteins is emphasised by the fact that mannose-binding rice lectin can interfere with development of the appressorium (Teraoka *et al.*, 1999).

One important component in the development of appressoria is a cell wall protein encoded by the *MPG1* gene. *MPG1* was first isolated as a pathogenicity determinant, specifically expressed during appressorium morphogenesis and *in planta* (Talbot *et al.*, 1993). *MPG1* was shown to encode a fungal hydrophobin, a class of proteins implicated in a number of fungal developmental processes, including, most notably, the formation of aerial hyphae (Talbot, 1999; Wösten *et al.*, 1999). Hydrophobins are morphogenetic proteins that are secreted by fungi and undergo polymerisation in response to air-water or hydrophobic surface interfaces (Wessels, 1997). Once polymerised, hydrophobins form amphipathic membranous structures with hydrophobic rodlets decorating one side. These rodlet layers coat the surfaces of spores and aerial hyphae of fungi, providing a hydrophobic surface that may be required for growth into the air and protection from subsequent desiccation (Wessels, 1997; Talbot, 1999). Hydrophobins may also play a role in spore dispersal and their abundance, variety and conservation among fungal species indicates that they perform a number of other functions in development (Kershaw and Talbot, 1998; Talbot, 1999).

The MPG1 hydrophobin was shown to self-assemble at the rice leaf surface, providing an amphipathic layer upon which subsequent appressorium development occurs (Talbot *et al.*, 1996). This polymer appears to serve as a morphogenetic sensor of an inductive surface for appressorium formation, because Δ*mpg1* mutants are reduced in their ability to form appressoria (Talbot *et al.*, 1993, 1996; Beckerman and Ebbole, 1996). The significance of the hydrophobin self-assembly process to appressorium development has been highlighted by the fact that diverse hydrophobins can substitute for MPG1p if expressed under control of the *MPG1* promoter in *M. grisea*. This shows that hydrophobin production is important for appressorium morphogenesis in *M. grisea*, and more significant than any particular (or unusual) characteristics of the MPG1 hydrophobin. Complementation of the Δ*mpg1* mutant phenotype has been observed by a large number of class I hydrophobin genes (Kershaw *et al.*, 1998).

1.4 Appressorium-mediated plant infection

The mechanism of appressorium differentiation in *M. grisea* has been well documented in a number of elegant cytological studies (Bourett and Howard, 1990; Howard and Valent, 1996). Briefly, conidial germination occurs from the apical and/or basal cell by elongation of a short germ tube. This step does not require preliminary attachment of the conidium to the substrate (Gilbert *et al.*, 1996), but some evidence suggests the involvement of plant cuticular waxes in promoting conidial germination (Hegde and Kolattukudy, 1997). Cuticular waxes may act by extracting lipophilic self-inhibitors that have been shown to be present on the conidium surface (Hegde and Kolattukudy, 1997). By 2–4 h after initial germination, tip growth of the germ tube arrests and is followed by flattening and hooking of the germ tube tip (Bourett and Howard, 1990). Mitosis occurs during germ tube elongation and one of the resulting daughter nuclei migrates into the incipient appressorium. A septation event then occurs, followed by synthesis of the specialised outer wall layers of the differentiating appressorium. The appressorium wall remaining in contact with the substratum possesses only a single cell wall layer and lacks chitin (Bourett and Howard, 1990). This region further differentiates into the appressorium pore during the cell's maturation. After 4–8 h of development, dihydroxynaphthalene (DHN)-melanin is deposited as a fibrillar and continuous layer between the plasma membrane and the appressorium cell wall, bound to chitin and, therefore, absent from the appressorium pore.

The appressorium pore, in turn, is surrounded by a ring, whose specific function is probably to seal the appressorium to the plant surface (Howard and Ferrari, 1989). A similar structure has been observed in the melanised appressoria of *Colletotrichum* species (Mendgen *et al.*, 1996). By 24–31 h, appressorium maturation and turgor generation occur. This phase is accompanied by loss of glycogen rosettes that are particularly abundant at the time of melanisation (Bourett and Howard, 1990). Meanwhile, the appressorium pore becomes covered by a bilayered pore wall overlay, through which a penetration peg emerges and extends, directing penetration through the cuticle. During this morphogenetic sequence, actin localises to the penetration peg and is probably essential for orchestration of germ tube growth, filasome-mediated secretion processes at the pore interface, and penetration peg extension (Howard, 1997). Appressorium-mediated penetration by *M. grisea* involves generation of a very large internal turgor pressure that has been estimated to be as high as 8 MPa, based on an incipient cytorrhysis assay (Howard *et al.*, 1991). This assay observes the number of appressoria collapsing during exposure to hyperosmotic concentrations of a solute. The concentration of solute causing collapse of appressoria is proportional to the internal concentration of solute and, therefore, the osmotic potential being created by the cell. In *M. grisea*, appressoria have a mean turgor of approximately 6 MPa using this assay (Howard *et al.*, 1991). Using a freezing point analysis technique, similar—though slightly lower—values were recorded, confirming that *M. grisea* appressoria produce significant pressures (Money and Howard, 1996).

Appressorium formation in *M. grisea* is similar in sequence and cell structure to the morphogenetic sequence in *Colletotrichum* species. This probably reflects a similar infection strategy in both species (see later in this section). In the latter, however, an appressorium cone structure within the infection cell and production of specific post-infection hyphal structures are quite distinct events (Mendgen *et al.*, 1996). Appressorium formation in the powdery mildew pathogen, *Erysiphe graminis*, the bean rust fungus, *Uromyces appendiculatus*, and other rust fungi differ appreciably from the sequence of events in *M. grisea*, and have been reviewed previously (Mendgen *et al.*, 1996; Howard, 1997; Staples and Hoch, 1997).

Knowledge regarding fungal metabolism during the pre-penetration phase is still fragmentary. Conidia germinate in pure water and there is no requirement for exogenous nutrients during appressorium morphogenesis. Energy for the process is, therefore, derived entirely from storage compounds within the conidium. At the onset of the infection process, trehalose and mannitol are the major carbohydrates stored in dormant conidia (Foster and Talbot, 1999), and are the most likely candidates to support the energy requirements for germ tube emergence and appressorium development (Xiao *et al.*, 1994b). Trehalose is rapidly broken down during germination (Foster and Talbot, unpublished results) and a mutation in a neutral trehalase-encoding gene (*PTH9*/*NTH1*) has been associated with a reduction in pathogenicity (Sweigard *et al.*, 1998). During germ tube elongation, internal glycerol levels rise dramatically, probably linked to membrane biosynthesis (De Jong *et al.*, 1997). After a sharp decline during appressorium differentiation, glycerol levels rise within appressoria to a maximum at 48 h after conidial germination, and concentrations of up to 3.2 M have been estimated (De Jong *et al.*, 1997). This concentration of glycerol is responsible for generating the large hydrostatic turgor produced by appressoria.

Theoretically, 3.2 M glycerol is sufficient to generate a turgor pressure of 8.7 MPa. Experimental observations using vapour pressure psychrometry have confirmed that the pressure generated by such a concentration of glycerol is at least 5.8 MPa (De Jong *et al.*, 1997). Pressure generated by appressoria is then translated into mechanical force, allowing the penetration peg to rupture the plant cuticle. Glycerol is well known as a compatible solute in fungi (an osmolyte) and is known to accumulate to very high concentrations in certain species in response to hyperosmotic stress (Yancey *et al.*, 1982). In *Saccharomyces cerevisiae*, for example, glycerol is formed from glycolysis-derived dihydroxyacetone-3-phosphate by reduction to glycerol-3-phosphate and subsequent dephosphorylation by a specific phosphatase (Nevoigt and Stahl, 1997). The route of glycerol biosynthesis within *M. grisea* appressoria is currently being investigated and the role of nicotinamide-adenine dinucleotide (NAD)-dependent glycerol-3-phosphate dehydrogenase is being directly tested by targeted gene replacement (Viaud and Talbot, unpublished). There are a number of possible origins for glycerol. Glycogen granules, which disappear in the appressorium cytoplasm during turgor generation, represent one potential source for glycerol biosynthesis. However, the amounts of glycogen rosettes, as estimated from the mean density of these bodies in electron micrograph sections, appear to be insufficient to generate such large concentrations of glycerol (Davis *et al.*, 1999). Trehalose, which is found in conidia in large concentrations, is likely to constitute an alternative precursor to glycerol (Foster and Talbot, unpublished), and recent evidence also indicates that lipid bodies, which are numerous in early appressorium development, may be another potential origin (Thines and Talbot, unpublished results). The complete elucidation of this biochemical pathway may prove to be pivotal in identifying a durable mechanism for rice blast control and in understanding the process of plant infection.

The significance of melanin in mediating mechanical infection by *M. grisea* appressoria is becoming increasingly apparent. DHN-melanin is a dark fungal pigment produced by polymerisation of 1,8-dihydroxynaphthalene. This monomer is obtained through a series of reactions involving polyketide synthesis, in which joining and cyclisation of acetate molecules occurs, followed by four subsequent steps alternating reduction and dehydration reactions. Three genes have been characterised, *ALB1, RSY1* and *BUF1*, which encode a polyketide synthase, a scytalone dehydratase and a polyhydroxynaphthalene reductase, respectively (Chumley and Valent, 1990; Vidal-Cros *et al.*, 1994; Howard and Valent, 1996; Motoyama *et al.*, 1998). There is some speculation that scytalone dehydratase can catalyse both steps of dehydration (Butler and Day, 1998). The role of melanin in the appressorium cell wall is to retard the efflux of glycerol from the appressorium, allowing hydrostatic turgor to be generated (De Jong *et al.*, 1997). A model explaining the role of melanin in glycerol accumulation and turgor generation has recently been described (Money, 1997), and takes into account the respective permeability of the melanised appressorium cell wall and of the plasma membrane. The appressorium wall represents a semipermeable layer, which allows free movement of water only and will retain larger molecules, such as glycerol, whilst the plasma membrane is relatively permeable to glycerol due to its phospholipid bilayer composition. In the model proposed by Money (1997), glycerol is prevented from leaking from the appressorium, due largely to the melanised cell wall, and is channelled back into the cytoplasm by transport proteins, such as H^+/glycerol symporters (Tamas *et al.*, 1999).

Melanin biosynthesis appears to play a similar role in appressorium-mediated penetration by *Colletotrichum lagenarium*, and genes encoding polyketide synthase, polyhydroxynaphthalene reductase and scytalone dehydratase have been isolated and shown to be required for appressorium-mediated penetration (Kubo *et al.*, 1996; Perpetua *et al.*, 1996; Takano *et al.*, 1997). These genes appear, therefore, to fulfil the same function as *M. grisea* melanin biosynthetic genes during plant infection. Interestingly, in *Alternaria alternata*, similar genes are present regulating melanin production, and two of these genes, *ALM* and *BRM2*, have been shown to be functional homologues of *M. grisea ALB1* and *BUF1*, respectively, restoring appressorium melanisation and pathogenicity to *alb1* and *buf1* mutants (Kawamura *et al.*, 1997). Despite the existence of appressoria in *A. alternata*, melanisation does not occur in this fungus during differentiation and is not relevant for host penetration. The distinct roles played by melanin biosynthetic genes in these species may reveal the possible route that evolution of the appressorium penetration process used by *M. grisea* and *C. lagenarium* has taken. It is possible, for example, that melanin pigmentation of conidia and hyphae arose for protection from ultraviolet light and desiccation and, therefore, functional homologues of the genes involved in this process are likely to be widespread among fungi. In a far smaller number of species, however, melanin biosynthesis has been co-opted into fulfilling a role in appressorium-mediated plant infection. In these fungi (such as *M. grisea* and *C. lagenarium*), the melanin biosynthetic genes can therefore be considered truly orthologous rather than simply functional homologues (as in *A. alternata*).

1.5 The plant response to infection: resistance and susceptibility

Appressoria are formed in *M. grisea* regardless of whether the host is resistant or susceptible to rice blast. Specificity, therefore, occurs during the primary infection of the host, although the plant appears to perceive the fungus prior to actual cuticle penetration (Xu *et al.*, 1998). Actin localisation and cytoplasm reorientation occur within plant epidermal cells during appressorium maturation but prior to penetration peg formation by the fungus. Whether appressorium development is required for perception of *M. grisea* by rice remains to be seen, but the growing number of mutants impaired in their ability to make infection structures should allow this to be tested directly.

Genetic studies of *M. grisea* have shown that a gene-for-gene interaction exists between the fungus and rice (for review see Laugé and De Wit, 1998). Rice cultivars have major dominant genes for blast resistance that have been introgressed from land races of rice and geographically distinct germplasm collections, and this forms the main method by which rice blast is controlled. In gene-for-gene interactions, the products of plant resistance genes recognise fungal proteins (directly or indirectly) that are encoded by avirulence genes. Fungal avirulence genes are genetically dominant and evolution of newly virulent forms occurs, therefore, by mutation or loss of avirulence gene products. The retention of avirulence genes in fungal species suggests that they have additional roles in pathogenesis or fitness, and simply encode proteins that plants have the ability to perceive when they carry an appropriate resistance gene (De Wit, 1992). Fungal avirulence genes have been isolated from only a small number of fungi so far, including *M. grisea*, the tomato pathogens, *Cladosporium fulvum* and *Phytophthora infestans*, and the barley scald fungus, *Rhynchosporium secalis* (Van den Ackerveken *et al.*, 1992; Joosten *et al.*, 1994; Rohe *et al.*, 1995; Kamoun *et al.*, 1998). In *M. grisea*, a large number of avirulence genes have been predicted, based on genetic studies, and are in the process of being cloned and characterised (Silué *et al.*, 1992a,b; Dioh *et al.*, 1996; Farman and Leong, 1998).

The first avirulence gene characterised from *M. grisea* was the *Avr2-YAMO* gene, which prevents infection of the rice cultivar Yashiro-Mochi. *Avr2-YAMO* was isolated by positional cloning after being found to map to the end of chromosome 1 of *M. grisea*. The gene is located in the sub-telomeric region and construction of a telomere-containing gene library was required for cloning. Perhaps due to its position, *Avr2-YAMO* appears to be subject to frequent mutations including deletions and insertions at the locus. *Avr2-YAMO* encodes a protein showing homology to a neutral Zn^{2+} protease (Valent, 1997), and the active site of this protease appears to be required for it to function as an avirulence gene. The putative enzymatic action of the *Avr2-YAMO* product is unusual compared to the avirulence genes identified in other pathogens. For example, Avr9 and Avr4 from *C. fulvum* are both cysteine-rich secreted proteins found in apoplastic fluids during infection of tomato. They may, therefore, be perceived extracellularly and the structure of the corresponding *Cf4* and *Cf9* resistance gene products is consistent with this (Jones and Jones, 1996; Laugé and De Wit, 1998).

Avirulence genes that control species-specificity have also been isolated from *M. grisea*. Although *M. grisea* can infect more than 50 species of grass, individual isolates of the fungus are normally restricted to a single or small number of species. The *PWL* gene family controls the ability of *M. grisea* to infect weeping lovegrass (*Eragrostis curvula*). *PWL1* prevents infection of weeping lovegrass and originates from isolates of *M. grisea* that are virulent on finger millet (*Eleusine coracana*). *PWL2* meanwhile is found in rice pathogenic strains of the fungus. When *PWL1* or *PWL2* are transformed into strains of *M. grisea* that normally infect weeping lovegrass, the resulting transformants are avirulent (Kang *et al.*, 1995; Sweigard *et al.*, 1995). This indicates that species-specificity works in a similar fashion to cultivar-specificity in *M. grisea*, by recognition of the products of dominant avirulence genes. *PWL2* encodes a 145 amino acid hydrophilic protein with no significant homologies. *PWL1* was isolated by homology to PWL2 and shares 75% amino acid identity. A number of other members of the PWL gene family have since been isolated based on homology. *PWL3* is from a finger millet pathogen and shows 51% identity with *PWL2*. *PWL4* originates from a weeping lovegrass pathogen and shows 72% identity with *PWL3* and 57% identity with *PWL2*. Interestingly, it was found that *PWL4* is non-functional as an avirulence gene (it was found in a weeping lovegrass pathogen) due to a promoter mutation. Expression of the *PWL4* open reading frame under control of the *PWL1* or *PWL2* promoter leads to transformants that are avirulent of weeping lovegrass (Kang *et al.*, 1995).

Recognition of *M. grisea* by resistant hosts, therefore, appears to occur rapidly and is the result of perception of fungal proteins encoded by avirulence genes. Whether these fungal proteins are pathogenicity factors, predominantly secreted proteins, or contribute to infection-related development is not yet clear and will depend on isolation of further avirulence genes (Dioh *et al.*, 1996; Farman and Leong, 1998). The site and timing of fungal perception by host plants is also an unresolved question and will require not only identification of rice blast resistance genes (Ronald, 1998) but also integration of the study of avirulence and resistance as initiated in the *C. fulvum*-tomato interaction (Laugé and De Wit, 1998).

1.6 Signal transduction pathways mediating plant infection

Several general transduction pathways have been implicated in the regulation of appressorium differentiation and function in *M. grisea*. Isolation of individual components of these pathways and targeted disruption of the corresponding genes has been used to test the involvement of signalling pathways in appressorium morphogenesis (for a full list of the pathogenicity genes isolated in *M. grisea* so far see Table 1.1). This information has been used to implicate several signal transduction pathways in appressorium formation that are best understood from the study of budding yeast (Banuett, 1998). These transduction pathways include the cyclic adenosine monophosphate (cAMP)/cAMP-dependent protein kinase A (PKA) pathway, the protein kinase C (PKC) pathway, Ca^{2+}/calmodulin signalling and two mitogen-activated protein kinase (MAPK) cascades. In budding and fission yeast, most of these pathways are

Table 1.1 Genes known to be involved in the pathogenicity of *Magnaporthe grisea*

Gene	Product	Pathogenicity	Phenotypes	Reference
MPG1	Hydrophobin	Reduced	Reduced conidiation Easily wetted	Talbot *et al.*, 1993, 1996
NPR1	Unknown	–	Nitrogen metabolism regulator	Lau and Hamer, 1996
NPR2	Unknown	–	As above	As above
CPKA/ PTH4	Protein kinase A	–	Penetration defective	Mitchell and Dean, 1995
MAGB	G protein α-subunit	–	No appressorium development	Liu and Dean, 1997
MAC1	Adenylate cyclase	–	No appressorium development	Choi and Dean, 1997
PMK1	MAPK	–	No appressorium development	Xu and Hamer, 1996
MPS1	MAPK	–	Penetration defective	Xu *et al.*, 1998
ABC1	ATP-efflux pump	Reduced	–	Urban *et al.*, 1999
APF1	Unknown	–	–	Silué *et al.*, 1998
APP1	Unknown	–	–	Zhu *et al.*, 1996
APP5	Unknown	–	–	Chun and Lee, 1999
CON1-7	Unknown	–	Conidial morphology mutants	Shi and Leung, 1995
SMO1	Unknown	Reduced	Conidial morphology	Hamer *et al.*, 1989b
PDE1	Unknown	–	Penetration defective	Balhadère *et al.*, 1999
PDE2	Unknown	–	As above	As above
IGD1	Unknown	Reduced	Invasive defect	As above
MET1	Unknown	Reduced	Methionine auxotroph	As above
GDE1	Unknown	Reduced on rice	–	As above
PTH1	*GRR1* homologue	Reduced	–	Sweigard *et al.*, 1998
PTH2	Carnitine acetyl transferase	–	–	As above
PTH3	Imidazole glycerol-P dehydratase	Reduced	Histidine auxotroph	As above
PTH8	Unknown yeast homologue	–	–	As above
PTH9	Trehalase	Reduced	–	As above
PTH10	Unknown	–	–	As above
ACR1	Conidiation negative regulator	Reduced	Conidial morphology	Lau and Hamer, 1998

Abbreviations: MAPK, mitogen-activated protein kinase; ATP, adenosine triphosphate.

at least partially controlled by the action of heterotrimeric G proteins and their coupled receptors (Thevelein, 1994; Banuett, 1998).

1.6.1 cAMP-mediated signalling

Cyclic AMP is a critical factor in appressorium morphogenesis in *M. grisea* (Lee and Dean, 1993). The first indication of the role of cAMP in appressorium development came from the observation that exogenous cAMP will induce appressorium formation by *M. grisea* even on non-inductive surfaces, such as glass and other hydrophilic plastic surfaces (Lee and Dean, 1993). Appressorium formation was similarly induced by the application of phosphodiesterase inhibitors and cAMP analogues and could even be made to occur from the tips of mature hyphae in fungal mycelium. More recently, cAMP has been shown to remediate appressorium-defective phenotypes of a number of mutants (Talbot *et al.*, 1996; Zhu *et al.*, 1996; Choi and Dean, 1997; Liu and Dean, 1997). Cyclic AMP is the central messenger molecule in the cAMP/PKA transduction pathway and acts by binding to the regulatory subunit of PKA, thus triggering release of the catalytic subunit of the enzyme. This constitutes the active PKA component and is responsible for phosphorylation-mediated activation of a range of proteins, including transcription factors (Walsh and Van Patten, 1994).

Genes encoding both the catalytic and the regulatory subunits of PKA, *CPKA* and *SUM1*, have been isolated (Mitchell and Dean, 1995; Xu and Hamer, 1996; Adachi and Hamer, 1998; Sweigard *et al.*, 1998). A targeted deletion of *CPKA* produced mutants that showed a clear delay in appressorium formation and were completely unable to cause disease (Mitchell and Dean, 1995; Xu *et al.*, 1997). This showed the importance of cAMP-mediated responses both for appressorium development and appressorium function. Consistent with this role, PKA activity has been measured in germlings and shown to correspond to appressorium differentiation (Kang *et al.*, 1999). Synthesis of cAMP by adenylate cyclase has been investigated by isolation and disruption of the *MAC1* adenylate cyclase gene (Choi and Dean, 1997; Adachi and Hamer, 1998). Δ*mac1* mutants were non-pathogenic and appressorium-deficient, although the mutant phenotype proved to be unstable in certain genetic backgrounds due to an extragenic suppressor mutation in the *SUM1* gene encoding the regulatory subunit of PKA (Adachi and Hamer, 1998). The dispensability of *CPKA* for appressorium formation, coupled with the Δ*mac1* mutant phenotype and its suppression by *SUM1* mutation, is consistent with more than one PKA catalytic activity being required for appressorium elaboration and function. There are, therefore, likely to be divergent cAMP signalling pathways for appressorium formation and turgor generation in *M. grisea* (Adachi and Hamer, 1998; Hamer and Talbot, 1998). Mutation of *MAC1* also causes a number of pleiotropic effects on growth and asexual development of *M. grisea*, which is consistent with the presence of divergent, partially redundant cAMP signalling pathways in *M. grisea* controlling appressorium morphogenesis, appressorium function, asexual morphogenesis and other developmental pathways (Adachi and Hamer, 1998).

The possibility of multiple, partially redundant cAMP signalling pathways in *M. grisea* is currently being tested, following recent isolation of the *MCH1* gene (Adachi *et al.*, 1999). This gene shares homology with the *S. cerevisiae SCH9* protein kinase gene and the *Schizosaccharomyces pombe SCK1* and *SCK2* protein kinase genes, all of which are structurally and functionally related to divergent PKA catalytic subunits (Toda *et al.*, 1988; Jin *et al.*, 1995; Fujita and Yamamoto, 1998). In budding yeast, yet another PKA-related protein kinase subunit has been identified, *YAK1* (Hartley *et al.*, 1994), providing a further possible contender for adaptation in *M. grisea* to a role in infection-related development. Among the proteins likely to be regulated by PKA phosphorylation during appressorium formation is the product of the *PTH9* neutral trehalase gene that contains a consensus PKA phosphorylation sequence and appears to be affected somewhat in enzyme activity in Δ*cpka* mutants (Foster and Talbot, unpublished observations).

1.6.2 MAPK signalling

Three MAPK-encoding genes have so far been identified in *M. grisea* and all have distinct roles in appressorium-mediated infection by the fungus. The first to be identified, and arguably the most significant in controlling pathogenic development in *M. grisea*, was the *PMK1* gene (Xu and Hamer, 1996). *PMK1* is a functional homologue of the budding yeast *FUS3/KSS1* MAPKs, which are involved in mating and pseudohyphal growth, respectively (Banuett, 1998). Δ*pmk1* null mutants are unable to differentiate appressoria and cannot cause rice blast disease symptoms, even when conidia are injected directly into rice leaf tissue. This indicates that *PMK1* is required not only for formation and function of appressorium, but that this is a prerequisite for all subsequent pathogenic development. This phenotype strongly contrasts with that of other appressorium-deficient mutants, such as the *buf1, alb1* and *rsy1* mutants and Δ*mpg1* mutants, all of which cause normal disease symptoms when conidia are injected into rice leaves or surfaces are abraded prior to infection (Chumley and Valent, 1990; Talbot *et al.*, 1993). The Δ*pmk1* mutant phenotype cannot be rescued by addition of exogenous cAMP, although some responsiveness to the secondary messenger can be observed when germinating conidia are treated on non-inductive hydrophilic surfaces (Xu and Hamer, 1996). The early stages of appressorium formation, including the swelling and flattening of germ tubes, can be observed after this treatment, although germ tubes always arrest growth prior to appressorium differentiation (Xu and Hamer, 1996). This suggests interconnection between the *PMK1* MAPK pathway and the cAMP pathway, with the MAPK cascade likely to act parallel with, or downstream of, the cAMP-dependent signalling pathway. The effect of the Δ*pmk1* null mutation strongly suggests that the gene is involved in control of growth polarity and cytoskeletal rearrangement during appressorium morphogenesis.

PMK1 is a functional homologue of the *S. cerevisiae FUS3/KSS1* MAPK genes, restoring mating ability to budding yeast when introduced and expressed under the *GAL1* promoter (Xu and Hamer, 1996). The *S. cerevisiae KSS1* gene is now known to be involved primarily in the pseudohyphal dimorphism shown by yeast upon exposure to certain environmental stresses. It seems likely that *PMK1* is functionally closely related to this gene, fulfilling a similar role in the control of polarity and a change in growth pattern. Consistent with this, although *PMK1* can complement the *fus3/kss1* mating defect in yeast it is dispensable for mating in *M. grisea* (Xu and Hamer, 1996). This observation indicates that *PMK1* does not operate in a directly orthologous pathway in *M.grisea* and may suggest existence of an additional MAPK (and corresponding cascade) associated with mating function. The relationship of *PMK1*, and perhaps other elements of the pheromone response pathway, to appressorium formation, may go some way to explain the inhibitory effect of yeast α-factor pheromone on appressorium development in *MAT 1–2* strains of *M. grisea* (Beckerman *et al.*, 1997). This remarkable observation suggests that pheromone signalling and, in particular, a pheromone receptor may be important in the control of appressorium morphogenesis by *M. grisea.* So far, only one gene encoding a putative receptor has been identified. This gene, called *PTH11*, was identified as a non-pathogenic REMI mutant and encodes a novel type of protein with multiple membrane-spanning domains. The gene seems to play an upstream role in signalling, although its precise relationship to elements in the cAMP and MAPK pathways has yet to be determined (DeZwaan *et al.*, 1999b).

The second MAPK gene identified in *M. grisea* is the *MPS1* gene. *MPS1* appears to control appressorium function among a number of other likely roles (Xu *et al.*, 1998). Δ*mps1* mutants are reduced in pathogenicity due to formation of non-functional appressoria. Mutants are also affected in their conidiation and fertility, and cultures become water-soaked after extensive growth. This indicates that MPS1 is required for appressorium penetration and related morphological processes, such as conidiation and ascospore generation. *MPS1* encodes a functional homologue of *S. cerevisiae SLT2/MPK1* (Xu *et al.*, 1998). In yeast, *SLT2* controls cell integrity in response to membrane stress and is activated partly through the PKC pathway and partly through the pheromone response pathway (Gustin *et al.*, 1998). It is possible that PKC signalling in *M. grisea* regulates the action of *MPS1* because biochemical studies using inhibitors and secondary messenger analogues have implicated PKC signalling in appressorium development (Thines *et al.*, 1997a,b). In mammalian cells, the PKC pathway is activated by diacylglycerols (DAGs) produced together with inositol-1,4,5-triphosphate (IP_3) as a result of conversion from phosphatidylinositol-4,5-biphosphate by phospholipase C in response to heterotrimeric G protein-mediated activation (Divecha and Irvine, 1995). In yeast, PKC is activated by the membrane-bound *RHO1*-encoded G protein and a putative mechanoreceptor (Banuett, 1998). The downstream targets of MPS1 phosphorylation and the relationship to IP_3-triggering of a potential Ca^{2+}/calmodulin pathway (Clapham, 1995) in *M. grisea* await further investigation. Recent biochemical evidence has suggested a role for calmodulin during conidial germination and appressorium induction (Lee and Lee, 1998; Liu and Kolattukudy, 1999). Characterisation of the gene encoding PKC and further analysis of calmodulin should clarify their roles in primary infection by *M. grisea*.

The third MAPK gene characterised from *M. grisea* is the *OSM1* gene, which regulates cellular turgor upon exposure to hyperosmotic stress. *OSM1* is a functional homologue of *HOG1* in *S. cerevisiae* and is able to complement a Δ*hog1* mutant, allowing it to grow under high salt concentrations (Dixon *et al.*, 1999). In yeast, HOG1 phosphorylation activates genes encoding enzymes leading to synthesis of glycerol, which acts as the principal compatible solute (Banuett, 1998). Because of this role, it seemed likely that *OSM1* might be the main regulator for appressorium turgor generation in *M. grisea*, carrying out the same process leading to glycerol accumulation. Somewhat surprisingly, this is not the case and Δ*osm1* mutants retain full pathogenicity and are able to accumulate glycerol normally. Instead, the action of *OSM1* during hyperosmotic stress centres on arabitol accumulation. Arabitol accumulates rapidly in *M. grisea* hyphae in response to hyperosmotic stress and is also able to accumulate in *M. grisea* appressoria when they are exposed to osmotic stress. *OSM1* does, however, perform a role in appressorium morphogenesis but here it appears to act as a negative regulator. Δ*osm1* mutants show production of multiple appressoria when conidia germinate and develop in the presence of hyperosmotic stress. This indicates that one role of OSM1 may be to negatively regulate the PMK1 MAPK during stressful conditions, thus preventing appressorium formation. The absence of *OSM1* in null mutants may, therefore, lead to multiple rounds of appressorium morphogenesis under such conditions, due to continued activation of PMK1, as shown in Figure 1.2 (Dixon *et al.*, 1999).

1.6.3 Heterotrimeric G-protein mediated signalling

Three genes encoding Gα subunits of heterotrimeric G-proteins have been cloned, *MAGA, MAGB* and *MAGC* (Liu and Dean, 1997). The products of these genes belong to the three major G-protein subgroups. *MAGA* belongs to group III and is related to mammalian Gαs that encode activators of adenylate cyclase (AC). *MAGB* belongs to group I and is related to Gα_i inhibitors of AC. *MAGC* belongs to group II, which has no mammalian counterpart indicative of

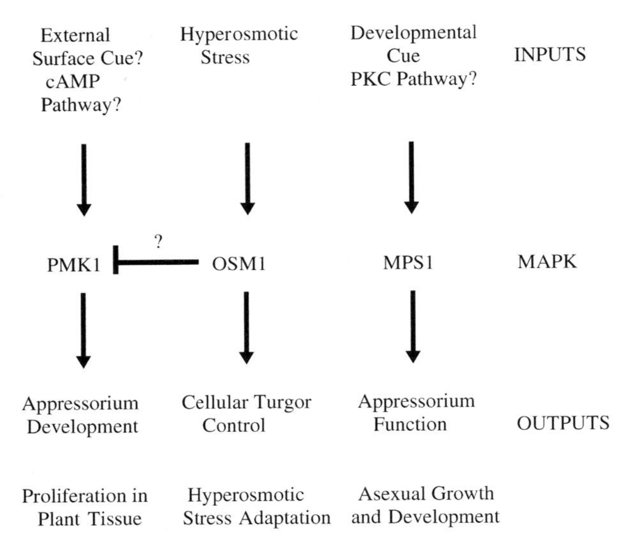

Figure 1.2 Model showing the mitogen-activated protein kinase (MAPK) signalling pathways in the rice blast fungus, *Magnaporthe grisea*. Three MAPK-encoding genes have been identified and characterised in *M. grisea*. *PMK1* regulates appressorium development and subsequent proliferation of the fungus in plant tissue. PMK1 may operate downstream (or parallel to) the cyclic adenosine monophosphate (cAMP) signalling pathway, which regulates appressorium formation (Xu and Hamer, 1996). *MPS1* regulates appressorium-mediated infection, potentially by controlling formation of the penetration peg. By analogy to *Saccharomyces cerevisiae*, MPS1 may act downstream of a protein kinase C (PKC) pathway (Xu *et al.*, 1998). *OSM1* regulates hyperosmotic stress adaptation and clearly acts in response to that stress. *OSM1* may also prevent the *PMK1* signalling pathway from operating during stressful environmental conditions.

their function (Bölker, 1998). *MAGB* is the only Gα subunit gene associated with pathogenicity (Liu and Dean, 1997), and this is somewhat surprising considering that it is likely to be an inhibitor of AC. The plausible explanation for the loss of appressorium development and pathogenicity in Δ*magB* transformants comes from the observation that both activated $G\alpha_i$ and Gβγ subunits can act as inhibitors of AC (Taussig *et al.*, 1993). This means that in Δ*magB* mutants, AC inhibition may occur constitutively due to the presence of a free Gβγ subunit (Bölker, 1998). This is consistent with the fact that the appressorium defect in Δ*magB* mutants can be overcome by addition of exogenous cAMP. Remediation of Δ*magB* mutant phenotypes by addition of hexadecanediol also highlights the fact that additional signalling mechanisms for appressorium development exist in *M. grisea* (Thines *et al.*, 1997a,b). Potential interactions between G-protein signalling and the *PTH11* receptor are currently being explored by Sweigard and co-workers, who have recently isolated two RAS homologues encoded by *RAS1* and *RAS2* (DeZwaan *et al.*, 1999a).

1.7 Future exploration of signalling mechanisms in *M. grisea*

A number of uncharacterised developmental mutants have been described in *M. grisea* with defects in appressorium formation. Cloning of *APP1-5* and the *APF1* locus (Zhu *et al.*, 1996; Silué *et al.*, 1998; Chun and Lee, 1999) will therefore be important in defining the regulatory mechanisms of infection structure formation in *M. grisea*. Additional valuable information will also result from characterising the substantial group of morphogenetic genes, such as *SMO1, ACR1* and *CON1-7*, identified in mutant screens for conidiogenesis and appressorium defects (Hamer *et al.*, 1989b; Shi and Leung, 1994, 1995; Lau and Hamer, 1998; Shi *et al.*, 1998). Significantly, all of these genes are involved in more than one morphogenetic process, either conidiogenesis, appressorium differentiation or formation of sexual reproductive structures. This is also true of the major regulatory genes defined previously (*PMK1, MPS1, CPKA, MAGB*) and highlights the similarity in these developmental processes, which all require departure from the hyphal growth pattern. Defining the signal transduction pathway required for appressorium formation is, therefore, unlikely to occur in the absence of an integrated study of the developmental pathways leading to conidium formation and perithecial development. In this context, defining epistatic relationships between the developmental mutants so far obtained, either through mutant screens or targeted replacement, should be an early priority (Shi *et al.*, 1998).

Another marked feature of the signalling pathways so far implicated in infection by *M. grisea* is the level of cooperation necessary for completion of appressorium development. For example, it seem that surface signals, a cAMP signal and cutin monomers are able to induce appressorium formation. However, mutations in key signalling steps, such as *MAGB*, can be circumvented by addition of a different signal (hexadecanediol), indicating that a number of different receptors and upstream signalling components are able to initiate appressorium development. Convergence of pathways appears to occur later, although at this time only *PMK1* can be confidently positioned as a central component in the developmental process (Xu and Hamer, 1996; Hamer and Talbot, 1998).

The organisation of signal transduction pathways in *M. grisea*, and the extent to which multiple signals converge to bring about infection structure formation, may be controlled by the presence of scaffold and adapter proteins that have been described in yeast (Herskowitz, 1995). In the pheromone response pathway, for example, STE5 appears to anchor the MAPK module together and is thought to provide a means by which MAPK pathways can be separately controlled and spatially limited. Another example is the mammalian synaptic AKAP79 anchoring protein (Faux and Scott, 1996). These proteins bring about association of enzymes into a complex and anchor the module to a specific cellular compartment. This creates an efficient means for controlling the phosphorylation state of a given protein in response to multiple intracellular signals (Faux and Scott, 1996). The presence of scaffold proteins in *M. grisea* might, therefore, constitute one mechanism by which signal transduction pathways for appressorium development, conidiogenesis and perithecial development (which contain many of the same components) may be separately regulated. The identification of such proteins and their corresponding genes may therefore be particularly informative.

The regulation of morphogenetic target genes has been studied extensively in developmental studies of *Neurospora crassa* and *Aspergillus nidulans*, and has been particularly useful in defining the importance of transcriptional regulators and corresponding upstream signalling pathways (for review see Adams *et al.*, 1998). Similar studies have not yet taken place in *M. grisea* but are likely to become increasingly important. In this context, the MPG1 hydrophobin gene may become an important tool for studying appressorium development because of its unusual pattern of expression (Talbot *et al.*, 1993) and importance in both appressorium formation and conidiation (Talbot *et al.*, 1996). The *MPG1* promoter displays an array of potential transcription factor binding sites, suggesting that it is regulated through the coordinated action of several factors. Putative GATA-like binding sites for *NUT1*, a functional homologue of *Aspergillus nidulans* AREA, the major nitrogen regulatory protein (Froeliger and Carpenter, 1996), are present in the *MPG1* promoter. This is consistent with elevated expression of MPG1 during nitrogen starvation stress. The promoter also contains consensus bristle response elements (BREs), which are potential binding sites for a putative *A. nidulans* brlA homologue. A *M. grisea* brlA homologue would be likely to regulate the conidiation-specific gene expression of *MPG1*, but might also be important for appressorium formation. Since *brlA*$^-$ mutants would be unlikely to produce conidia, defining a role for such a gene in appressorium formation would be difficult and may, in fact, only be possible by such a promoter analysis. Preliminary *MPG1* promoter analysis using the green fluorescent protein reporter (Kershaw *et al.*, 1998) has confirmed the importance of *NUT1*-mediated regulation during nitrogen starvation stress and also the requirement for the *NPR1* and *NPR2* loci for high level *MPG1* expression (Lau and Hamer, 1996; Soanes and Talbot, unpublished). Interestingly, reporter gene studies have also revealed previously unforeseen roles for *CPKA* and *PMK1* in regulating *MPG1* expression (Soanes and Talbot, unpublished), highlighting the utility of this type of analysis (see Bell-Pederson *et al.*, 1992).

1.8 Conclusions

Recent studies in *M. grisea* have begun to define important processes that bring about plant infection. These include strong attachment of conidia to the leaf surface, rapid germination without the need for exogenous nutrients, correct perception of the leaf surface using extracellular cell wall components, including a hydrophobin, and subsequent production of melanin-pigmented infection cells. Once formed, appressoria generate turgor by accumulation of intracellular glycerol, which is maintained by the melanised cell wall, and force a narrow penetration peg into the leaf cuticle. Production of appressoria and their subsequent activity is controlled by components of cAMP and MAPK signalling pathways, which respond to more than one external signal and probably regulate very large numbers of genes. The critical question now is how widespread are these features likely to be among more diverse plant pathogenic fungi?

While developmental details and methods of plant infection differ markedly among plant pathogens, preliminary evidence suggests that underlying signalling processes may be conserved. Preliminary studies to address this have shown, for example, that homologues of the *PMK1* MAPK gene are required for pathogenicity in *Botrytis cinerea* and *Colletotrichum lagenarium* (Takano *et al.*, 1997; Zheng *et al.*, 1999). Furthermore, cAMP signalling for plant infection is important for pathogens as diverse as *Ustilago maydis* and *Erysiphe graminis* (Durrenberger *et al.*, 1998; Oliver *et al.*, 1999). Studying infection mechanisms in detail in *M. grisea* may, therefore, prove to be of considerable value in learning about fungal pathogenesis.

The advent of functional genomic approaches to the study of fungal biology is likely to accelerate discovery of pathogenicity determinants, as the bottleneck in identifying and cloning genes is effectively removed. The new challenge will, therefore, be in integrating information concerning the actions of individual genes to form a picture of the whole process of plant infection. This will require a multidisciplinary approach encompassing genetics, biochemistry and cell biology techniques to explore the orchestrated action of gene sets during pathogenesis. The newly emerging discipline of bioinformatics will also be vital in order to assimilate, interrogate and gain meaning from the large volume of information that is likely to be generated in the next few years.

References

Adachi, K. and Hamer, J.E. (1998) Divergent cAMP signalling pathways regulate growth and pathogenesis in the rice blast fungus, *Magnaporthe grisea*. *Plant Cell*, **10** 1361-73.

Adachi, K., Urban, M. and Hamer, J.E. (1999) *MCH1*, a gene encoding Sch9 homolog kinase of *Magnaporthe grisea*. *Fungal Genet. Newslett.*, **46** (Suppl.) 73.

Adams, T.H., Wieser, J.K. and Yu, J.H. (1998) Asexual sporulation in *Aspergillus nidulans*. *Microbiol. Mol. Biol. Rev.*, **62** 35-54.

Balhadère, P.V., Foster, A.J. and Talbot, N.J. (1999) Identification of pathogenicity mutants of the rice blast fungus, *Magnaporthe grisea*, by insertional mutagenesis. *Mol. Plant-Microbe Interact.*, **2** 129-42.

Banuett, F. (1998) Signalling in the yeasts: an informational cascade with links to the filamentous fungi. *Microbiol. Mol. Biol. Rev.*, **62** 249-74.

Beckerman, J.L. and Ebbole, D.J. (1996) *MPG1*, a gene encoding a fungal hydrophobin of *Magnaporthe grisea*, is involved in surface recognition. *Mol. Plant-Microbe Interact.*, **9** 450-56.

Beckerman, J.L., Naider, F. and Ebbole, D.J. (1997) Inhibition of pathogenicity of the rice blast fungus by *Saccharomyces cerevisiae* α-factor. *Science*, **276** 1116-19.

Bell-Pederson, D., Dunlap, J.C. and Loros, J.J. (1992) Distinct *cis*-acting elements mediate clock, light and developmental regulation of the *Neurospora crassa EAS* (*ccg-2*) gene. *Mol. Cell Biol.*, **16** 513-21.

Bölker, M. (1998) Sex and crime: heterotrimeric G proteins in fungal mating and pathogenesis. *Fungal Genet. Biol.*, **25** 143-56.

Bourett, T.M. and Howard, R.J. (1990) *In vitro* development of penetration structures in the rice blast fungus, *Magnaporthe grisea*. *Can. J. Bot.*, **68** 329-42.

Butler, M.J. and Day, A.W. (1998) Fungal melanins: a review. *Can. J. Microbiol.*, **44** 1115-36.

Chida, T. and Sisler, H.D. (1987) Restoration of appressorial penetration ability by melanin precursors in *Pyricularia oryzae* treated with antipenetrants and in melanin-deficient mutants. *J. Pesticide Sci.*, **12** 49-55.

Choi, W. and Dean, R.A. (1997) The adenylate cyclase gene, *MAC1*, of *Magnaporthe grisea* controls appressorium formation and other aspects of growth and development. *Plant Cell*, **9** 1973-83.

Chumley, F.G. and Valent, B. (1990) Genetic analysis of melanin deficient, non-pathogenic mutants of *Magnaporthe grisea*. *Mol. Plant-Microbe Interact.*, **3** 135-43.

Chun, S.J. and Lee, Y.-H. (1999) Genetic analysis of a mutation on appressorium formation in *Magnaporthe grisea*. *FEMS Microbiol. Lett.*, **173** 133-37.

Clapham, D.E. (1995) Calcium signalling. *Cell*, **80** 259-68.

Davis, D.J., Burlak, C. and Money, N.P. (2000) Biochemical and biomechanical aspects of appressorial development of *Magnaporthe grisea*, in *Progress in Rice Blast Research* (eds. N.J. Talbot, M.-H. Lebrun, D. Tharreau and J.-L. Notteghem), Springer-Verlag, Berlin (in press).

Dean, R.A. (1997) Signal pathways and appressorium morphogenesis. *Annu. Rev. Phytopathol.*, **35** 211-34.

De Jong, J.C., McCormack, B.J., Smirnoff, N. and Talbot, N.J. (1997) Glycerol generates turgor in rice blast. *Nature*, **389** 244-45.

De Wit, P.J.G.M. (1992) Molecular characterization of gene-for-gene systems in plant-fungus interactions and the application of avirulence genes in control of plant pathogens. *Annu. Rev. Phytopathol.*, **30** 391-418.

DeZwaan, T.M., Carroll, A.M. and Sweigard, J.A. (1999a) The role of ras and G-alpha homologs in PTH11-mediated signalling in *Magnaporthe grisea*. *Fungal Genet. Newslett.*, **46** (Suppl.) 102.

DeZwaan, T.M., Carroll, A.M., Valent, B. and Sweigard, J.A. (1999b) The *PTH11* gene encodes a novel upstream component of pathogenicity signalling in the rice blast fungus. *Fungal Genet. Newslett.*, **46** (Suppl.) 102.

Diaz-Perez, S.V., Crouch, V.W. and Orbach, M.J. (1996) Construction and characterization of a *Magnaporthe grisea* bacterial artificial chromosome library. *Fungal Genet. Biol.*, **20** 280-88.

Dioh, W., Tharreau, D., Gomez, R., Roumen, E., Orbach, M., Notteghem, J.-L. and Lebrun, M.-H. (1996) Mapping avirulence genes in the rice blast fungus, *Magnaporthe grisea*, in *Rice Genetics III* (ed. G.S. Khush), IRRI, Philippines, pp. 916-20.

Divecha, N. and Irvine, R.F. (1995) Phospholipid signalling. *Cell*, **80** 269-78.

Dixon, K.P., Xu, J.-R., Smirnoff, N. and Talbot, N.J. (1999) Independent signalling pathways regulate cellular turgor generation during osmotic stress and appressorium-mediated plant infection by *Magnaporthe grisea*. *Plant Cell*, **11** 2045-58.

Durrenberger, F., Wong, K. and Kronstad, J.W. (1998) Identification of a cAMP-dependent protein kinase catalytic subunit required for virulence and morphogenesis in *Ustilago maydis*. *Proc. Natl. Acad. Sci. USA*, **95** 5684-89.

Ebata, Y., Yamamoto, H. and Uchiyama, T. (1998) Chemical composition of the glue from appressoria of *Magnaporthe grisea*. *Biosci. Biotechnol. Biochem.*, **62** 672-74.

Farman, M.L. and Leong, S.A. (1998) Chromosome walking to the *AVR1-CO39* avirulence gene of *Magnaporthe grisea*: discrepancy between the physical and genetic maps. *Genetics*, **150** 1049-58.

Faux, M.C. and Scott, J.D. (1996) More on target with protein phosphorylation: conferring specificity by location. *Trends Biochem. Sci.*, **21** 312-15.

Foster, A.J. and Talbot, N.J. (2000) The role of carbohydrates in the pathogenicity of the rice blast fungus, *Magnaporthe grisea*, in *Progress in Rice Blast Research* (eds. N.J. Talbot, M.-H. Lebrun, D. Tharreau and J.-L. Notteghem), Springer-Verlag, Berlin (in press).

Froeliger, E.H. and Carpenter, B.E. (1996) *NUT1*, a major nitrogen regulatory gene in *Magnaporthe grisea*, is dispensable for pathogenicity. *Mol. Gen. Genet.*, **251** 647-56.

Fujita, M. and Yamamoto, M. (1998) *S. pombe sck2* (+), a second homologue of *S. cerevisiae SCH9* in fission yeast, encodes a putative protein kinase closely related to PKA in function. *Curr. Genet.*, **33** 248-54.

Gilbert, R.D., Johnson, A.M. and Dean, R.A. (1996) Chemical signals responsible for appressorium formation in the rice blast fungus. *Physiol. Mol. Plant Pathol.*, **48** 335-46.

Gustin, M.C., Albertyn, J., Alexander, M. and Davenport, K. (1998) MAP kinase pathways in the yeast *Saccharomyces cerevisiae. Microbiol. Mol. Biol. Rev.*, **62** 1264-300.

Hamer, J.E. and Holden, D.W. (1997) Linking approaches in the study of fungal pathogenesis: a commentary. *Fungal Genet. Biol.*, **21** 11-16.

Hamer, J.E. and Talbot, N.J. (1998) Infection-related development in the rice blast fungus, *Magnaporthe grisea. Curr. Opin. Microbiol.*, **1** 693-98.

Hamer, J.E., Howard, R.J., Chumley, F.G. and Valent, B. (1988) A mechanism for surface attachment of spores of a plant pathogenic fungus. *Science*, **239** 288-90.

Hamer, J.E., Farrall, L., Orbach, M.J., Valent, B. and Chumley, F.G. (1989a) Host species-specific conservation of a family of repeated DNA sequences in the genome of a fungal plant pathogen. *Proc. Natl. Acad. Sci. USA*, **86** 9981-85.

Hamer, J.E., Valent, B. and Chumley, F.G. (1989b) Mutations at the *SMO* locus affect the shapes of divers cell types in the rice blast fungus. *Genetics*, **122** 351-61.

Hartley, A.D., Ward, M.P. and Garrett, S. (1994) The YAK1 protein kinase of *Saccharomyces cerevisiae* moderates thermotolerance and inhibits growth by an SCH9 protein-kinase independent mechanism. *Genetics*, **136** 465-74.

Hayashi, N. and Naito, H. (1994) Genome mapping of *Magnaporthe grisea*, in *Rice Blast Disease* (eds. R.S. Zeigler, S.A. Leong and P.S. Teng), CABI, Wallingford, p. 590.

Hegde, Y. and Kolattukudy, P.E. (1997) Cuticular waxes relieve self-inhibition of germination and appressorium formation by the conidia of *Magnaporthe grisea. Physiol. Mol. Plant Pathol.*, **51** 75-84.

Herskowitz, I. (1995) MAP kinases in yeast: for mating and more. *Cell*, **80** 187-97.

Howard, R.J. (1997) Breaching the outer barriers—cuticle and cell wall penetration, in *The Mycota V Part B Plant Relationships* (eds. G.C. Carroll and P. Tudzynski), Springer-Verlag, Berlin, Heidelberg, pp. 43-60.

Howard, R.J. and Ferrari, M.A. (1989) Role of melanin in appressorium formation. *Exp. Mycol.*, **13** 403-18.

Howard, R.J. and Valent, B. (1996) Breaking and entering—host penetration by the fungal rice blast pathogen, *Magnaporthe grisea. Annu. Rev. Microbiol.*, **50** 491-512.

Howard, R.J., Ferrari, M.A., Roach, D.H. and Money, N.P. (1991) Penetration of hard substrates by a fungus employing enormous turgor pressures. *Proc. Natl. Acad. Sci. USA*, **88** 11281-84.

Jelitto, T.C., Page, H.A. and Read, N.D. (1994) Role of external signals in regulating the pre-penetration phase of infection by the rice blast fungus, *Magnaporthe grisea. Planta*, **194** 471-77.

Jin, M., Fujita, M., Culley, B.M., Apolinario, E., Yamamoto, M., Maundrell, K. and Hoffman, C.S. (1995) *sck1*, a high copy number suppressor of defects in the cAMP-dependent protein kinase pathway fission yeast, encodes a protein homologous to the *Saccharomyces cerevisiae* SCH9 kinase. *Genetics*, **140** 457-67.

Jones, D.A. and Jones, J.D.G. (1996) The role of leucine-rich repeat proteins in plant defences. *Adv. Bot. Res.*, **24** 91-167.

Joosten, M.H.A.J., Cozijnsen, T.J. and De Wit, P.J.G.M. (1994) Host resistance to a fungal tomato pathogen lost by a single base-pair change in an avirulence gene. *Nature*, **367** 384-86.

Kamoun, S., Van West, P., Vleeshouwers, V.G.A.A., De Groot, K.E. and Govers, F. (1998) Resistance of *Nicotiana benthamiana* to *Phytophthora infestans* is mediated by the recognition of the elicitor protein, *INF1*. *Plant Cell*, **10** 1413-26.

Kang, S., Sweigard, J.A. and Valent, B. (1995) The *PWL* host specificity gene family in the blast fungus *Magnaporthe grisea*. *Mol. Plant-Microbe Interact.*, **8** 939-48.

Kang, S.H., Khang, C.H. and Lee, Y.-H. (1999) Regulation of cAMP-dependent protein kinase during appressorium formation in *Magnaporthe grisea*. *FEMS Microbiol. Lett.*, **170** 419-23.

Kawamura, C., Moriwaki, J., Kimura, N., Fujita, Y., Fuji, S., Hirano, T., Koizumi, S. and Tsuge, T. (1997) The melanin biosynthesis genes of *Alternaria alternata* can restore pathogenicity of the melanin-deficient mutants of *Magnaporthe grisea*. *Mol. Plant-Microbe Interact.*, **4** 446-53.

Kershaw, M.J. and Talbot, N.J. (1998) Hydrophobins and repellents: proteins with fundamental roles in fungal morphogenesis. *Fungal Genet. Biol.*, **23** 18-33.

Kershaw, M.J., Wakley, G.E. and Talbot, N.J. (1998) Complementation of the *Mpg1* mutant phenotype in *Magnaporthe grisea* reveals functional relationships between fungal hydrophobins. *EMBO J.*, **17** 3838-49.

Kronstad, J.W., De Maria, A., Funnell, D., Laidlaw, R.D., Lee, N., Moniz de Sá, M. and Ramesh, M. (1998) Signalling via cAMP in fungi: interconnections with mitogen-activated protein kinase pathways. *Arch. Microbiol.*, **170** 395-404.

Kubo, Y., Takano, Y., Endo, N., Yasuda, N., Tajima, S. and Furusawa, I. (1996) Cloning and structural analysis of the melanin biosynthesis gene, *SCD1*, encoding scytalone dehydratase in *Colletotrichum lagenarium*. *Appl. Environ. Microbiol.*, **62** 4340-44.

Lau, G.W. and Hamer, J.E. (1996) Regulatory genes controlling *MPG1* expression and pathogenicity in the rice blast fungus, *Magnaporthe grisea*. *Plant Cell*, **8** 771-81.

Lau, G.W. and Hamer, J.E. (1998) *Acropetal*: a genetic locus required for conidiophore architecture and pathogenicity in the rice blast fungus. *Fungal Genet. Biol.*, **24** 228-39.

Laugé, R. and De Wit, P.J.G.M. (1998) Fungal avirulence genes: structure and possible functions. *Fungal Genet. Biol.*, **24** 285-97.

Lee, Y.-H. and Dean, R.A. (1993) cAMP regulates infection structure formation in the plant pathogenic fungus, *Magnaporthe grisea*. *Plant Cell*, **5** 693-700.

Lee, Y.-H. and Dean, R.A. (1994) Hydrophobicity of contact surface induces appressorium formation in *Magnaporthe grisea*. *FEMS Microbiol. Lett.*, **115** 71-75.

Lee, S.C. and Lee, Y.-H. (1998) Calcium/calmodulin-dependent signalling for appressorium formation in the plant pathogenic fungus, *Magnaporthe grisea*. *Mols. Cells*, **8** 698-704.

Leung, H., Lehtinen, U., Karjalainen, R., Skinner, D.Z., Tooley, P.W., Leong, S.A. and Ellingboe, A.H. (1990) Transformation of the rice blast fungus, *Magnaporthe grisea*, to hygromycin B resistance. *Curr. Genet.*, **17** 409-11.

Liu, S. and Dean, R.A. (1997) G protein α subunit genes control growth, development and pathogenicity of *Magnaporthe grisea*. *Mol. Plant-Microbe Interact.*, **10** 1075-86.

Liu, Z.M. and Kolattukudy, P.E. (1999) Early expression of the calmodulin gene, which precedes appressorium formation in *Magnaporthe grisea*, is inhibited by self-inhibitors and requires surface attachment. *J. Bacteriol.*, **181** 3571-77.

Mendgen, K., Hahn, M. and Deising, H. (1996) Morphogenesis and mechanisms of penetration by plant pathogenic fungi. *Annu. Rev. Phytopathol.*, **34** 367-86.

Mitchell, T.K. and Dean, R.A. (1995) The cAMP-dependent protein kinase catalytic subunit is required for appressorium formation and pathogenesis by the rice blast fungus, *Magnaporthe grisea*. *Plant Cell*, **7** 1869-78.

Money, N.P. (1995) Turgor pressure and the mechanics of fungal penetration. *Can. J. Bot.*, **73** (Suppl. 1), S96-S102.

Money, N.P. (1997) Mechanism linking cellular pigmentation and pathogenicity in rice blast disease. *Fungal Genet. Biol.*, **22** 151-52.

Money, N.P. and Howard, R.J. (1996) Confirmation of a link between fungal pigmentation, turgor pressure and pathogenicity using a new method of turgor measurement. *Fungal Genet. Biol.*, **20** 217-27.

Motoyama, T., Imanishi, K. and Yamaguchi, I. (1998) cDNA cloning, expression and mutagenesis of scytalone dehydratase needed for pathogenicity of the rice blast fungus, *Pyricularia oryzae. Biosci. Biotechnol. Biochem.*, **62** 564-66.

Nevoigt, E. and Stahl, U. (1997) Osmoregulation and glycerol metabolism in the yeast, *Saccharomyces cerevisiae. FEMS Microbiol. Rev.*, **21** 231-41.

Oliver, R.P., Kinane, J., Bindslev, L., Dalvin, S., Rasmussen, S.W., Rouster, J., Giese, H., Talbot, N.J. and Kershaw, M.J. (1999) Acquisition and analysis of an EST database of barley mildew: signal transduction pathways. *Fungal Genet. Newslett.*, **46** 86 (Appendix).

Ou, S.H. (1985) *Rice Diseases*, Kew, Surrey, Commonwealth Mycological Institute, C.A.B.I, pp. 109-201.

Parsons, K.A., Chumley, F.G. and Valent, B. (1987) Genetic transformation of the fungal pathogen responsible for rice blast disease. *Proc. Natl. Acad. Sci. USA*, **84** 4161-65.

Perpetua, N.S., Kubo, Y., Yasuda, N., Takano, Y. and Furusawa, I. (1996) Cloning and characterization of a melanin biosynthetic *THR1* reductase gene essential for appressorial penetration of *Colletotrichum lagenarium. Mol. Plant-Microbe Interact.*, **9** 323-29.

Rohe, M., Gierlach, A., Hermann, H., Hahn, M., Schmidt, B., Rosahl, S. and Knogge, W. (1995) The race-specific elicitor, *NIP1*, from the barley pathogen, *Rhynchosporium secalis*, determines avirulence on host plants of the *Rrs1* resistance genotype. *EMBO J.*, **14** 4168-77.

Ronald, P.C. (1998) Resistance gene evolution. *Curr. Opin. Plant Biol.*, **1** 294-98.

Shi, Z. and Leung, H. (1994) Genetic analysis and rapid mapping of a sporulation mutation in *Magnaporthe grisea. Mol. Plant-Microbe Interact.*, **7** 113-20.

Shi, Z. and Leung, H. (1995) Genetic analysis of sporulation in *Magnaporthe grisea* by chemical and insertional mutagenesis. *Mol. Plant-Microbe Interact.*, **8** 949-59.

Shi, Z., Christian, D. and Leung, H. (1998) Interactions between spore morphogenetic mutations affect cell types, sporulation and pathogenesis in *Magnaporthe grisea. Mol. Plant-Microbe Interact.*, **11** 1199-207.

Silué, D., Notteghem, J.-L. and Tharreau, D. (1992a) Evidence for a gene-for-gene relationship in the *Oryza sativa-Magnaporthe grisea* pathosystem. *Phytopathology*, **82** 577-80.

Silué, D., Tharreau, D. and Notteghem, J.-L. (1992b) Identification of *Magnaporthe grisea* avirulence genes to seven rice cultivars. *Phytopathology*, **82** 1462-67.

Silué, D., Tharreau, D., Talbot, N.J., Clergeot, P.-H., Notteghem, J.-L. and Lebrun, M.-H. (1998) Identification and characterization of *apf1*$^-$, a non-pathogenic mutant of the rice blast fungus, *Magnaporthe* grisea, which is unable to differentiate appressoria. *Physiol. Mol. Plant Pathol.*, **53** 239-51.

Smith, J.R. and Leong, S.A. (1994) Mapping of a *Magnaporthe grisea* locus affecting rice (*Oryza sativa*) cultivar specificity. *Theor. Appl. Genet.*, **88** 901-908.

Staples, R.C. and Hoch, H.C. (1997) Physical and chemical cues for spore germination and appressorium formation by fungal pathogens, in *The Mycota V Part B Plant Relationships* (eds. G.C. Carroll and P. Tudzynski), Springer-Verlag, Berlin, Heidelberg, pp. 27-40.

Sweigard, J.A., Chumley, F.G. and Valent, B. (1992) Disruption of a *Magnaporthe grisea* cutinase gene. *Mol. Gen. Genet.*, **232** 183-90.

Sweigard, J.A., Valent, B., Orbach, M.J., Walter, A.M. and Rafalski, A. (1993) Genetic map of the rice blast fungus, *Magnaporthe grisea* (n = 7), in *Genetic Maps*, Edn. 6 (ed. S.J. O'Brien), Cold Spring Harbor Laboratory Press, Cold Spring Harbor, New York, pp. 3.112-3.114.

Sweigard, J.A., Carroll, A.M., Kang, S., Farrall, L., Chumley, F.G. and Valent, B. (1995) Identification, cloning and characterization of *PWL2*, a gene for host species specificity in the rice blast fungus. *Plant Cell*, **7** 1221-33.

Sweigard, J.A., Carroll, A.M., Farrall, L., Chumley, F.G. and Valent, B. (1998) *Magnaporthe grisea* pathogenicity genes obtained through insertional mutagenesis. *Mol. Plant-Microbe Interact.*, **11** 404-12.

Takano, Y., Kubo, Y., Kawamura, C., Tsuge, T. and Furusawa, I. (1997) The *Alternaria alternata* melanin biosynthesis gene restores appressorial melanization and penetration of cellulose membranes in the melanin-deficient albino mutant of *Colletotrichum lagenarium. Fungal Genet. Biol.*, **21** 131-40.

Talbot, N.J. (1999) Fungal biology—coming up for air and sporulation. *Nature*, **398** 295-96.

Talbot, N.J., Ebbole, D.J. and Hamer, J.E. (1993) Identification and characterization of *MPG1*, a gene involved in pathogenicity from the rice blast fungus, *Magnaporthe grisea. Plant Cell*, **5** 1575-90.

Talbot, N.J., Kershaw, M.J., Wakley, G.E., de Vries, O.M.H., Wessels, J.G.H. and Hamer, J.E. (1996) *MPG1* encodes a fungal hydrophobin involved in surface interactions during infection-related development of *Magnaporthe grisea. Plant Cell*, **8** 985-99.

Tamas, M.J., Luyten, K., Sutherland, F.C., Hernandez, A., Albertyn, J., Valadi, H., Li, H., Prior, B.A., Kilian, S.G., Ramos, J., Gustafsson, L., Thevelein, J.M. and Hohmann, S. (1999) Fps1p controls the accumulation and release of the compatible solute glycerol in yeast osmoregulation. *Mol. Microbiol.*, **31** 1087-104.

Taussig, R., Iniguezlluhi, J.A. and Gilman, A.G. (1993) Inhibition of adenylyl-cyclase by G(i-alpha). *Science*, **261** 218-21.

Teraoka, T., Nagaoka, M., Hirano, K., Takahashi, H. and Hosokawa, D. (2000) Possible role of concanavalin A-binding glycoprotein secreted from germinating conidia of *Magnaporthe grisea*, in *Progress in Rice Blast Research* (eds. N.J. Talbot, M.-H. Lebrun, D. Tharreau and J.-L. Notteghem), Springer-Verlag, Berlin (in press).

Thevelein, J.M. (1994) Signal transduction in yeast. *Yeast*, **10** 1753-90.

Thines, E., Eilbert, F., Sterner, O. and Anke, H. (1997a) Glisoprenin A, an inhibitor of the signal transduction pathway leading to appressorium formation in germinating conidia of *Magnaporthe grisea* on hydrophobic surfaces. *FEMS Microbiol. Lett.*, **151** 219-24.

Thines, E., Eilbert, F., Sterner, O. and Anke, H. (1997b) Signal transduction leading to appressorium formation in germinating conidia of *Magnaporthe grisea*: effects of second messengers diacylglycerol, ceramides and sphingomyelin. *FEMS Microbiol. Lett.*, **156** 91-94.

Toda, T., Cameron, S., Sass, P. and Wigler, M. (1988) *SCH9*, a gene of *Saccharomyces cerevisiae* that encodes a protein distinct from, but functionally and structurally related to, cAMP-dependent protein kinase catalytic subunits. *Genes Dev.*, **2** 517-27.

Uchiyama, T. and Okuyama, K. (1990) Participation of *Oryza sativa* leaf wax in appressorium formation by *Pyricularia oryzae. Phytochemistry*, **29** 91-92.

Urban, M., Bhargava, T. and Hamer, J.E. (1999) An ATP-driven efflux pump is a novel pathogenicity factor in rice blast disease. *EMBO J.*, **18** 512-21.

Valent, B. (1997) The rice blast fungus, *Magnaporthe grisea*, in *The Mycota V Part B Plant Relationships* (eds. G.C. Carroll and P. Tudzynski), Springer-Verlag, Berlin, Heidelberg, pp. 37-54.

Valent, B., Crawford, M.S., Weaver, C.G. and Chumley, F.G. (1986) Genetic studies of fertility and pathogenicity in *Magnaporthe grisea. Iowa State J. Res.*, **60** 569-94.

Valent, B., Farrall, L. and Chumley, F.G. (1991) *Magnaporthe grisea* genes for pathogenicity and virulence identified through a series of backcrosses. *Genetics*, **127** 87-101.

Van den Ackerveken, G.F.J.M., Van Kan, J.A.L. and De Wit, P.J.G.M. (1992) Molecular analysis of the avirulence gene, *avr9*, fully supports the gene-for-gene hypothesis. *Plant J.*, **2** 359-66.

Vidal-Cros, A., Viviani, F., Labesse, G., Boccara, M. and Gaudry, M. (1994) Polyhydroxynaphthalene reductase involved in melanin biosynthesis in *Magnaporthe grisea. Eur. J. Biochem.*, **219** 985-92.

Walsh, D.A. and Van Patten, S.M. (1994) Multiple pathway signal transduction by the cAMP-dependent protein kinase *FASEB J.*, **8** 1227-36.

Wessels, J.G.H. (1997) Hydrophobins: proteins that change the nature of the fungal surface. *Adv. Microbiol. Physiol.*, **38** 1-45.

Wösten, H.A.B., van Wetter, M.A., Lugones, L.G., van der Mei, H.C., Busscher, H.J. and Wessels, J.G.H. (1999) How a fungus escapes the water to grow into the air. *Curr. Biol.*, **9** 85-88.

Xiao, J., Ohshima, A., Kamakura, T., Ishiyama, T. and Yamaguchi, I. (1994a) Extracellular glycoprotein(s) associated with cellular differentiation in *Magnaporthe grisea. Mol. Plant-Microbe Interact.*, **5** 639-44.

Xiao, J., Watanabe, T., Kamakura, T., Ohshima, A. and Yamaguchi, I. (1994b) Studies on cellular differentiation of *Magnaporthe grisea*: physicochemical aspects of substratum surfaces in relation to appressorium formation. *Physiol. Mol. Plant Pathol.*, **44** 227-36.

Xu, J.-R. and Hamer, J.E. (1996) MAP kinase and cAMP signalling regulate infection structure formation and pathogenic growth in the rice blast fungus, *Magnaporthe grisea. Genes Dev.*, **10** 2696-706.

Xu, J.-R., Urban, M., Sweigard, J.A. and Hamer, J.E. (1997) The *CPKA* gene of *Magnaporthe grisea* is essential for appressorial penetration. *Mol. Plant-Microbe Interact.*, **10** 187-94.

Xu, J.-R., Staiger, C.J. and Hamer, J.E. (1998) Inactivation of the mitogen-activated protein kinase, Mps1, from the rice blast fungus prevents penetration of host cells but allows activation of plant defense responses. *Proc. Natl. Acad. Sci. USA*, **95** 12713-18.

Yancey, P.H., Clark, M.E., Hand, S.C., Bowlus, R.D. and Somero, G.N. (1982) Living with water stress: evolution of osmolyte systems. *Science*, **217** 1214-22.

Yoder, O.C. and Turgeon, B.G. (1996) Molecular-genetic evaluation of fungal molecules for roles in pathogenesis to plants. *J. Genet.*, **75** 425-40.

Zeigler, R.S. (1998) Recombination in *Magnaporthe grisea. Annu. Rev. Phytopathol.*, **36** 249-75.

Zheng, L., Campbell, M., Lam, S. and Xu, J.R. (1999) The *bmp1* kinase gene is essential for pathogenicity in *Botrytis cinerea. Fungal Genet. Newslett.*, **46** (Appendix) p. 72.

Zhu, H., Whitehead, D.S., Lee, Y.-H. and Dean, R.A. (1996) Genetic analysis of developmental mutants and rapid chromosome mapping of *APP1*, a gene required for appressorium formation in *Magnaporthe grisea. Mol. Plant-Microbe Interact.*, **9** 767-74.

Zhu, H., Choi, S.D., Johnston, A.K., Wing, R.A. and Dean, R.A. (1997) A large-insert (130 kbp) bacterial artificial chromosome library of the rice blast fungus, *Magnaporthe grisea*: genome analysis, contig assembly and gene cloning. *Fungal Genet. Biol.*, **21** 337-47.

2 Bacterial pathogenicity

Elisabeth Huguet

2.1 Introduction

Bacterial diseases of plants are most frequent and severe in moist and warm climates. This is due, in part, to the mode of dissemination of bacteria, which primarily involves water. The spattering effect of rain, for example, enables the transmission of the pathogens from the soil, a possible source of inoculum, to plants, and from one plant to another. Bacteria penetrate inside the plant through natural openings, such as stomata or wounds, and typically multiply in the intercellular space or in the xylem (apoplast). Bacterial phytopathogens cause a wide variety of disease symptoms on many different host plants (Agrios, 1988). The best-studied bacterial phytopathogens are from the Gram-negative genera, *Agrobacterium, Erwinia, Pseudomonas, Ralstonia* and *Xanthomonas*. Some species have been further subdivided into pathovars (pv.) to distinguish strains differing in the plant species they infect, i.e. *Pseudomonas syringae* pv. *phaseolicola* causes halo blight of bean, whilst *P. syringae* pv. *tomato* is the causal agent of bacterial speck of tomato. This review focuses on the mechanisms employed by various species of Gram-negative bacteria to become successful pathogens. The bacteria are all faced with the same challenges: to suppress or avoid the host plants' defences; and to modify the local plant environment to satisfy their nutritional needs. However, the strategies employed to reach this goal differ enormously. The Gram-positive phytopathogens, *Clavibacter* spp. and *Streptomyces* spp., are not discussed in this chapter.

An overview of the different aspects of pathogenicity of Gram-negative phytopathogenic bacteria is presented before focusing on the strategy adopted by the biotrophic necrogenic bacteria. Research on these latter pathogens has benefited from the finding that the mechanisms of infection are similar to those of many mammalian bacterial pathogens.

2.2 To become a successful phytopathogen

2.2.1 Agrobacterium and plant transformation—a world apart

2.2.1.1 Disease symptoms

The soil bacterium, *Agrobacterium tumefaciens*, is the causal agent of crown gall, which affects many woody and herbaceous dicotyledonous plants. The disease is characterised by the formation of tumours, often just below the soil surface, at the 'crown' of the plant. Plants with tumours grow poorly and have reduced yields. *Agrobacterium rhizogenes*, the hairy root agent, also causes hyperplasias on host plants.

2.2.1.2 Overview of the infection process—a unique strategy

The mode of infection of *Agrobacterium* holds a unique position in plant-bacterial interactions, as it involves inter-kingdom transformation. This section focuses on the infection processes of *Agrobacterium tumefaciens* that harbour the tumour-inducing (Ti) plasmid. Following attachment of the bacteria to the plant cells, and sensing of plant signals at the sites of infection, part of the Ti plasmid, the transferred DNA (T-DNA) is transported into the plant cell and into the nucleus. Once inside the nucleus, the T-DNA integrates into the plant genome.

The T-DNA harbours two sets of genes that are expressed in the transformed plant cell. The first set encodes enzymes involved in the synthesis of plant growth hormones, the expression of which leads to unconstrained cell growth and division to produce the tumour. The second set of genes encodes enzymes that synthesise opines, which are a major source of carbon and nitrogen for *Agrobacterium*. Agrobacteria have been classified according to the type of opines specified by the T-DNA. To utilise opines as a nutrient source, *Agrobacterium* requires specific enzymes; and because these enzymes are encoded in the Ti plasmid, practically no other microorganism can metabolise opines, creating a favourable biological niche specific for *Agrobacterium*.

2.2.1.3 Genes and mechanisms involved in the voyage of the T-DNA

Processing and transfer of the T-DNA are mediated by proteins encoded by the *vir* genes, organised in seven major loci on the Ti-plasmid. *A. tumefaciens vir* mutants are non-pathogenic. The T-DNA is delineated by two imperfect direct repeats, known as the T-DNA borders, which delimit the transferred segment. It has been proposed that T-DNA processing and movement into the plant cell occur via mechanisms that are reminiscent of bacterial conjugation. Besides functional analogies, there are sequence similarities between the Vir proteins and proteins involved in transfer of protein-DNA complexes from a donor to a recipient cell, described as type IV secretion, which is different from the bacterial type I, II and III secretion/transfer systems (see sections 2.2.2.1 and 2.2.4.5) (Salmond, 1994). Type IV systems may use common mechanisms to transfer substrates of varying nature (Christie, 1997). For example, in *Bordetella pertussis*, the operon required for secretion of pertussis toxin encodes nine proteins that show homology to those of plasmid conjugation systems (Weiss *et al.*, 1993).

The *vir* genes are tightly regulated; they are transcriptionally activated by the VirA/VirG two-component regulatory system, in response to plant signal molecules, such as specific phenolic compounds and sugars that can be released by wounded plant cells (Winans *et al.*, 1994). *vir*-gene-induced *Agrobacterium* cells generate a linear single-stranded (ss) copy of the T-DNA region (T-strand) (Stachel *et al.*, 1986). The T-strand is thought to be transferred out of the bacterium and into the plant cell as a protein-nucleic acid complex (Howard and Citovsky, 1990), which is composed of one molecule of VirD2, an endonuclease, co-valently attached to the T-strand (Herrera-Estrella *et al.*, 1988), and several hundred molecules of VirE2, a ssDNA binding molecule, that coat the DNA (Citovsky *et al.*, 1988, 1989). The transfer of the T-complex across the two bacterial membranes and the plant cytoplasmic membrane requires ten products of the *virB* operon and *virD4* (Christie, 1997). Cell fractionation and immunoelectron microscopy have localised most *virB* products and VirD4 in the bacterial membranes, with many proteins distributed equally between inner and outer membranes in accord with a transmembrane structure held together by protein-protein interactions (Christie, 1997). Interestingly, Fullner *et al.* (1996) observed by electron microscopy that *Agrobacterium* produced pili of 3.8 nm in diameter. Pili assembly required *virA/virG* as well as *virB* genes and *virD4*. Candidate proteins to be structural components of the pilus are a processed form of VirB1, VirB1*, which has been shown to be secreted (Baron *et al.*, 1997), and VirB2 and VirB5 because of sequence homology with the pilin of the *Escherichia coli* F plasmid and a putative pilin subunit encoded by the IncN plasmid, pKM101 (Shirasu *et al.*, 1994; Fullner *et al.*, 1996). In fact, recent evidence shows that a 7.2-kDa processed form of VirB2 is the major subunit of the pilus (Lai and Kado, 1998). In the natural infection process, the pili may contact plant cells and mediate formation of a mating bridge by other *virB* products.

The nuclear import of the T-strand is probably mediated by its associated proteins, VirD2 and VirE2. VirD2 contains a functional nuclear localisation signal (NLS), which conforms to the bipartite consensus motif (Howard *et al.*, 1992). Its biological relevance has been shown by the observation that *Agrobacterium* tumorigenicity is reduced in NLS deletion mutants of VirD2 (Shurvinton *et al.*, 1992). However, VirD2 may not be the sole mediator of T-complex nuclear uptake, as VirE2 also harbours two functional NLS sequences that are both required for nuclear localisation (Citovsky *et al.*, 1992). It is a matter of debate as to whether VirE2 function is limited to the protection of the T-strand or not (Sheng and Citovsky, 1996).

The final step of the T-DNA transfer process is the integration of foreign DNA into the host chromosome. The mechanism is still not fully understood but the integration is believed to occur via illegitimate recombination (Gheysen *et al.*, 1991). Since T-DNA does not encode enzymatic activities enabling integration, insertion into the plant genome must be mediated by proteins imported from the infecting bacterium and/or by plant cell factors (Sheng and Citovsky, 1996).

To summarise, *Agrobacterium* creates a very intimate relationship with host plant cells. The bacterium incites the plant to synthesise specific nutrients that enable it to multiply. *Agrobacterium* is now widely used to produce transgenic plants expressing genes of interest, since the coding region of the wild-type T-DNA, between the T-DNA borders, can be replaced by any sequence without affecting its transfer to the plant (Sheng and Citovsky, 1996).

2.2.2 Factors involved in aggressiveness

The capacity of a microbe to cause disease, whatever the intensity of the symptoms observed, is defined as pathogenicity. Virulence or aggressiveness, on the other hand, determines the severity of disease symptoms. Bacterial strains with mutations in genes that are directly or indirectly involved in the synthesis of virulence factors develop attenuated and/or delayed symptoms. Cell-wall-degrading enzymes, exopolysaccharides and toxins have been shown to be involved in the aggressiveness of bacterial pathogens.

2.2.2.1 Extracellular plant cell-wall-degrading enzymes

Necrotrophic bacterial pathogens, such as *Erwinia chrysanthemi* and *Erwinia carotovora*, attack living plant tissues in the field or in storage and cause macerating diseases (soft rots). The bacteria multiply in the intercellular spaces, where they produce large quantities of extracellular plant cell-wall-degrading enzymes, which results in disorganisation of the plant cell wall and softening of tissues. Although the spectacular symptoms suggest that these pathogens have adopted a 'brute-force' strategy, in fact the whole infection process is dependent on environmental conditions, with the synthesis and secretion of the different enzymes subject to refined regulation.

To start with, despite the fact that *E. chrysanthemi* and *E. carotovora* induce similar symptoms, they do not secrete exactly the same cocktail of enzymes. *E. chrysanthemi*, for example, synthesises pectinases (pectin methylesterases, pectate lyases, pectine lyase, polygalacturonase), cellulases, proteases and a phospholipase (Collmer and Keen, 1986). Among all these enzymes, the pectinases and, in particular, the pectate lyases (Pels) have a predominant role in the soft rot symptoms (Collmer and Keen, 1986). In *E. chrysanthemi*, five major isoenzymes, PelA–PelE, have been identified (Bertheau *et al.*, 1984; Ried and Collmer, 1986). Mutations in individual *pel* genes have indicated that none of them are essential for maceration (Ried and Collmer, 1988; Beaulieu *et al.*, 1993). Unexpectedly, a deletion of all *pel* genes failed to eliminate tissue maceration, which enabled the identification of a second set of Pels, thought to be synthesised only in the plant (Ried and Collmer, 1988; Beaulieu *et al.*, 1993). Interestingly, Pel isoenzymes differ in their relative contribution to maceration and systemic invasion of plants (Barras *et al.*, 1987; Boccara *et al.*, 1988), and, furthermore, the Pels vary in importance in different hosts and may contribute to the wide host range of *E. chrysanthemi* (Beaulieu *et al.*, 1993).

The depolymerisation of the plant cell wall and catabolism of the derived residues is one means by which these bacteria obtain nutrients. In *E. chrysanthemi*, certain pectate degradation products act by inducing transcription of genes involved in pectinolysis (Nasser *et al.*, 1991), including genes encoding the Out secretion machinery (Condemine *et al.*, 1992). The Out system is a type II secretion system, through which certain enzymes, for example the pectinases (Pels, pectin methylesterases, polygalacturonases) and cellulases, are secreted (Andro *et al.*, 1984; Chatterjee *et al.*, 1985; Thurn and Chatterjee, 1985; Ji *et al.*, 1987). Proteins that are secreted by type II secretion systems harbour a cleavable N-terminal signal peptide and secretion is dependent on the *sec* gene products (Pugsley, 1993).

In addition, the pectinase genes are regulated by various environmental conditions, such as plant extracts, temperature, anaerobiosis, iron limitation, osmolarity and nitrogen starvation, which are mediated by several regulatory circuits (Barras *et al.*, 1994; Hugouvieux-Cotte-Pattat *et al.*, 1996). Interestingly, distinct *pel* genes respond differently to the same signals and are differentially regulated depending on the host plant, illustrating once more the finely-tuned regulation behind soft rot diseases. Moreover, environmental stimuli also affect the expression of cellulases and proteases, which permits the coordinate synthesis of all plant cell-wall-degrading enzymes. Proteases are secreted via a different route to that of other enzymes: by a type I secretion system, exemplified by the *E. coli* haemolysin secretion system. Here, secreted proteins carry a C-terminal secretion motif and are thought to be secreted into the extracellular milieu in one step (Wandersman, 1996).

The finding that pectic oligomers with a high degree of polymerisation activate certain plant defence mechanisms (Ryan and Farmer, 1991), whilst short pectic oligomers preferentially induce the pathogen's pectinases, highlights the high-wire on which *Erwinia* walks. The outcome of the interaction, disease or resistance, will most likely be influenced by the oligomers surrounding the bacteria, and therefore also by the catabolic efficiency of the enzymes, which are tightly regulated by environmental conditions.

2.2.2.2 Toxins

Toxins can be virulence determinants for phytopathogenic bacteria, and this has been particularly well studied among the pathovars of *Pseudomonas syringae* (Gross, 1991). Whereas some fungal plant pathogens synthesise host-specific toxins that are essential for pathogenicity (Walton, 1996), bacterial toxins show no host specificity, are typically not required for bacterial growth and contribute by enhancing disease symptoms.

Different types of toxins, usually peptides, are produced by *P. syringae* pathovars. The toxins, syringomycin, tagetitoxin and phaseolotoxin, are produced exclusively by *P. s.* pv. *syringae, P. s.* pv. *tagetis* and *P. s.* pv. *phaseolicola*, respectively, whereas coronatine and tabtoxin can be synthesised by several pathovars, including *P. s.* pv. *tomato* and *P. s.* pv. *tabaci*, respectively (Gross, 1991). The actual role and mode of action of these toxins is not completely understood. Syringomycin is thought to disrupt plant cellular signalling by formation of transmembrane pores that are permeable to cations (Hutchison *et al.*, 1995). Pores would benefit the bacteria by releasing cellular metabolites into the intercellular spaces of host tissues. Coronatine is the only bacterial toxin for which there is evidence that production leads to higher bacterial populations (Bender *et al.*, 1987). Coronatine may play a role during the early stages of infection by suppressing the activation of defence-related genes (Mittal and Davis, 1995). On the other hand, it has been suggested that it acts, at least in part, by mimicking methyl-jasmonate, which regulates wound-inducible defence responses in plants (Feys *et al.*, 1994; Weiler *et al.*, 1994; Palmer and Bender, 1995). The role of tabtoxin is unknown and, curiously, strains in the field spontaneously delete tabtoxin biosynthetic genes without loss of pathogenicity (Willis *et al.*, 1991). In general, many of these toxins display wide-spectrum antibiotic activity, which suggests that they also act by reducing microbial competition during the epiphytic life cycle and intracellular colonisation of the plant (Gross, 1991).

2.2.2.3 Exopolysaccharide production

Many bacteria, including phytopathogens, produce extracellular polysaccharides (EPSs), which surround the bacteria as a capsule or as fluidal slime. EPS has been shown in *Ralstonia solanacearum* (Kao *et al.*, 1992; Denny, 1995), *Erwinia amylovora* (Steinberger and Beer, 1988; Bernhard *et al.*, 1993) and *Erwinia stewartii* (Coplin and Majerczak, 1990) to contribute to wilt or water-soaking symptoms, without being absolutely required for pathogenicity. EPS may provide a selective advantage for bacteria by protecting them from stressful environmental conditions, such as desiccation. In pathogenicity, the presence of EPS is thought to reduce contact with toxic molecules, impede activation of plant defence signals and sustain production of disease symptoms (Denny, 1995).

2.2.3 Quorum-sensing and bacterial pathogenesis

Quorum-sensing is a communication mechanism by which bacteria regulate the expression of certain genes in response to their population density. The bacteria sense population size through the concentration of diffusible signal molecules, named autoinducers, that they produce themselves. Only at high cell densities, or in a confined environment, will the autoinducers reach a threshold concentration required for gene activation of products that are presumably beneficial to the bacteria in that particular habitat.

Quorum-sensing was first described in the marine symbiont, *Vibrio fischeri* (Nealson *et al.*, 1970), in which the autoinducer (an *N*-acyl homoserine lactone) regulates the expression of bioluminescent *lux* genes in a cell-density-dependent manner. LuxR, a trancriptional activator, requires a threshold concentration of *V. fischeri* autoinducer, a diffusible product of *luxI*, to activate transcription of the *lux* operon (Fuqua *et al.*, 1996). Autoinducers and LuxI/R homologues have since been described in many bacteria, including Gram-negative plant-associated bacteria (Gray, 1997; Cha *et al.*, 1998). It is thought that plant pathogens or symbiotic microorganisms form microcolonies within the host, where high bacterial densities can indeed be achieved (Fuqua *et al.*, 1996). In the best studied systems, the autoinducers are *N*-acylated derivatives of L-homoserine lactone. So far, five plant pathogens, *Erwinia carotovora* (Jones *et al.*, 1993; Pirhonen *et al.*, 1993), *E. stewartii* (Beck von Bodman and Farrand, 1995), *R. solanacearum* (Clough *et al.*, 1997), *Xanthomonas campestris* pv. *campestris* (Barber *et al.*, 1997) and *E. chrysanthemi* (Nasser *et al.*, 1998; Reverchon *et al.*, 1998) have been shown to regulate genes involved in pathogenicity by quorum-sensing mechanisms. Similar systems also regulate Ti-plasmid conjugal transfer in *A. tumefaciens* (Piper *et al.*, 1993).

In the case of a pectolytic attack by *Erwinia* spp., it is interesting to note that such 'quorum-sensing' control presumably withholds pathogen attack until the population density is large enough to overcome the host defences that it is likely to trigger (Jones *et al.*, 1993; Pirhonen *et al.*, 1993).

As more data have been acquired, it has become apparent that quorum-sensing regulated functions are extremely complex, in that they are not only limited to systems involving LuxI/R homologues, and that additional global regulatory mechanisms overlap and also influence responses (Gray, 1997). In *R. solanacearum*, for example, the acyl-homoserine lactone-dependent autoinduction system: 1) is regulated by a higher level autoinducer system that is responsive to 3-hydroxypalmitic acid methyl ester; and 2) requires a homologue of the alternative sigma factor, RpoS (Clough *et al.*, 1997; Flavier *et al.*, 1997, 1998). Recently, Reverchon *et al.* (1998) have begun to integrate the ExpI/ExpR quorum-sensing system of *E. chrysanthemi* into the multiple regulatory networks controlling extracellular enzyme production. In this system, several levels of control occur via cyclic AMP receptor protein (CRP), a global regulator of sugar metabolism, the PecS repressor, which regulates pectinase and cellulase production, and the RsmA protein, which suppresses synthesis of extracellular enzymes (Reverchon *et al.*, 1998).

The presence of related quorum-sensing systems among different bacteria opens up the possibility of interspecific cross-talk. Furthermore, bacterial autoinducers may also be perceived by the host plant. This is supported by the finding that human epithelial cells produce interleukin-8 in response to *Pseudomonas aeruginosa* autoinducer (DiMango *et al.*, 1995). Unravelling the role(s) autoinducers may play as a means of communication between bacterial pathogens and as signal molecules to the host plants is an exciting field for new research.

2.2.4 hrp *genes*

2.2.4.1 Isolation of genes involved in basic pathogenicity

A random mutagenesis approach has led to the isolation of genes involved in basic pathogenicity of Gram-negative, biotrophic, necrogenic bacteria. The isolated non-pathogenic mutants have been tested for their capacity to grow on minimal medium to ensure no basic housekeeping genes have been affected. The genes were designated *hrp* (hypersensitive reaction and pathogenicity) because of their pleiotropic mutant phenotype. These *hrp* mutants are not only unable to cause disease or multiply in a susceptible host plant (compatible interaction) but have also lost the capacity to induce the macroscopically visible defence response, the hypersensitive reaction (HR), on a resistant host or non-host plant (incompatible interaction). This rapid defence response involves localised plant cell death, often associated with production of active oxygen species, phytoalexin biosynthesis, cell wall reinforcement and release of inhibitory proteins (Dixon *et al.*, 1994; Greenberg *et al.*, 1994; Hammond-Kosack and Jones, 1996). The HR coincides with prevention of pathogen multiplication and colonisation, thus halting disease. The HR is macroscopically visible, as confluent necrosis, only when bacterial cells are infiltrated into the plant tissue at high inoculum densities (i.e. 10^6–10^8 colony-forming units (CFU)/ml). At lower inoculum levels, in natural infections for example, individual plant cells die in response to individual bacteria (Klement, 1982).

hrp genes have been isolated from *P. syringae* pathovars (Lindgren *et al.*, 1986, 1988; Huang *et al.*, 1988), *R. solanacearum* (formerly *Burkholderia*), and *Pseudomonas* (Boucher *et al.*, 1987), *X. campestris* pathovars (Arlat *et al.*, 1991; Bonas *et al.*, 1991), and from *E. amylovora* (Steinberger and Beer, 1988; Barny *et al.*, 1990; Walters *et al.*, 1990; Bauer and Beer, 1991). *hrp*-homologous genes have been isolated in *E. stewartii*, and designated *wts*; mutations in *wts* genes result in loss of the water-soaking (disease) phenotype (Frederick *et al.*, 1993). *E. chrysanthemi* also contains *hrp*-homologues that are not essential for maceration but may play a role in the early stages of pathogenesis (Bauer *et al.*, 1994). Interestingly, proteins showing homology to *hrp* products have also been identified in the symbiotic bacteria, *Rhizobium fredii* and species NGR234 (Meinhardt *et al.*, 1993; Freiberg *et al.*, 1997). In *R. fredii*, the genes concerned are involved in host specificity, since a mutation in these genes leads to host-range extension of strain USDA257 (Meinhardt *et al.*, 1993).

2.2.4.2 Host-specificity determinants in bacterial phytopathogens

Host plants are often able to recognise certain isolates of a given pathogen. Flor was first to describe, in the flax-flax rust fungus interaction, the 'gene-for-gene concept' that governs host specificity in plant-microbe interactions (Flor, 1956). His observation was that for each plant resistance gene (*R*) there is a corresponding avirulence (*Avr*) gene in the pathogen. Only if the two matching genes are present does resistance (for example, elicitation of the HR) occur. Flor's hypothesis has been extended to plant-bacterial, viral and nematode pathogenic interactions, and has been confirmed by the cloning of several *R* and *Avr* genes (De Wit, 1997).

More than 30 bacterial *Avr* genes have been cloned from *P. syringae* pathovars and *Xanthomonas* species (Dangl, 1994; Leach and White, 1996). Although *Avr* genes encode, *a priori*, for products that impede pathogen multiplication, understanding *Avr* gene function has helped to improve our knowledge of the mechanisms involved in pathogenicity. Firstly, *Avr* gene function is dependent on *hrp* gene function, since *hrp* mutants are unable to elicit the HR. For example, *X. campestris* pv. *vesicatoria* is the causal agent of bacterial spot disease on pepper and tomato plants. Strains carrying the *avrBs3* avirulence gene are specifically recognised by pepper plants harbouring the *Bs3* resistance gene (Bonas *et al.*, 1989), and *AvrBs3* function is dependent on *hrp* genes (Knoop *et al.*, 1991). Secondly, since *Avr* genes have been maintained in pathogen populations, they must present an advantage, and many research studies are now focusing on unravelling the biological role of *Avr* genes in pathogens (see sections 2.3.4.2 and 2.4.3.3).

2.2.4.3 hrp *gene organisation*

Group 1 and Group 2 hrp *gene organisation.* *hrp* genes have been extensively characterised in *P. s.* pv. *syringae*, *E. amylovora*, *R. solanacearum* and *X. c.* pv. *vesicatoria* (Alfano and Collmer, 1997). Most *hrp* genes in these bacteria are clustered in 25-kb regions on the chromosome; in *R. solanacearum* they are on a mega-plasmid. The *hrp* gene clusters can be grouped, based on sequence similarity, operon structures and regulatory systems, into Group 1 (*P. s.* pv. *syringae* and *E. amylovora*) and Group 2 (*R. solanacearum* and *X. c.* pv. *vesicatoria*) (see Figure 2.1) (Alfano and Collmer, 1997; He, 1997). Eleven *hrp* genes are conserved in all *hrp* clusters. A number of genes are characteristic of Group 1 or Group 2, and certain genes show no homology and appear to be unique to each system (see Figure 2.1). The organisation of operons is also characteristic of each group.

2.2.4.4 hrp *gene regulation*

Expression of *hrp* genes is tightly regulated and dependent on environmental conditions: most genes are suppressed in complex medium but are induced *in planta*, and in certain minimal media modulated by the carbon and nitrogen source (Huynh *et al.*, 1989; Arlat *et al.*, 1992; Rahme *et al.*, 1992; Schulte and Bonas, 1992; Wei *et al.*, 1992b; Xiao *et al.*, 1992). The *in vitro* conditions are thought to mimic physiological conditions encountered by the bacteria during infection. The regulatory circuits that activate the *hrp* genes are distinct in Group 1 and Group 2 *hrp* gene clusters.

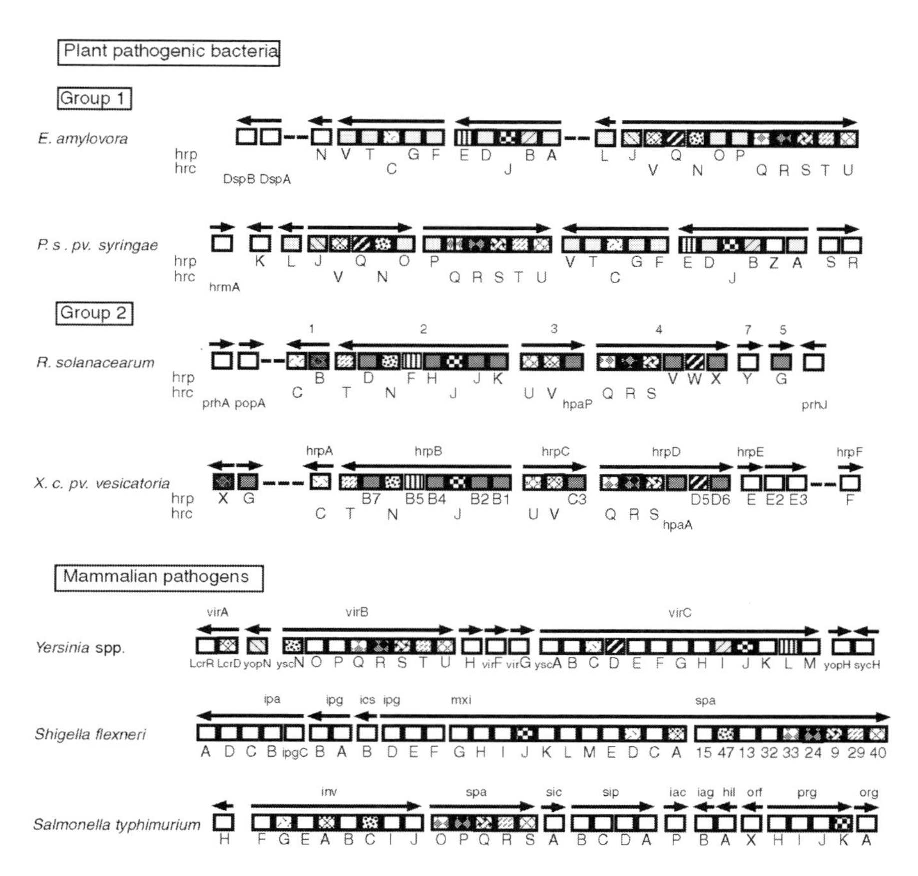

Figure 2.1 Genetic organization of genes encoding proteins common to plant pathogenic bacteria (*hrp* gene clusters) and to animal bacterial pathogens. Transcription units and relative positions of the genes are presented. Genes that encode related proteins are represented by the same pattern. HrpD5 and HrpB5 homologues (in the *Xanthomonas campestris* pv. *vesicatoria* nomenclature) display only weak similarity. For the plant pathogens, genes that are conserved among Group 1 (*Pseudomonas syringae* pv. *syringae* and *Erwinia amylovora*) pathogens are indicated in light grey. Genes that are conserved among Group 2 (*Ralstonia solanacearum* and *X. c.* pv. *vesicatoria*) pathogens are indicated in dark grey. Genes that are unique to each plant pathogen are indicated in white. Two genes from the Group 1 plant pathogens encode proteins that are related to *Yersinia* (*hrpB* and *hrpJ* of *E. amylovora*). Two dashed lines in between genes indicate they are within 4 kb, or indicate gaps in the reported sequence of the *hrp* gene cluster. Three dashed lines indicate the genes are elsewhere on the chromosome.

Group 1—P. syringae *and* E. amylovora. Expression of *P. syringae hrp* genes is environmentally regulated in response to carbon and nitrogen sources, pH and osmotic conditions (Lindgren *et al.*, 1989; Rahme *et al.*, 1991, 1992; Xiao *et al.*, 1992). In *E. amylovora*, carbon and nitrogen sources are important as well as temperature (Wei *et al.*, 1992b). The Hrp regulatory cascades of

P. syringae and *E. amylovora* appear to be similar. The present section focuses on *P. syringae*, since this system has been more extensively studied.

In *P. syringae*, *hrp* gene activation depends on HrpR and HrpS. These proteins are 60% identical to each other and belong to a class of σ54 enhancer-binding proteins. HrpR and HrpS are homologous to response regulators of two-component systems, like NifA and XylR (Grimm and Panopoulos, 1989; Rahme *et al.*, 1991; Xiao *et al.*, 1994), and have been shown to specifically activate the *hrpL* promoter. *hrpL* encodes an alternative sigma factor, which stimulates expression of *hrp*, *avr* genes and *hrmA* (Fellay *et al.*, 1991; Xiao *et al.*, 1994; Leach and White, 1996). HrpL is believed to recognise a conserved sequence motif in *hrp* and *avr* promoters, the '*hrp* box' (GGAACCNA) (Huynh *et al.*, 1989; Fellay *et al.*, 1991; Innes *et al.*, 1993; Salmeron and Staskawicz, 1993; Shen and Keen, 1993). In *E. amylovora*, *hrpL* also controls the expression of the *dsp* (disease-specific) region (Gaudriault *et al.*, 1997; Bogdanove *et al.*, 1998b). The mechanism by which HrpR, HrpS and HrpL perceive and transduce environmental signals remains to be elucidated.

Group 2—R. solanacearum *and* X. c. *pv.* vesicatoria. Two key regulators, named *hrpG*, have been identified in *X. c.* pv. *vesicatoria* and *R. solanacearum* (Wengelnik *et al.*, 1996b; Brito *et al.*, 1999). HrpG of *X. c.* pv. *vesicatoria* and *R. solanacearum* are 36% identical and show similarity to response-regulator proteins of bacterial two-component systems, in particular to the subgroup containing OmpR, PhoB of *Escherichia coli* and VirG of *A. tumefaciens*, but not HrpR and HrpS of *P. syringae* (Wengelnik *et al.*, 1996b; Brito *et al.*, 1999). *hrpG* controls the expression of *hrpXv* and *hrpA* in *X. c.* pv. *vesicatoria* (Wengelnik *et al.*, 1996b) and the expression of *hrpB* in *R. solanacearum* (Brito *et al.*, 1999).

Most *hrp* genes of *X. c.* pv. *vesicatoria* and *R. solanacearum* are regulated by HrpXv and HrpB, respectively. HrpXv and HrpB are 40% identical, functional homologues and are members of the AraC- and XylS-family of regulatory proteins (Genin *et al.*, 1992; Wengelnik and Bonas, 1996), which also includes the VirF proteins of the bacterial mammalian pathogens, *Yersinia enterocolitica* and *Shigella flexneri* (Sakai *et al.*, 1988; Adler *et al.*, 1989; Cornelis *et al.*, 1989). In *R. solanacearum, hrpB* positively regulates four out of five *hrp* transcription units (units 1, 2, 3 and 4) as well as genes situated in a 16-kb region to the left of the *hrp* gene cluster, including *popA* (Genin *et al.*, 1992; Arlat *et al.*, 1994; Marenda *et al.*, 1998). In *X. c.* pv. *vesicatoria*, *hrpXv* regulates five out of the six *hrp* operons (*hrpB, hrpC, hrpD, hrpE* and *hrpF*) (Wengelnik and Bonas, 1996). Inspection of upstream sequences of *hrpXv*- and *hrpB*-regulated loci in *X. c.* pv. *vesicatoria* and *R. solanacearum* revealed a conserved motif, designated the PIP-box (TTCGC-N15-TTCGC), which might be involved in HrpXv-, HrpB-mediated activation (Fenselau and Bonas, 1995; Wengelnik and Bonas, 1996).

In contrast to *P. syringae*, *E. amylovora* and *R. solanacearum*, the *hrp* regulatory genes of *X. c.* pv. *vesicatoria* are not located in or adjacent to the large gene cluster (Wengelnik *et al.*, 1996b).

2.2.4.5 The Hrp type III secretion system

Nine Hrp products are homologous to components of type III virulence protein secretion systems described in Gram-negative bacterial pathogens of mammals, such as *Yersinia* (Ysc/Lcr proteins), *Shigella* (Inv/Spa), *Salmonella* (Inv/Spa) spp. and enteropathogenic *E. coli* (Sep) (Menard *et al.*, 1996; Collazo and Galan, 1997a; Cornelis and Wolf-Watz, 1997; Donnenberg *et al.*, 1997). Type III secretion has also been shown to be involved in flagella biosynthesis in *Salmonella*, *E. coli* and in the Gram-positive bacterium, *Bacillus subtilis* (Macnab, 1996) (see Figure 2.1 and Table 2.1).

Table 2.1 Homologous proteins involved in type III secretion in plant and mammalian pathogens and in flagella biosynthesis

Plant pathogens	*Yersinia*	*Shigella*	*Salmonella*	Flagella proteins
HrcC	YscC	MxiD	InvG	
HrcJ	YscJ	MxiJ	PrgK	FliF
HrcN	YscN	Spa47	InvC	FliL
HrcQ	YscQ	Spa33	SpaO	FliN (FliY)
HrcR	YscR	Spa24	SpaP	FliP
HrcS	YscS	Spa9	SpaQ	FliQ
HrcT	YscT	Spa29	SpaR	FliR
HrcU	YscU	Spa40	SpaS	FlhB
HrcV	YscV	MxiA	InvA	FlhA
HrpB5*	YscL			
HrpD5*	YscD			

*: *Xanthomonas campestris* pv. *vesicatoria* nomenclature.

These exciting homologies suggested that *hrp* genes would also be involved in type III protein secretion. Subsequently, proteins secreted by the Hrp type III pathway were identified: harpins, PopA, HrpA, DspA and HrpW (Wei *et al.*, 1992a; He *et al.*, 1993; Arlat *et al.*, 1994; Bogdanove *et al.*, 1996b, 1998a; Gaudriault *et al.*, 1997, 1998; Roine *et al.*, 1997). The nine conserved *hrp* genes have been renamed *hrc* (*hrp*-conserved), followed by a capital letter of the corresponding *Yersinia ysc* (*Yersinia* secretion) homologue (Bogdanove *et al.*, 1996a).

The next two sections focus on the features that animal and plant pathogens have in common regarding type III secretion, but also on the specificities, in part imposed by the plant environment, of the Hrp pathway and its substrates.

2.3 Features of the Hrp type III pathway that are conserved with type III secretion systems in animal pathogens

The Hrp type III pathway shares characteristics with type III secretion systems of mammalian pathogens: both from a structural point of view, i.e. certain proteins (the Hrc proteins) that are part of the secretion apparatus are conserved; and also from a functional point of view, the features of secreted proteins and their final destination in the eukaryotic host.

2.3.1 What is known about the role of the Hrc proteins in the Hrp secretion apparatus?

The Hrc proteins are predicted, and in some cases have been demonstrated, to be essential structural components of the Hrp pathway. In animal pathogens, non-polar mutagenesis has shown which genes are required for secretion through the type III system in *Yersinia* spp. and *S. flexneri*, and for the type III-dependent invasion phenotype in *Salmonella* (Allaoui *et al.*, 1992, 1993, 1994, 1995a and b; Sasakawa *et al.*, 1993; Bergman *et al.*, 1994; Kaniga *et al.*, 1994; Woestyn *et al.*, 1994; Plano and Straley, 1995; Galan, 1996; Koster *et al.*, 1997). An overview of what is known, or predicted, for the function of the Hrc proteins is presented here, based on results obtained in plant and animal pathogens.

Five proteins, HrcR, HrcS, HrcT, HrcU and HrcV, are predicted to be inner membrane proteins, since they harbour potential transmembrane domains and carry no signal peptide (Fenselau *et al.*, 1992; Huang *et al.*, 1995; Van Gijsegem *et al.*, 1995; Bogdanove *et al.*, 1996b; Huguet *et al.*, 1998; O. Rossier and U. Bonas, unpublished). In *P. s.* pv. *syringae*, the secreted harpin protein was found to accumulate in the cytoplasm of non-polar *hrcU* mutants, and not in the periplasmic fraction or the extracellular milieu. This suggests that *hrcU* is involved in secretion of proteins from the cytoplasm to the periplasm (Charkowski *et al.*, 1997). In *Yersinia*, YscU-, LcrD- (HrcV homologue) and YscR-phoA fusions have enabled prediction of the topology of these proteins in the inner membrane (Plano *et al.*, 1991; Allaoui *et al.*, 1994; Fields *et al.*, 1994).

HrcQ proteins are predicted to be hydrophilic (Huang *et al.*, 1995; Van Gijsegem *et al.*, 1995; Bogdanove *et al.*, 1996b). The *Salmonella* spp. homologue, SpaO, is secreted via the Inv/Spa type III secretion system (Li *et al.*, 1995; Collazo and Galan, 1996). A mutation in *spaO* prevents the export of other proteins secreted by this pathway (Collazo and Galan, 1996). In the plant pathogens, it is not known whether HrcQ is secreted and whether this protein is required for the secretion of others.

HrcC contains an N-terminal signal peptide and has been localised in *X. c.* pv. *vesicatoria*, to the bacterial outer membrane, where it is proposed to form a multimer (Wengelnik *et al.*, 1996a). *P. syringae* and *R. solanacearum* require

HrcC for secretion of harpin and PopA, respectively (He *et al.*, 1993; Arlat *et al.*, 1994). Furthermore, in *P. s.* pv. *syringae*, a non-polar *hrcC* mutant accumulates harpin in both the cytoplasm and the periplasm, indicating that *hrcC* is involved in secretion across the outer membrane (Charkowski *et al.*, 1997).

HrcC has a unique status—not only does it show similarity with components involved in type III secretion systems but it is also related to components of other export pathways, such as the type II pathway, release and assembly of filamentous phages and DNA uptake (Genin and Boucher, 1994; Russel, 1994). Members of this family of proteins, now designated secretins, display sequence homology in their C-terminal halves, but the N-terminal regions are only conserved between proteins of related secretion systems (Martin *et al.*, 1993a; Genin and Boucher, 1994). Secretins of the type II pathway, such as PulD (*Klebsiella oxytoca*) and OutD (*E. chrysanthemi*), as well as the pIV filamentous phage protein, have all been localised in the bacterial outer membrane, where they form large multimers (Kazmierczak *et al.*, 1994; Hardie *et al.*, 1996a; Shevchik *et al.*, 1997). The prediction that secretins form large channels in the outer membrane has recently been supported by the purification of the YscC complex of *Yersinia enterocolitica* (Koster *et al.*, 1997). Electron microscopy has revealed that purified YscC complexes form ring-shaped structures of 20 nm, with an apparent central pore (Koster *et al.*, 1997). It is therefore proposed that all secretins, including the HrcC proteins of phytopathogenic bacteria, also form large channels in the outer membrane, which enable the secretion of proteins to the extracellular milieu.

HrcJ homologues in plant pathogenic bacteria and their counterparts in animal pathogens are predicted to be lipoproteins, with a possible outer membrane localisation (Allaoui *et al.*, 1992; Fenselau *et al.*, 1992; Huang *et al.*, 1995; Van Gijsegem *et al.*, 1995). For YscJ in *Y. enterocolitica* (Michiels *et al.*, 1991) and an MxiJ-PhoA fusion in *S. flexneri* (Allaoui *et al.*, 1992), fatty acylation of the proteins has been demonstrated using (^{3}H)-palmitate. It is not clear what role HrcJ plays in the secretion apparatus. In type II secretion, PulD requires a small lipoprotein, PulS, for protection against proteolytic degradation and for targeting to the outer membrane (Hardie *et al.*, 1996b). In *Y. enterocolitica* a similar role might be played by VirG, a 14-kDa lipoprotein predicted to be in the outer membrane (Allaoui *et al.*, 1995b; Koster *et al.*, 1997). YscJ, MxiJ and HrcJ also show similarity in their N-terminal region to a protein involved in flagellum biogenesis, FliF (Huang *et al.*, 1995; Van Gijsegem *et al.*, 1995). In *Salmonella typhimurium*, FliF is an inner membrane protein that forms the membrane and supramembrane (MS)-ring. Careful localisation studies may provide clues to the function of HrcJ homologues.

The HrcN proteins are members of a family of proteins with homology to the β-subunit of the F_0F_1 adenosine triphosphatase (ATPase). All homologues contain the consensus nucleotide-binding motifs (boxes A and B) described by Walker and co-workers (1982) (Fenselau *et al.*, 1992; Huang *et al.*, 1995;

Van Gijsegem *et al.*, 1995; Bogdanove *et al.*, 1996b). The ability to hydrolyse ATP has been demonstrated for the *S. typhimurium* InvC (Eichelberg *et al.*, 1994). HrcN proteins are therefore postulated to be ATPases that energise the Hrp type III pathway, either by helping in the assembly of the secretion apparatus, or in the actual transport of proteins, or both.

It has recently emerged that the type III apparatus may exist as a supramolecular structure that spans the inner and outer membranes of bacteria. Supramolecular 'needle-shaped' structures, isolated from *S. typhimurium*, were shown to contain at least three type III secretion components: InvG (secretin family, HrcC homologue); PrgK (lipoprotein, HrcJ homologue); and PrgH (also a lipoprotein) (Kubori *et al.*, 1998). Although data are accumulating on the localisation and function of individual components of type III secretion systems, little is known about how the proteins interact and how they are organised in the bacterial envelope. Cross-linking experiments will be useful to address these points. In addition, it is necessary to construct non-polar mutations in all *hrc* and *hrp* genes in order to evaluate their role in secretion.

2.3.2 The Hrp secretion system—involved in translocation?

Features of type III secretion systems in animal pathogens are that: 1) secreted proteins lack a classical, cleavable signal peptide at their N-terminus (the pathway is *sec*-independent); 2) secretion is induced after contact with the host cell; and 3) certain secreted proteins, the 'effector proteins', are not only secreted but translocated (transported across the eukaryotic cell plasma membrane) into the eukaryotic host cell (Cornelis and Wolf-Watz, 1997). The Hrp pathway functions as a protein secretion system, and recent indirect evidence suggests that it is also capable of translocating proteins into the plant cell (Gopalan *et al.*, 1996; for a review, see Bonas and Van den Ackerveken, 1997).

2.3.3 Protein secretion via the type III pathway

2.3.3.1 Characteristics of secreted proteins in animal pathogens

Yersinia spp. are pathogenic to rodents and humans: *Y. pestis* is the causal agent of bubonic plague; *Y. pseudotuberculosis* causes adenitis and septicaemia; and *Y. enterocolitica* causes gastrointestinal diseases of variable severity. All three species resist the non-specific immune response, and in particular they are resistant to phagocytosis by macrophages and polymorphonuclear leukocytes. This capacity depends upon the plasmid-encoded Yop virulon, composed of the Ysc type III secretion system, and 12 (so far) secreted proteins, designated Yops (*Yersinia* outer membrane proteins). In *S. flexneri*, the causal agent of bacillary dysentery in humans, Ipa (invasion plasmid antigens) proteins are secreted via the type III pathway encoded by the *mxi* and *spa* genes. The Ipas are essential virulence determinants required for bacterial entry into epithelial cells. Yops and Ipas do not harbour an N-terminal signal peptide and secretion has been shown to be independent of the general export pathway (Cornelis, 1994; Parsot, 1994).

In *Yersinia*, secretion of Yops is controlled both by positive and negative regulatory elements. The first level of regulation is temperature: expression of genes involved in Yop synthesis and secretion only occurs at 37°C. Thermoregulation is coordinated by a transcriptional activator, VirF. A second level of regulation is mediated by Ca^{2+}, which *in vitro* represses *yop* operons. Therefore, at 37°C in the absence of Ca^{2+}, *Yersinia* secrete Yops (Cornelis and Wolf-Watz, 1997). In *Shigella*, the Ipa virulence proteins are present as a pool in the bacterial cytoplasm, and Ipa secretion can be triggered and enhanced by the addition of foetal bovine serum or Congo red to the medium (Menard *et al.*, 1994; Parsot *et al.*, 1995).

2.3.3.2 Proteins secreted via the Hrp pathway

To identify proteins secreted through the Hrp pathway, laboratories initially analysed bacterial culture supernatants for proteins that were 'active' on plants, and dependent on *hrp* genes for their secretion in the extracellular milieu. These experiments led to the isolation of proteins designated harpins in *E. amylovora* and *P. s.* pv. *syringae*, and PopA in *R. solanacearum* (Wei *et al.*, 1992a; He *et al.*, 1993; Arlat *et al.*, 1994). These secreted proteins share no sequence homology but have characteristics in common: 1) they are glycine-rich proteins that lack cysteine; 2) they are heat-stable; and 3) they possess HR-elicitor activity when infiltrated into the leaves of tobacco and several other plants. Since then, other proteins have been identified in culture supernatants of bacteria grown in *hrp*-inducing medium, such as the HrpA protein of *P. s.* pv. *tomato* (Roine *et al.*, 1997), HrpW of *E. amylovora* (Gaudriault *et al.*, 1998) and DspA (also called DspE; Bogdanove *et al.*, 1998a), an essential pathogenicity factor of *E. amylovora* (Gaudriault *et al.*, 1997; Bogdanove *et al.*, 1998a). None of the proteins secreted by the Hrp pathway contains a characteristic signal peptide. Interestingly, the genes that encode the secreted proteins are located either in the *hrp* gene clusters or in regions flanking the clusters (see Figure 2.1). The harpins of *E. amylovora* and *P. s.* pv. *syringae* are encoded by *hrpN* and *hrpZ*, respectively. Properties, and putative functions of these proteins are discussed in section 2.4.2.

Avirulence proteins were also thought to be good candidates to be secreted via the Hrp pathway, since their function depends on *hrp* genes (Dangl, 1994; Leach and White, 1996). For *avr* genes of *P. syringae* pathovars one explanation for this dependency could be that their expression is dependent on Hrp regulatory factors (Leach and White, 1996). However, expression of *P. syringae avr* genes from constitutive promoters still required a functional *hrp* system to induce the HR, and in *Xanthomonas*, *avr* genes appear to be constitutively expressed (Knoop *et al.*, 1991; U. Bonas, unpublished).

A particularly interesting example is the *avrBs3* gene of *X. c.* pv. *vesicatoria*, which encodes a protein of 122 kDa (Bonas *et al.*, 1989). *avrBs3* was the first identified member of a large gene family widely distributed in *Xanthomonas* pathovars, which includes *avrXa10* of *X. oryzae* pv. *oryzae* and *pthA* of *X. citri* (Leach and White, 1996). A striking feature of the protein is the presence of 17.5 copies of nearly identical 34 amino acid repeats in the internal portion of the protein (Bonas *et al.*, 1989). The repetitive region of *avrBs3* determines race specificity, in the sense that deletion of repeat units generates new avirulence specificities, and unmasks undiscovered resistance genes in pepper and tomato (Herbers *et al.*, 1992). For example, *X. c.* pv. *vesicatoria* strains harbouring *avrBs3*Δ*rep16* (which lacks four repeats) are no longer recognised by pepper genotype *Bs3* but by pepper genotype *bs3* (Herbers *et al.*, 1992). The fact that changes in the repeat region can give rise to new specificities, strongly suggests that the AvrBs3 protein is the elicitor that is secreted and recognised by the plant. Despite this hypothesis, Hrp-dependent secretion of AvrBs3 has not been observed (Knoop *et al.*, 1991; Young *et al.*, 1994; Brown *et al.*, 1995) until very recently (O. Rossier, K. Hahn and U. Bonas, unpublished). Furthermore, infiltration of purified Avr proteins (e.g. AvrBs3 of *Xanthomonas* or AvrB of *P. syringae*) into the plant intercellular space did not induce the HR (Gopalan *et al.*, 1996; Van den Ackerveken *et al.*, 1996). This paradoxical situation was clarified by the finding that in animal pathogens certain secreted proteins are, in fact, translocated into the eukaryotic host cell.

2.3.4 Type III pathways and translocation of proteins into the host cell

2.3.4.1 Type III pathways in animal pathogens deliver effector proteins into the host cell

The first formal demonstration of type III-dependent translocation of proteins was obtained in *Yersinia* spp. (Cornelis and Wolf-Watz, 1997). It was shown that extracellular adherent *Yersinia* induce a cytotoxic effect on Helen Lake (HeLa) cells, which is due, in part, to the action of YopE (Forsberg and Wolf-Watz, 1988; Rosqvist *et al.*, 1990). However, YopE had no cytotoxic effect unless it was microinjected into HeLa cells, indicating that the target of this bacterial protein is inside the eukaryotic host cell (Rosqvist *et al.*, 1991).

A first approach to demonstrating YopE translocation consisted of analysing different fractions of monolayers of HeLa cells infected with *Yersinia*. YopE could only be detected in the cytoplasmic fraction of the infected eukaryotic cells and not in the tissue culture medium (Rosqvist *et al.*, 1994). Furthermore, immunofluorescence and confocal laser-scanning microscopy showed that YopE was present in the perinuclear region of *Yersinia*-infected HeLa cells (Rosqvist *et al.*, 1994). A second approach was based on a reporter enzyme strategy (Sory and Cornelis, 1994), where YopE was fused to the calmodulin-dependent adenylate cyclase of *Bordetella pertussis*. Because bacteria do not produce calmodulin, adenylate cyclase activity, i.e. accumulation of cyclic adenosine monophosphate (cAMP), would indicate that the fusion protein is in the cytoplasmic compartment of *Yersinia*-infected HeLa cells. Infection of monolayers of HeLa cells by the recombinant *Yersinia*, expressing the YopE-cya fusion, did indeed result in a significant increase in cAMP (Sory and Cornelis, 1994).

Other Yops have since been shown to be translocated into the host cell, such as: YopH, a protein tyrosine phosphatase (Persson *et al.*, 1995; Sory *et al.*, 1995); YopM (Boland *et al.*, 1996); YpkA/YopO a serine/threonine kinase (Hakansson *et al.*, 1996a); and YopP (Mills *et al.*, 1997). YopH acts on tyrosine-phosphorylated proteins of macrophages, thus inhibiting bacterial uptake and the oxidative burst (Bliska and Black, 1995; Andersson *et al.*, 1996), presumably by dephosphorylating key proteins involved in signal transduction. YpkA is also thought to interfere with signal-transduction pathways of the eukaryotic cell.

Targeting of type III secreted effector proteins to the eukaryotic host cell has also been demonstrated for other animal pathogens, such as: *S. flexneri* (IpaB) (Chen *et al.*, 1996); *Salmonella dublin* (SopE and SopB) (Wood *et al.*, 1996; Galyov *et al.*, 1997); *S. typhimurium* (SipB, SipC and SptB) (Collazo and Galan, 1997b; Fu and Galan, 1998); and enteropathogenic *E. coli* (EPEC) (EspB) (Knutton *et al.*, 1998; Wolff *et al.*, 1998); and therefore seems to be a general pathogenicity mechanism. The translocation process requires bacterial attachment to the host cell. For example, in *Yersinia*, Yop effectors are translocated into eukaryotic cells only if contact has been established between the pathogen and the target cell (Rosqvist *et al.*, 1994; Sory and Cornelis, 1994; Persson *et al.*, 1995; Sory *et al.*, 1995; Hakansson *et al.*, 1996b). Furthermore, contact increases the rate of transcription of the *yop* virulence genes (Pettersson *et al.*, 1996). The mechanism by which the 'contact signal' is transduced is not completely clear. The outer membrane protein, YopN, is believed to play a role, since *yopN* mutants express and secrete Yops in the absence of target cells (Forsberg *et al.*, 1991; Rosqvist *et al.*, 1994). A negative regulatory loop is involved in repressing Yop expression in the absence of contact; and secretion of one or several negative regulators is believed to trigger *yop* expression after contact with the host cell (Cornelis and Wolf-Watz, 1997).

Contact-dependent secretion has been demonstrated for *Shigella* and *Salmonella* (Menard *et al.*, 1994; Watarai *et al.*, 1995; Zierler and Galan, 1995). Since, in these pathogens, the proteins which will be secreted already exist as a pool, contact appears to trigger the opening of the secretion channel. Interestingly, certain bacterial culture media, which probably mimic contact with the host-cell, have enabled laboratories to obtain high levels of secretion *in vitro* for these animal pathogens (i.e. Ca^{2+} chelation for *Yersinia* and Congo red for *Shigella*).

Translocation of effector virulence factors into host cells is also dependent on proteins that act as 'translocators'. For example, in *Yersinia* spp., translocation of effector Yop proteins is dependent on YopB and YopD (Rosqvist *et al.*, 1994; Sory and Cornelis, 1994; Boland *et al.*, 1996; Hakansson *et al.*, 1996b). YopB and YopD are secreted proteins that are thought to be translocators involved in delivering the Yop 'effectors' across the host cell plasma membrane. YopB shows similarity to members of the RTX family of α-haemolysins and leukotoxins, and has been shown to cause haemolysis of erythrocytes, suggesting this protein may form a pore through which effector Yops are translocated (Hakansson *et al.*, 1996b).

2.3.4.2 Evidence for translocation of proteins by plant pathogenic bacteria

The discoveries in the animal field led many laboratories working on plant-bacterial interactions to the hypothesis that certain proteins, for example Avr proteins, might be translocated, via the Hrp pathway, into the plant cell. A major breakthrough was the finding that Avr proteins are recognised within the plant cell. So far, this has been shown for AvrB, AvrPto and AvrRpt2, AvrPphE.R2 from *P. syringae* pathovars and AvrBs3 protein of *X. c.* pv. *vesicatoria*. In the plant, *avr* gene expression was performed either by *Agrobacterium*-mediated transient expression (Tang *et al.*, 1996; Van den Ackerveken *et al.*, 1996; Stevens *et al.*, 1998), biolistic transient expression (Gopalan *et al.*, 1996; Leister *et al.*, 1996), or by stable transformation of plants (Gopalan *et al.*, 1996). The present section focuses on *Agrobacterium*-mediated transient expression of the *X. c.* pv. *vesicatoria avrBs3* gene. The *avr* gene was cloned between the T-DNA borders of a binary plant transformation vector, under the control of a strong promoter, the CaMV35S promoter, active in plants. An *A. tumefaciens* strain harbouring such a construct was induced for T-DNA transfer, and infiltrated into susceptible and resistant plants. Transformed cells should then express the *avr* gene. In the resistant pepper plant, ECW-30R, carrying the *Bs3* resistance gene, transient expression of *avrBs3* resulted in the induction of confluent HR. In susceptible plants, no reaction was observed, indicating that the recognition is genotype-specific (Van den Ackerveken *et al.*, 1996). These experiments were further validated by expressing the derivative of *avrBs3*, *avrBs3Δrep16*, which induced the 'inverse specificity', i.e. HR on *bs3* pepper plants and not on *Bs3* plants (Van den Ackerveken *et al.*, 1996).

Since Avr proteins depend on *hrp* genes for their function, the Hrp pathway is postulated to be involved in their secretion and translocation into the host cell (Bonas and Van den Ackerveken, 1997). However, the actual transfer of Avr proteins from the bacterial pathogens to the host cell has not been formally demonstrated, and the mechanisms for the translocation process are completely unknown.

The rapid, strong and clearly visible phenotype (HR induction) associated with *avr* gene function has enabled plant phytopathologists to demonstrate that certain Avr proteins are recognised inside the plant cell. It is probable that other avirulence and also virulence proteins follow the same route, since disease induction is also dependent on *hrp* genes. Intracellular action of individual virulence factors might be difficult to prove, since disease is probably the result of the action of several factors and not of one single protein, as in the case of *avr-R* gene recognition. As discussed previously, *avr* genes may play a role in pathogenicity. In this context, the disease specific *dspA* gene of *E. amylovora*, flanking the *hrp* gene cluster, is a good example. DspA is homologous to AvrE of *P. s.* pv. *tomato*. The expression of *avrE* in *P. s.* pv. *glycinea* renders virulent strains avirulent on certain cultivars of soybean (Lorang and Keen, 1995). In *Erwinia*, *dspA* is required for pathogenicity on the host plant but not for HR elicitation on tobacco (Gaudriault *et al.*, 1997; Bogdanove *et al.*, 1998b). DspA was shown to be secreted *in vitro* by the Hrp pathway (Gaudriault *et al.*, 1997; Bogdanove *et al.*, 1998a). Excitingly, Bogdanove *et al.*, 1998a demonstrated that *P. s.* pv. *glycinea* harbouring DspA was avirulent on soybean and that *avrE* could partially restore pathogenicity to *E. amylovora dspA* mutant strains. It will now be interesting to determine whether DspA and AvrE function inside the host cell.

2.3.4.3 Evidence for contact-dependent secretion in plant pathogens?

The fact that avirulence proteins have been difficult to detect in culture supernatants of *hrp*-induced bacteria suggests that the conditions used so far have not been sufficient to allow 'enhanced' secretion. Recent findings for *R. solanacearum* suggest that contact with the host cell may also be important for plant pathogens. Indeed, co-culture of *R. solanacearum* with *Arabidopsis* or tomato cell suspensions induced a much higher level of *hrp* gene transcription than the *hrp*-inducing minimal medium, i.e. 20 times higher (Marenda *et al.*, 1998).

This gene induction in co-culture is controlled by *prh* genes (plant regulatory *hrp*) specifically involved in transduction of plant-cell-derived signal(s). *prhA*, which codes for a putative outer membrane receptor with homology to TonB-dependent siderophore receptors, is postulated to be a receptor for plant signals (Marenda *et al.*, 1998). Sensing of plant-derived signals would then be transduced to PrhJ, which shares homology to regulatory proteins of the LuxR/UhpA family of transcriptional activators. PrhJ, in turn, activates *hrpG*, which leads to *hrp* gene activation (Brito *et al.*, 1999). The fact that no diffusible inducing activity could be found in the cell-conditioned medium led to the suggestion that *hrp* expression might be due to contact with the plant cells (Marenda *et al.*, 1998). Furthermore, *hrpJ* of *E. amylovora* encodes a homologue of the *Yersinia* YopN (Bogdanove *et al.*, 1996b). It will be interesting to determine whether *hrpJ* is also involved in sensing contact with the plant cell.

2.4 Specific aspects of type III protein secretion in plant pathogenic bacteria

Although the basic mechanisms of pathogenicity, involving type III protein secretion, appear to be conserved among Gram-negative bacterial pathogens, there are unique features associated with plant pathogens. Likewise, among mammalian pathogens, each bacterium has its unique attributes, due to differences in epidemiology and the final outcome: invasion, neutralisation or killing of the target cell.

The plant pathogen has to adapt to and modify the plant environment, and avoid plant defence responses, in order to multiply. Furthermore, if translocation of proteins is involved in the infection process (see section 2.3.4.2), there is one major obstacle that has to be overcome to gain access to the plant cell plasma membrane, that is, the plant cell wall. The plant cell wall is an extracellular matrix on the external surface of the plant cell. Although animal cells also have extracellular matrix components on their surface as part of the cell coat or glycocalyx, the plant cell wall is generally thicker and stronger (i.e. from 0.1 to several μm in thickness). The wall is composed of cellulose fibres embedded in a highly cross-linked matrix of polysaccharides (predominantly hemicellulose and pectin) and protein (mainly hydroxyproline-rich glycoproteins). Although the structure appears to be rigid, the wall contains pores of 3.2–5.2 nm in diameter, or more, depending on the cell type.

This section focuses on aspects of the type III secretion system that appear to be characteristic of phytopathogenic bacteria.

2.4.1 Proteins that are specific to plant phytopathogenic bacteria

As shown in Figure 2.1, certain genes in, or associated with, the *hrp* gene clusters encode proteins that are either unique to plant pathogens (genes in light grey for Group 1 and in dark grey for Group 2 phytopathogens), or are present only in one particular plant pathogen (indicated in white). Some of these proteins, which are not conserved in animal pathogens, have been shown to be regulators (i.e. HrpR, HrpS, HrpL in *P. s.* pv. *syringae* and *E. amylovora*, HrpB and HrpG in *R. solanacearum*, and HrpX and HrpG in *X. c.* pv. *vesicatoria*), and others are secreted via the Hrp pathway (i.e. PopA in *R. solanacearum*, DspA in *E. amylovora*) (see section 2.4.2). The roles of the remaining 'plant specific' genes are unknown, but they may encode elements of the secretion or translocation machinery that are adapted to the plant environment, or proteins that are secreted or translocated via the Hrp pathway.

It is now necessary to determine the mutant phenotype associated with each gene, in order to evaluate the importance of a given gene in the interaction with the host plant and in type III secretion. Non-polar mutagenesis of the *hrpD* operon in *X. c.* pv. *vesicatoria* has revealed that one gene, named *hpaA* (for *hrp*-associated), is specifically required for disease development. *hpaA* mutants are affected in pathogenicity but retain, in part, the ability to induce the host-specific hypersensitive reaction, suggesting that the *hpaA* product acts as a virulence factor (Huguet *et al.*, 1998).

2.4.2 *Proteins that are secreted via the Hrp pathway*

2.4.2.1 *Harpins, PopA and HrpW*

The harpins of *E. amylovora* and *P. s.* pv. *syringae* are non-specific HR elicitors, whereas PopA may represent a host-specificity factor because the isolated protein elicits the HR selectively on those plants in which *R. solanacearum* also elicits the HR (Wei *et al.*, 1992a; He *et al.*, 1993; Arlat *et al.*, 1994; Preston *et al.*, 1995). Harpins of *E. amylovora* and *P. s.* pv. *syringae* have been shown to cause extracellular alkalisation and potassium efflux in suspension-cultured tobacco cells, responses normally associated with disease resistance (Wei *et al.*, 1992a; Hoyos *et al.*, 1996). The harpin protein of *P. s.* pv. *syringae* is postulated to bind to cell walls of intact plant cells, since it is unable to bind to protoplasts or to induce HR-associated responses in protoplasts (Hoyos *et al.*, 1996).

The role of these proteins in pathogenicity is not entirely clear. *R. solanacearum popA* mutants are still pathogenic (Arlat *et al.*, 1994), and *P. s.* pv. *syringae hrpZ* (codes for harpin) mutants are either slightly or not affected in virulence (Alfano *et al.*, 1996), whereas *E. amylovora hrpN* mutants are reported to be more significantly reduced in virulence or are non-pathogenic (Wei *et al.*, 1992a; Barny, 1995). Analysis of various *hrpZ* gene deletions under various *hrp* inducing conditions suggests that harpin may be required for avirulence protein translocation into host cells (Gopalan *et al.*, 1996; Pirhonen *et al.*, 1996). It has, therefore, been suggested that harpin plays a role in modulation of protein secretion, by the modification of the plant cell wall during transport of avirulence or virulence effector proteins (He, 1997).

HrpW shares structural similarities with HrpN of *E. amylovora* and PopA of *R. solanacearum*. Interestingly, the C-terminal region of HrpW is homologous to class III pectate lyases. The role of *HrpW* is unknown. *hrpW* mutants are not affected in pathogenicity; however, they elicit the HR response on tobacco at lower inoculum densities than a wild-type strain (Gaudriault *et al.*, 1998).

2.4.2.2 *HrpA and the Hrp pilus*

Intriguingly, the *P. s.* pv. *tomato* HrpA protein is associated with a macromolecular pilus-like structure, designated 'the Hrp pilus', which has a diameter of 8 nm and a length of 2 μm or more (Roine *et al.*, 1997). A non-polar *hrpA* mutant is an *hrp* mutant, suggesting that the formation of the Hrp pili is essential for both compatible and incompatible interactions with the plant, and therefore also for presumed translocation of proteins into the plant cell. Interestingly, *A. tumefaciens*, which delivers macromolecules into plant cells, also forms pili

(Fullner *et al.*, 1996; see also section 2.2.1.3). In general, these structures may traverse the plant cell wall and allow contact with the host cell plasma membrane. Whether pili are involved in adhesion or form an actual tube permitting transit of secreted proteins is unknown, and an exciting question for future research.

Macromolecular structures associated with type III secretion systems in animal pathogens have also been identified. In *S. typhimurium*, filamentous appendages of 60 nm in diameter have been observed on the surface of the bacteria after contact with the host cell. The production of these 'invasomes' is transient and under the control of the type III secretion system (Ginocchio *et al.*, 1994). Similarly, stimulation of type III secretion in *Shigella* leads to the appearance of sheet-like structures *in vitro*, of which the secreted IpaB and IpaC proteins are constituents (Parsot *et al.*, 1995). Excitingly, enteropathogenic *E. coli* also produce filamentous structures, composed of EspA, that are secreted via a type III pathway. These structures appear during the early stages of infection. Immunoelectron microscopy has shown that the structures form a physical bridge between the bacteria and the infected eukaryotic cell (Knutton *et al.*, 1998). Interestingly, the filaments which have a diameter of 50 nm in immunofluorescence studies, when observed under negative staining in electron microscopy appear to be composed of a substructure of 7–8 nm in diameter. The role of these type III surface structures in protein translocation or adhesion, or both, awaits further experiments.

2.4.2.3 The disease-specific protein, DspA

The *dsp* region of *E. amylovora*, which is separated from the *hrp* gene cluster by a 4.0-kb region, is required for pathogenicity but not for HR elicitation (Gaudriault *et al.*, 1997; Bogdanove *et al.*, 1998b). DspA has been shown to encode a 190-kDa protein that is secreted into the extracellular milieu in an Hrp-dependent manner. The DspB protein, encoded downstream of *dspA*, is required for DspA secretion. It is postulated that DspB is a chaperone of DspA, since it has certain characteristics in common with the *Yersinia* Ysc chaperones: it is a small, acidic protein (Gaudriault *et al.*, 1997). This is the first evidence of the existence of chaperones in the Hrp secretion system.

The role of DspA in pathogenicity is unknown. Whether *in vivo* DspA is secreted or translocated is the next question to be answered. The homology of DspA to AvrE, and the fact that some avirulence proteins have been shown to function inside the host cell, suggests that DspA is translocated.

2.4.3 Plant perception of bacterial proteins

2.4.3.1 Resistance genes and signalling cascades

Many laboratories are interested in understanding the molecular basis of plant disease resistance for fundamental and obvious agronomic purposes. The simplest model for the molecular basis of the gene-for-gene interaction predicts physical interaction between the corresponding *avr* and *R* gene products. By using the yeast two-hybrid system, it was shown that the *P. s.* pv. *tomato* AvrPto protein indeed interacts with its cognate *R* gene product, Pto, from tomato (Scofield *et al.*, 1996; Tang *et al.*, 1996). Pto shows similarity to serine-threonine protein kinases (Martin *et al.*, 1993b), and exhibits protein kinase catalytic activity *in vitro* (Loh and Martin, 1995; Rommens *et al.*, 1995). Several Pto-interacting plant proteins have been isolated; some resemble transcription factors and are capable of binding to a DNA sequence that is found in promoters of genes encoding pathogenesis-related proteins (Bent, 1996; Zhou *et al.*, 1997). Taken together, these results suggest that direct interaction between AvrPto and Pto mediates a signal transduction cascade that leads to activation of plant defence-related genes. Direct interaction between other *avr* and *R* gene products is probable, since most plant *R* genes cloned contain leucine-rich-repeat regions (LRRs), which mediate protein-protein interactions (Kobe and Deisenhofer, 1994; Bent, 1996; Baker *et al.*, 1997; Zhou *et al.*, 1997).

2.4.3.2 AvrBs3 contains nuclear localisation signals

AvrBs3 of *X. c.* pv. *vesicatoria*, and other members of the AvrBs3 family (Leach and White, 1996), have a particular feature: they contain nuclear localisation signals (NLSs) in their C-terminus. The C-terminal region of AvrBs3 has been shown to target a cytoplasmic reporter-protein β-glucuronidase (GUS) to the plant cell nucleus (Van den Ackerveken *et al.*, 1996). Furthermore, the presence of at least one out of two NLSs appears to be important for full AvrBs3 activity: specific induction of the HR on *Bs3* pepper plants; and for nuclear localisation activity. Together, these results suggest that AvrBs3 is recognised inside the plant cell and that nuclear factors are involved in this perception (Van den Ackerveken *et al.*, 1996).

2.4.3.3 What could be the fate of virulence proteins?

A role in virulence (i.e. pathogen fitness or symptom formation) has been described for a few *avr* genes, such as: *avrBs2* of *X. c.* pv. *vesicatoria* (Kearney and Staskawicz, 1990; Swords *et al.*, 1996); *avrE* and *avrA* of *P. s.* pv. *tomato* (Lorang *et al.*, 1994); *avrRpm1* from *P. s.* pv. *maculicola* (Ritter and Dangl, 1995); *pthA* from *X. citri* (Swarup *et al.*, 1992); *avrB6* from *X. c.* pv. *malvacearum* (De Feyter *et al.*, 1993); and *avrXa7* from *X. oryzae* pv. *oryzae* (Leach and White, 1996). It will be interesting to determine the fate of the corresponding avirulence proteins in compatible interactions, and, more generally, the fate of predicted, or so far unknown, virulence factors. Interestingly, the HpaA protein of *X. c.* pv. *vesicatoria* contains two functional nuclear localisation signals, which are important for the interaction with the plant. HpaA may therefore be an effector 'virulence factor' that is translocated into the host cell (Huguet *et al.*, 1998). Virulence proteins may be involved, as certain Yop effector proteins, in the interference of signal transduction pathways of the eukaryotic host, e.g. by disrupting plant gene expression or diverting plant gene expression to the pathogen's advantage.

2.4.4 *Hrp-dependent suppression of host defences?*

Although understanding of *hrp* gene function has progressed over the last few years, it is not known how bacteria establish a relationship with plants that allows them to multiply and cause disease. It has been suggested that *hrp* mutants are unable to secrete factors that would allow release of nutrients from the host cell (Mansfield *et al.*, 1994). However, there are indications that *hrp* mutants are unable to multiply in the plant tissue because they have lost the ability to suppress non-specific host defences.

2.4.4.1 *Evidence for non-specific host defence responses and suppression in compatible interactions*

Jakobek and Lindgren (1993) reported that during the incompatible interaction between bean and *P. s.* pv. *tabaci* there is activation of genes encoding phenylalanine ammonia lyase and chitinase and there is accumulation of phytoalexins, processes that are generally associated with plant defence. Surprisingly, *hrp* mutants, heat-killed bacteria or non-pathogenic bacteria (i.e. *E. coli*), which are not able to induce the HR, also induced transcript and phytoalexin accumulation, and with the same temporal pattern. These results suggested that non-specific defence reactions are induced in the plant in response to bacteria. In the compatible interaction between *P. s.* pv. *phaseolicola* and bean, no transcript accumulation was visible. Furthermore, prior infiltration with *P. s.* pv. *phaseolicola* suppressed the typical transcript accumulation and phytoalexin accumulation that occurs in bean after infiltration with *P. s.* pv. *tabaci*, or *hrp* mutants or *E. coli* (Jakobek and Lindgren, 1993; Jakobek *et al.*, 1993). This suppression was not due to the production of the toxin phaseolotoxin by *P. s.* pv. *Phaseolicola*; however, the nature of the suppressor has not been determined (Jakobek *et al.*, 1993).

2.4.4.2 *Evidence for hrp-dependent suppression of non-specific host defences*

Electron microscopy has been used to examine responses of pepper leaf mesophyll cells to different strains of *X. c.* pv. *vesicatoria*. *hrp* mutants caused localised modification of the cell wall associated with deposition of papillae in adjacent cells. Similar deposits were visible after infiltration with the saprophytic *X. campestris* strain, T55 (which lacks the *hrp* gene cluster), or with heat-killed bacteria, but not with live pathogenic strains (Brown *et al.*, 1995). The results suggest that the localised cell wall modifications and papillae represent a defence response of plant cells to *X. c.* pv. *vesicatoria*, and that suppression of these reactions requires functional *hrp* genes (Brown *et al.*, 1995). The *hrp* gene, or other genes, which encode the suppressor activity have not been identified.

2.5 Concluding remarks

Based on the results obtained for *hrp* genes, and analogies with the mammalian bacterial pathogens, a model for a possible sequence of events after infection by a biotrophic plant pathogen is presented in Figure 2.2. In this model: 1) Bacteria enter the plant and come into contact with the cell wall; environmental signals (i.e. plant signals) or contact, trigger(s) *hrp* gene expression. 2) This results in the formation of Hrp secretion units (H) and pili (P); pili may be involved in establishing a link with the plant cell plasma membrane. 3) In the susceptible plant, virulence factors (V) are secreted into the apoplast, and possibly translocated into the host cell; some factors may disrupt the integrity

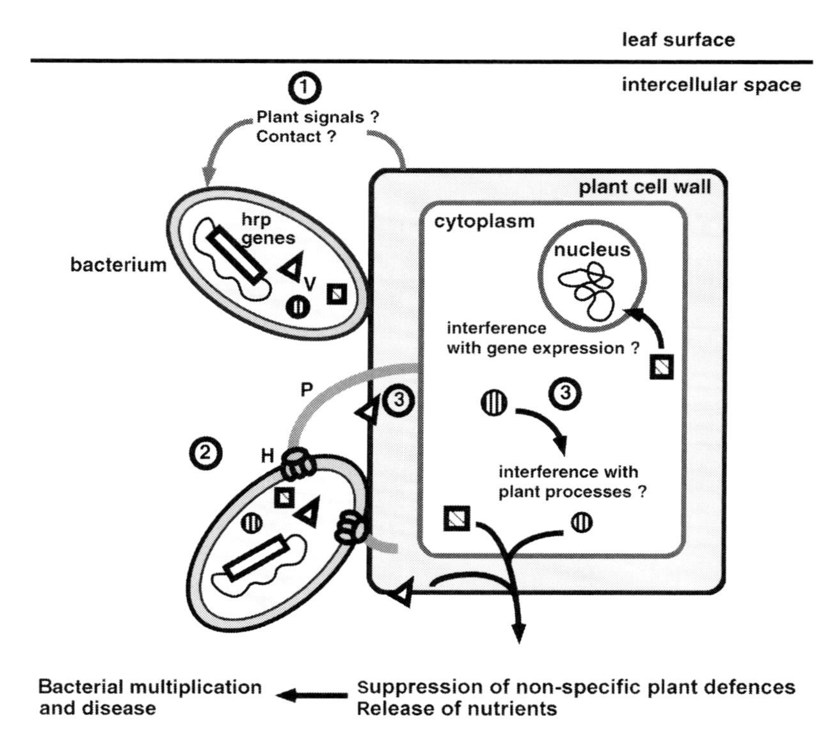

Figure 2.2 Model of the infection process of biotrophic bacterial plant pathogens. Abbreviations: H, Hrp pathway; P, Hrp pilus; V, virulence factors. See text for details.

of the cell wall, thereby facilitating access to the plant cell and releasing nutrients. Translocated products may interfere with plant intracellular processes or may act directly on gene transcription. The global action of these virulence factors is to inactivate non-specific host defences and allow the release of nutrients.

In the case of a resistant plant, the plant has 'learned' to recognise certain proteins (avirulence proteins) translocated by the bacteria. Avr protein recognition results in activation of a signalling cascade, which in turn switches on the activation of plant defence-related genes, resulting in the HR.

Figure 2.2 should only be viewed as a working hypothesis, since there are still many black boxes at all levels of the infection process. It is not known how the bacteria transfer effector proteins into the plant cell. The proteins could be 'injected' or secreted by the bacteria to be taken up by the plant cell by endocytosis. Pili may be directly involved in transport, or may act by facilitating contact and adhesion to the cell. One challenge for future research is to identify new secreted virulence factors, and to determine how and in which compartment these proteins act.

Interestingly, as our knowledge of plant bacterial pathogenesis is advanced, more and more parallels are established between the different infection strategies. Globally, bacterial virulence genes are subject to tight environmental control. *Agrobacterium* is faced with the challenge of transferring macromolecules into the host cell, and into the plant nucleus. As with previously described virulence factors (extracellular cell-wall-degrading enzymes, toxins), so it will now be necessary to determine the mode of action of *hrp*-dependent virulence factors. The recent finding that certain symbiotic bacteria also encode a type III secretion system will probably provide a new dimension to type III secretion systems (Freiberg *et al.*, 1997). Finally, progress in understanding type III-dependent secretion and pathogenicity in mammalian pathogens has already inspired the plant pathologist, and will most probably continue to do so.

Acknowledgements

The author would like to thank Ulla Bonas, Ombeline Rossier, Christian Boucher and John Mansfield for critical reading of this manuscript.

References

Adler, B., Sasakawa, C., Tobe, T., Makino, S., Komatsu, K. and Yoshikawa, M. (1989) A dual transcriptional activation system for the 230 kb plasmid genes coding for virulence-associated antigens of *Shigella flexneri*. *Mol. Microbiol.*, **3** 627-35.

Agrios, G.N. (1988) *Plant Pathology*, Third Edition, Academic Press Inc., New York.

Alfano, J.R. and Collmer, A. (1997) The Type III (Hrp) secretion pathway of plant pathogenic bacteria: trafficking Harpins, Avr proteins and death. *J. Bacteriol.*, **179** 5655-62.

Alfano, J.R., Bauer, D.W., Milos, T.M. and Collmer, A. (1996) Analysis of the role of the *Pseudomonas syringae* pv. *syringae* HrpZ harpin in elicitation of the hypersensitive response in tobacco using functionally non-polar *hrpZ* deletion mutations, truncated HrpZ fragments and *hrmA* mutations. *Mol. Microbiol.*, **19** 715-28.

Allaoui, A., Sansonetti, P.J. and Parsot, C. (1992) MxiJ, a lipoprotein involved in secretion of *Shigella* Ipa invasins, is homologous to YscJ, a secretion factor of the *Yersinia* Yop proteins. *J. Bacteriol.*, **174** 7661-69.

Allaoui, A., Sansonetti, P.J. and Parsot, C. (1993) MxiD, an outer membrane protein necessary for the secretion of the *Shigella flexneri* Ipa invasins. *Mol. Microbiol.*, **7** 59-68.

Allaoui, A., Woestyn, S., Sluiters, C. and Cornelis, G.R. (1994) YscU, a *Yersinia enterocolitica* inner membrane protein involved in Yop secretion. *J. Bacteriol.*, **176** 4534-42.

Allaoui, A., Sansonetti, P.J., Menard, R., Barzu, S., Mounier, J., Phalipon, A. and Parsot, C. (1995a) MxiG, a membrane protein required for secretion of *Shigella* spp. Ipa invasins: involvement in entry into epithelial cells and in intercellular dissemination. *Mol. Microbiol.*, **17** 461-70.

Allaoui, A., Schulte, R. and Cornelis, G.R. (1995b) Mutational analysis of the *Yersinia enterocolitica virC* operon: characterization of yscE, F, G, I, J, K required for Yop secretion and yscH encoding YopR. *Mol. Microbiol.*, **18** 343-55.

Andersson, K., Carballeira, N., Magnusson, K.E., Persson, C., Stendahl, O., Wolf-Watz, H. and Fallman, M. (1996) YopH of *Yersinia pseudotuberculosis* interrupts early phosphotyrosine signalling associated with phagocytosis. *Mol. Microbiol.*, **20** 1057-69.

Andro, T., Chambost, J.P., Kotoujansky, A., Cattaneo, J., Bertheau, Y., Barras, F., van Gijsegem, F. and Coleno, A. (1984) Mutants of *Erwinia chrysanthemi* defective in secretion of pectinase and cellulase. *J. Bacteriol.*, **160** 1199-203.

Arlat, M., Gough, C.L., Barber, C.E., Boucher, C. and Daniels, M.J. (1991) *Xanthomonas campestris* contains a cluster of *hrp* genes related to the larger *hrp* cluster of *Pseudomonas solanacearum*. *Mol. Plant-Microbe Interact.*, **4** 593-601.

Arlat, M., Gough, C.L., Zischek, C., Barberis, P.A., Trigalet, A. and Boucher, C.A. (1992) Transcriptional organization and expression of the large *hrp* gene cluster of *Pseudomonas solanacearum*. *Mol. Plant-Microbe Interact.*, **5** 187-93.

Arlat, M., van Gijsegem, F., Huet, J.C., Pernollet, J.C. and Boucher, C.A. (1994) PopA1, a protein which induces a hypersensitivity-like response on specific *Petunia* genotypes, is secreted via the Hrp pathway of *Pseudomonas solanacearum*. *EMBO J.*, **13** 543-53.

Baker, B., Zambryski, P., Staskawicz, B. and Dinesh-Kumar, S.P. (1997) Signalling in plant-microbe interactions. *Science*, **276** 726-33.

Barber, C.E., Tang, J.L., Feng, J.X., Pan, M.Q., Wilson, T.J.G., Slater, H., Dow, J.M., Williams, P. and Daniels, M.J. (1997) A novel regulatory system required for pathogenicity of *Xanthomonas campestris* is mediated by a small diffusible signal molecule. *Mol. Microbiol.*, **24** 555-66.

Barny, M.A. (1995) *Erwinia amylovora hrpN* mutants, blocked in harpin synthesis, express a reduced virulence on host plants and elicit variable hypersensitive reactions on tobacco. *Eur. J. Plant Pathol.*, **101** 333-40.

Barny, M.A., Guinebretière, M.H., Marçais, B., Coissac, E., Paulin, J.P. and Laurent, J. (1990) Cloning of a large gene cluster involved in *Erwinia amylovora* CFBP1430 virulence. *Mol. Microbiol.*, **4** 777-86.

Baron, C., Llosa, M., Zhou, S. and Zambryski, P. C. (1997) VirB1, a component of the T-complex transfer machinery of *Agrobacterium tumefaciens*, is processed to a C-terminal secreted product, VirB1. *J. Bacteriol.*, **179** 1203-10.

Barras, F., Thurn, K.K. and Chatterjee, A.K. (1987) Resolution of four pectate lyase structural genes of *Erwinia chrysanthemi* (EC16) and characterization of the enzymes produced in *Escherichia coli*. *Mol. Gen. Genet.*, **209** 319-25.

Barras, F., van Gijsegem, F. and Chatterjee, A.K. (1994) Extracellular enzymes and pathogenesis of soft rot *Erwinia*. *Annu. Rev. Phytopathol.*, **32** 201-34.

Bauer, D.W. and Beer, S.V. (1991) Further characterization of an *hrp* gene cluster of *Erwinia amylovora*. *Mol. Plant-Microbe Interact.*, **4** 493-99.

Bauer, D.W., Bogdanove, A.J., Beer, S.V. and Collmer, A. (1994) *Erwinia chrysanthemi hrp* genes and their involvement in soft rot pathogenesis and elicitation of the hypersensitive response. *Mol. Plant-Microbe Interact.*, **7** 573-81.

Beaulieu, C., Boccara, M. and van Gijsegem, F. (1993) Pathogenic behaviour of pectinase-defective *Erwinia chrysanthemi* mutants on different plants. Mol. *Plant-Microbe Interact.*, **6** 197-202.

Beck von Bodman, S. and Farrand, S.K. (1995) Capsular polysaccharide biosynthesis and pathogenicity of *Erwinia stewartii* require induction by an *N*-acylhomoserine lactone autoinducer. *J. Bacteriol.*, **177** 5000-5008.

Bender, C.L., Stone, H.E., Sims, J.J. and Cooksey, D.A. (1987) Reduced pathogen fitness of *Pseudomonas syringae* pv. *tomato* Tn*5* insertions defective in coronatine production. *Physiol. Mol. Plant Pathol.*, **30** 273-83.

Bent, A.F. (1996) Plant disease resistance genes: function meets structure. *Plant Cell*, **8** 1757-71.

Bergman, T., Erickson, K., Galyov, E., Persson, C. and Wolf-Watz, H. (1994) The *lcrB* (*yscN/U*) gene cluster of *Yersinia pseudotuberculosis* is involved in Yop secretion and shows high homology to the *spa* gene clusters of *Shigella flexneri* and *Salmonella typhimurium*. *J. Bacteriol.*, **176** 2619-26.

Bernhard, T., Coplin, D.L. and Geider, K. (1993) A gene cluster for amylovoran synthesis in *Erwinia amylovora*: characterization and relation to *cps* genes in *Erwinia stewartii*. *Mol. Gen. Genet.*, **239** 158-68.

Bertheau, Y., Madgidi-Hervan, E., Kotoujansky, A., Nguyen-The, C., Andro, T. and Coleno, A. (1984) Detection of depolymerase isoenzymes after electrophoresis or electrofocusing, or in titration curves. *Analytical Biochem.*, **139** 383-89.

Bliska, J.B. and Black, D.S. (1995) Inhibition of the Fc receptor-mediated oxidative burst in macrophages by the *Yersinia pseudotuberculosis* tyrosine phosphatase. *Infect. Immun.*, **63** 681-85.

Boccara, M., Diolez, A., Rouve, M. and Kotoujansky, A. (1988) The role of the individual pectate lyases of *Erwinia chrysanthemi* strain 3937 in pathogenicity on *Saintpaulia* plants. *Physiol. Mol. Plant Pathol.*, **33** 95-104.

Bogdanove, A., Beer, S.V., Bonas, U., Boucher, C.A., Collmer, A., Coplin, D.L., Cornelis, G.R., Huang, H.-C., Hutcheson, S.W., Panopoulos, N.J. and van Gijsegem, F. (1996a) Unified nomenclature for broadly conserved *hrp* genes of phytopathogenic bacteria. *Mol. Microbiol.*, **20** 681-83.

Bogdanove, A.J., Wei, Z.M., Zhao, L. and Beer, S.V. (1996b) *Erwinia amylovora* secretes harpin via a type III pathway and contains a homolog of *yopN* of *Yersinia* spp. *J. Bacteriol.*, **178** 1720-30.

Bogdanove, A.J., Bauer, D.W. and Beer, S.V. (1998a) *Erwinia amylovora* secretes DspE, a pathogenicity factor and functional AvrE homolog, through the Hrp (type III secretion) pathway. *J. Bacteriol.*, **180** 2244-47.

Bogdanove, A.J., Kim, J.F., Wei, Z., Kolchinsky, P., Charkowski, A.O., Conlin, A.K., Collmer, A. and Beer, S.V. (1998b) Homology and functional similarity of an *hrp*-linked pathogenicity locus, *dspEF*, of *Erwinia amylovora* and the avirulence locus, *avrE*, of *Pseudomonas syringae* pathovar *tomato*. *Proc. Natl. Acad. Sci. USA*, **95** 1325-30.

Boland, A., Sory, M.P., Iriarte, M., Kerbourch, C., Wattiau, P. and Cornelis, G.R. (1996) Status of YopM and YopN in the *Yersinia* Yop virulon: YopM of *Y. enterocolitica* is internalized inside the cytosol of PU5-1.8 macrophages by the YopB, D, N delivery apparatus. *EMBO J.*, **15** 5191-201.

Bonas, U. and Van den Ackerveken, G. (1997) Recognition of bacterial avirulence proteins occurs inside the plant cell: a general phenomenon in resistance to bacterial diseases. *Plant J.*, **12** 1-7.

Bonas, U., Stall, R.E. and Staskawicz, B. (1989) Genetic and structural characterization of the avirulence gene, *avrBs3*, from *Xanthomonas campestris* pv. *vesicatoria*. *Mol. Gen. Genet.*, **218** 127-36.

Bonas, U., Schulte, R., Fenselau, S., Minsavage, G.V. and Staskawicz, B.J. (1991) Isolation of a gene cluster from *Xanthomonas campestris* pv. *vesicatoria* that determines pathogenicity and the hypersensitive response on pepper and tomato. *Mol. Plant-Microbe Interact.*, **4** 81-88.

Boucher, C.A., van Gijsegem, F., Barberis, P.A., Arlat, M. and Zischek, C. (1987) *Pseudomonas solanacearum* genes controlling both pathogenicity on tomato and hypersensitivity on tobacco are clustered. *J. Bacteriol.*, **169** 5626-32.

Brito, B., Marenda, M., Barberis, P., Boucher, C. and Genin, S. (1999) *prhJ* and *hrpG*, two new components of the plant signal-dependent regulatory cascade controlled by PrhA in *Ralstonia solanacearum*. *Mol. Microbiol.*, **31** 237-51.

Brown, I., Mansfield, J. and Bonas, U. (1995) *hrp* genes in *Xanthomonas campestris* pv. *vesicatoria* determine ability to suppress papillae deposition in pepper mesophyll cells. *Mol. Plant-Microbe Interact.*, **8** 825-36.

Cha, C., Gao, P., Chen, Y-C., Shaw, P.D. and Farrand S.K. (1998) Production of acyl-homoserine lactone quorum-sensing signals by Gram-negative plant-associated bacteria. *Mol. Plant-Microbe Interact.*, **11** 1119-29.

Charkowski, A.O., Huang, H.-C. and Collmer, A. (1997) Altered localization of HrpZ in *Pseudomonas syringae* pv. *syringae hrp* mutants suggests that different components of the Type III secretion pathway control protein translocation across the inner and outer membranes of Gram-negative bacteria. *J. Bacteriol.*, **179** 3866-74.

Chatterjee, A.K., Ross, L.M., McEvoy, J.L. and Thurn, K.K. (1985) pULB113, an RP4:mini-Mu plasmid, mediates chromosomal mobilisation and R-prime formation in *Erwinia amylovora, Erwinia chrysanthemi* and subspecies of *Erwinia carotovora*. *Appl. Environ. Microbiol.*, **50** 1-9.

Chen, Y., Smith, M.R., Thirumalai, K. and Zychlinsky, A. (1996) A bacterial invasin induces macrophage apoptosis by binding directly to ICE. *EMBO J.*, **15** 3853-60.

Christie, P.J. (1997) *Agrobacterium tumefaciens* T-complex transport apparatus: a paradigm for a new family of multifunctional transporters in eubacteria. *J. Bacteriol.*, **179** 3085-94.

Citovsky, V., De Vos, G. and Zambryski, P. (1988) Single-stranded DNA binding protein encoded by the *virE* locus of *Agrobacterium tumefaciens*. *Science*, **240** 501-504.

Citovsky, V., Wong, M.L. and Zambryski, P. (1989) Cooperative interaction of *Agrobacterium* VirE2 protein with single-stranded DNA: implications for the T-DNA transfer process. *Proc. Natl. Acad. Sci. USA*, **86** 1193-97.

Citovsky, V., Zupan, J., Warnick, D. and Zambryski, P. (1992) Nuclear localization of *Agrobacterium* VirE2 protein in plant cells. *Science*, **256** 1802-1805.

Clough, S.J., Lee, K.E., Schell, M.A. and Denny, T.P. (1997) A two-component system in *Ralstonia solanacearum* modulates production of *phcA*-regulated virulence factors in response to 3-hydroxypalmitic acid methyl ester. *J. Bacteriol.*, **179** 3639-48.

Collazo, C.M. and Galan, J.E. (1996) Requirement for exported proteins in secretion through the invasion-associated type III system of *Salmonella typhimurium*. *Infect. Immun.*, **64** 3524-31.

Collazo, C.M. and Galan, J.E. (1997a) The invasion-associated type III system of *Salmonella typhimurium* directs the translocation of Sip proteins into the host cell. *Mol. Microbiol.*, **24** 747-56.

Collazo, C.M. and Galan, J.E. (1997b) The invasion-associated type-III protein secretion system in *Salmonella*: a review. *Gene*, **192** 51-59.

Collmer, A. and Keen, N.T. (1986) The role of pectic enzymes in plant pathogenesis. *Annu. Rev. Phytopathol.*, **24** 383-409.

Condemine, G., Dorel, C., Hugouvieux-Cotte-Pattat, N. and Robert-Baudouy, J. (1992) Some of the Out genes involved in the secretion of pectate lyases in *Erwinia chrysanthemi* are regulated by kdgR. *Mol. Microbiol.*, **6** 3199-211.

Coplin, D.L. and Majerczak, D.R. (1990) Extracellular polysaccharide genes in *Erwinia stewartii* directed mutagenesis and complementation analysis. *Mol. Plant-Microbe Interact.*, **3** 286-92.

Cornelis, G.R. (1994) *Yersinia* pathogenicity factors. *Curr. Topics Microbiol. Immunol.*, **192** 243-63.

Cornelis, G.R. and Wolf-Watz, H. (1997) The *Yersinia* Yop virulon: a bacterial system for subverting eukaryotic cells. *Mol. Microbiol.*, **23** 861-67.

Cornelis, G., Sluiters, C., de Rouvroit, C.L. and Michiels, T. (1989) Homology between *virF*, the transcriptional activator of the *Yersinia* virulence regulon, and AraC, the *Escherichia coli* arabinose operon regulator. *J. Bacteriol.*, **171** 254-62.

Dangl, J.L. (1994) The enigmatic avirulence genes of phytopathogenic bacteria. *Curr. Topics Microbiol. Immunol.*, **192** 99-118.

De Feyter, R., Yang, Y. and Gabriel, D.W. (1993) Gene-for-gene interactions between cotton *R* genes and *Xanthomonas campestris* pv. *malvacearum avr* genes. *Mol. Plant-Microbe Interact.*, **6** 225-37.

De Wit, J.G.M. (1997) Pathogen avirulence and plant resistance: a key role for recognition. *Trends Plant Sci.*, **2** 452-57.

Denny, T.P. (1995) Involvement of bacterial polysaccharide in plant pathogenesis. *Annu. Rev. Phytopathol.*, **33** 173-97.

DiMango, E., Zar, H.J., Bryan, R. and Prince A. (1995) Diverse *Pseudomonas aeruginosa* gene products stimulate respiratory epithelial cells to produce interleukin-8. *J. Clin. Invest.*, **96** 2204-10.

Dixon, R.A., Harrison, M.J. and Lamb, C.J. (1994) Early events in the activation of plant defense responses. *Annu. Rev. Phytopathol.*, **32** 479-501.

Donnenberg, M.S., Kaper, J.B. and Finlay, B.B. (1997) Interactions between enteropathogenic *Escherichia coli* and host epithelial cells. *Trends Microbiol.*, **5** 109-14.

Eichelberg, K., Ginocchio, C. and Gla, J.E. (1994) Molecular and functional characterization of the *Salmonella typhimurium* invasion genes, *invB* and *invC*: homology of InvC to the F_0F_1 ATPase family of proteins. *J. Bacteriol.*, **176** 4501-10.

Fellay, R., Rahme, L.G., Mindrinos, M.N., Frederick, R.D., Pisi, A. and Panopoulos, N.J. (1991) Genes and signals controlling the *Pseudomonas syringae* pv. *phaseolicola*-plant interaction, in *Advances in Molecular Genetics of Plant-Microbe Interactions* (eds. H. Hennecke and D.P.S. Veram), Kluwer Academic Publishers, Dordrecht, pp. 45-52.

Fenselau, S. and Bonas, U. (1995) Sequence and expression analysis of the *hrpB* pathogenicity operon of *Xanthomonas campestris* pv. *vesicatoria*, which encodes eight proteins with similarity to components of the Hrp, Ysc, Spa and Fli secretion systems. *Mol. Plant-Microbe Interact.*, **8** 845-54.

Fenselau, S., Balbo, I. and Bonas, U. (1992) Determinants of pathogenicity in *Xanthomonas campestris* pv. *vesicatoria* are related to proteins involved in secretion in bacterial pathogens of animals. *Mol. Plant-Microbe Interact.*, **5** 390-96.

Feys, B.J.F., Benedetti, C.E., Penfold, C.N. and Turner, J.G. (1994) *Arabidopsis* mutants selected for resistance to the phytotoxin, coronatine, are male-sterile, insensitive to methyl jasmonate and resistant to a bacterial pathogen. *Mol. Plant-Microbe Interact.*, **6** 751-59.

Fields, K.A., Plano, G.V. and Straley, S.C. (1994) A low-Ca^{2+} response (LCR) secretion (*ysc*) locus lies within the *lcrB* region of the LCR plasmid in *Yersinia pestis*. *J. Bacteriol.*, **176** 569-79.

Flavier, A.B., Ganova-Raeva, L.M., Schell, M.A. and Denny, T.P. (1997) Hierarchical autoinduction in *Ralstonia solanacearum*: control of acyl-homoserine lactone production by a novel autoregulatory system responsive to 3-hydroxypalmitic acid methyl ester. *J. Bacteriol.*, **179** 7089-97.

Flavier, A.B., Schell, M.A. and Denny, T.P. (1998) An RpoS (sigmaS) homologue regulates acyl-homoserine lactone-dependent autoinduction in *Ralstonia solanacearum*. *Mol. Microbiol.*, **28** 475-86.

Flor, H.H. (1956) The complementary genetic systems in flax and flax rust. *Adv. Genet.*, **8** 29-54.

Forsberg, A. and Wolf-Watz, H. (1988) The virulence protein, YopE, of *Yersinia pseudotuberculosis* is regulated at transcriptional level by plasmid-plB1-encoded trans-acting elements controlled by temperature and calcium. *Mol. Microbiol.*, **2** 121-33.

Forsberg, A., Viitanen, A.M., Skurnik, M. and Wolf-Watz, H. (1991) The surface-located YopN protein is involved in calcium signal transduction in *Yersinia pseudotuberculosis*. *Mol. Microbiol.*, **5** 977-86.

Frederick, R.D., Majerczak, D.R. and Coplin, D.L. (1993) *Erwinia stewartii* WtsA, a positive regulator of pathogenicity gene expression, is similar to *Pseudomonas syringae* pv. *phaseolicola* HrpS. *Mol. Microbiol.*, **9** 477-85.

Freiberg, C., Fellay, R., Bairoch, A., Broughton, W.J., Rosenthal, A. and Perret, X. (1997) Molecular basis of symbiosis between *Rhizobium* and legumes. *Nature*, **387** 394-401.

Fu, Y. and Galan, J.E. (1998) The *Salmonella typhimurium* tyrosine phosphatase SptP is translocated into the host cells and disrupts the actin cytoskeleton. *Mol. Microbiol.*, **27** 359-68.

Fullner, K.J., Lara, J.C. and Nester, E.W. (1996) Pilus assembly by *Agrobacterium* T-DNA transfer genes. *Science*, **273** 1107-1109.

Fuqua, C., Winans, S.C. and Greenberg, E.P. (1996) Census and consensus in bacterial ecosystems: the LuxR-LuxI family of quorum-sensing transcriptional regulators. *Annu. Rev. Microbiol.*, **50** 727-51.

Galan, J.E. (1996) Molecular genetic bases of *Salmonella* entry into host cells. *Mol. Microbiol.*, **20** 263-71.

Galyov, E.E., Wood, M.W., Rosqvist, R., Mullan, P.B., Watson, P.R., Hedges, S. and Wallis, T.S. (1997) A secreted effector protein of *Salmonella dublin* is translocated into eukaryotic cells and mediates inflammation and fluid secretion in infected ileal mucosa. *Mol. Microbiol.*, **25** 903-12.

Gaudriault, S., Malandrin, L., Paulin, J.-P. and Barny, M.-A. (1997) DspA, an essential pathogenicity factor of *Erwinia amylovora* showing homology with AvrE of *Pseudomonas syringae*, is secreted via the Hrp secretion pathway in a DspB-dependent way. *Mol. Microbiol.*, **26** 1057-69.

Gaudriault, S., Brisset, M.N. and Barny, M.-A. (1998) HrpW of *Erwinia amylovora*, a new Hrp-secreted protein. *FEBS Lett.*, **29** 224-28.

Genin, S. and Boucher, C.A. (1994) A superfamily of proteins involved in different secretion pathways in Gram-negative bacteria: modular structure and specificity of the N-terminal domain. *Mol. Gen. Genet.*, **243** 112-18.

Genin, S., Gough, C.L., Zischek, C. and Boucher, C.A. (1992) Evidence that the *hrpB* gene encodes a positive regulator of pathogenicity genes from *Pseudomonas solanacearum*. *Mol. Microbiol.*, **6** 3065-76.

Gheysen, G., Villarroel, R. and Van Montagu, M. (1991) Illegitimate recombination in plants: a model for T-DNA integration. *Genes Dev.*, **5** 287-97.

Ginocchio, C.C., Olmsted, S.B., Wells, C.L. and Galan, J.E. (1994) Contact with epithelial cells induces the formation of surface appendages on *Salmonella typhimurium*. *Cell*, **76** 717-24.

Gopalan, S., Bauer, D.W., Alfano, J.A., Loniello, A.O., He, S.Y. and Collmer, A. (1996) Expression of the *Pseudomonas syringae* avirulence protein, AvrB, in plant cells alleviates its dependence on the hypersensitive response and pathogenicity (Hrp) secretion system in eliciting genotype-specific hypersensitive cell death. *Plant Cell*, **8** 1095-105.

Greenberg, J.T., Guo, A., Klessig, D.F. and Ausubel, F.M. (1994) Programmed cell death in plants: a pathogen-triggered response activated coordinately with multiple defense functions. *Cell*, **77** 551-64.

Gray, K.M. (1997) Intercellular communication and group behavior in bacteria. *Trends Microbiol.*, **5** 184-87.

Grimm, C. and Panopoulos, N.J. (1989) The predicted protein product of a pathogenicity locus from *Pseudomonas syringae* pv. *phaseolicola* is homologous to a highly conserved domain of several procaryotic regulatory proteins. *J. Bacteriol.*, **171** 5031-38.

Gross, D.C. (1991) Molecular and genetic analysis of toxin production by pathovars of *Pseudomonas syringae*. *Annu. Rev. Phytopathol.*, **29** 247-78.

Hakansson, S., Galyov, E.E., Rosqvist, R. and Wolf-Watz, H. (1996a) The *Yersinia* YpkA Ser/Thr kinase is translocated and subsequently targeted to the inner surface of the HeLa cell plasma membrane. *Mol. Microbiol.*, **20** 593-603.

Hakansson, S., Schesser, K., Persson, C., Galyov, E.E., Rosqvist, R., Homble, F. and Wolf-Watz, H. (1996b) The YopB protein of *Yersinia pseudotuberculosis* is essential for the translocation of Yop effector proteins across the target cell plasma membrane and displays a contact-dependent membrane disrupting activity. *EMBO J.*, **15** 5812-23.

Hammond-Kosack, K.E. and Jones, J.D.G. (1996) Resistance gene-dependent plant defense responses. *Plant Cell*, **8** 1773-91.

Hardie, K.R., Lory, S. and Pugsley, A.P. (1996a) Insertion of an outer membrane protein in *Escherichia coli* requires a chaperone-like protein. *EMBO J.*, **15** 978-88.

Hardie, K.R., Seydel, A., Guilvout, I. and Pugsley, A.P. (1996b) The secretin-specific, chaperone-like protein of the general secretory pathway: separation of proteolytic protection and piloting functions. *Mol. Microbiol.*, **22** 967-76.

He, S.Y. (1997) Hrp-controlled interkingdom protein transport: learning from flagellar assembly? *Trends Microbiol.*, **5** 489-95.

He, S.Y., Huang, H.C. and Collmer, A. (1993) *Pseudomonas syringae* pv. *syringae* harpinPss: a protein that is secreted via the Hrp pathway and elicits the hypersensitive response in plants. *Cell*, **73** 1255-66.

Herbers, K., Conrads-Strauch, J. and Bonas, U. (1992) Race-specificity of plant resistance to bacterial spot disease determined by repetitive motifs in a bacterial avirulence protein. *Nature*, **356** 172-74.

Herrera-Estrella, A., Chen, Z.M., Van Montagu, M. and Wang, K. (1988) VirD proteins of *Agrobacterium tumefaciens* are required for the formation of a co-valent DNA-protein complex at the 5′ terminus of T-strand molecules. *EMBO J.*, **7** 4055-62.

Howard, E.A. and Citovsky, V. (1990) The emerging structure of the *Agrobacterium* T-DNA transfer complex. *Bioessays*, **12** 103-108.

Howard, E.A., Zupan, J.R., Citovsky, V. and Zambryski, P.C. (1992) The VirD2 protein of *A. tumefaciens* contains a C-terminal bipartite nuclear localization signal: implications for nuclear uptake of DNA in plant cells. *Cell*, **68** 109-18.

Hoyos, M.E., Stanley, C.M., He, S.Y., Pike, S., Pu, X.-A. and Novacky, A. (1996) The interaction of HarpinPss with plant cell walls. *Mol. Plant-Microbe Interact.*, **9** 608-16.

Huang, H.C., Schuurink, R., Denny, T.P., Atkinson, M.M., Baker, C.J., Yucel, I., Hutcheson, S.W. and Collmer, A. (1988) Molecular cloning of a *Pseudomonas syringae* pv. *syringae* gene cluster that enables *Pseudomonas fluorescens* to elicit the hypersensitive response in tobacco plants. *J. Bacteriol.*, **170** 4748-56.

Huang, H.C., Lin, R.H., Chang, C.J., Collmer, A. and Deng, W.L. (1995) The complete *hrp* gene cluster of *Pseudomonas syringae* pv. *syringae* 61 includes two blocks of genes required for harpinPss secretion that are arranged colinearly with *Yersinia ysc* homologs. *Mol. Plant-Microbe Interact.*, **8** 733-46.

Hugouvieux-Cotte-Pattat, N., Condemine, G., Nasser, W. and Reverchon, S. (1996) Regulation of pectinolysis in *Erwinia chrysanthemi*. *Annu. Rev. Microbiol.*, **50** 213-57.

Huguet, E., Hahn, K., Wengelnik, K. and Bonas, U. (1998) *hpaA* mutants of *Xanthomonas campestris* pv. *vesicatoria* are affected in pathogenicity but retain the ability to induce host-specific hypersensitive reaction. *Mol. Microbiol.*, **29** 1379-90.

Hutchison, M.L., Tester, M.A. and Gross, D.C. (1995) Role of biosurfactant and ion channel-forming activities of syringomycin in transmembrane ion flux: a model for the mechanism of action in the plant-pathogen interaction. *Mol. Plant-Microbe Interact.*, **8** 610-20.

Huynh, T.V., Dahlbeck, D. and Staskawicz, B.J. (1989) Bacterial blight of soybean: regulation of a pathogen gene determining host cultivar specificity. *Science*, **245** 1374-77.

Innes, R.W., Bent, A.F., Kunkel, B.N., Bisgrove, S.R. and Staskawicz, B.J. (1993) Molecular analysis of avirulence gene, *avrRpt2*, and identification of a putative regulatory sequence common to all known *Pseudomonas syringae* avirulence genes. *J. Bacteriol.*, **175** 4859-69.

Jakobek, J.L. and Lindgren, P.B. (1993) Generalized induction of defense responses in bean is not correlated with the induction of the hypersensitive reaction. *Plant Cell*, **5** 49-56.

Jakobek, J.L., Smith, J.A. and Lindgren, P.B. (1993) Suppression of bean defense responses by *Pseudomonas syringae*. *Plant Cell*, **5** 57-63.

Ji, J., Hugouvieux-Cotte-Pattat, N. and Robert-Baudouy, J. (1987) Use of Mu-*lac* insertions to study the secretion of pectate lyases by *Erwinia chrysanthemi*. *J. Gen. Microbiol.*, **133** 793-802.

Jones, S., Yu, B., Bainton, N.J., Birdsall, M., Bycroft, B.W., Chhabra, R.R., Cox, A.J.R., Golby, P., Reeves, P.J., Stephens, S., Winson, M.K., Salmond, G.P.C., Stewart, G.S.A.B. and Williams, P. (1993) The *lux* autoinducer regulates the production of exoenzyme virulence determinants in *Erwinia carotovora* and *Pseudomonas aeruginosa*. *EMBO J.*, **12** 2477-82.

Kaniga, K., Bossio, J.C. and Galan, J.E. (1994) The *Salmonella typhimurium* invasion genes, *invF* and *invG*, encode homologues of the AraC and PulD family of proteins. *Mol. Microbiol.*, **13** 555-68.

Kao, C.C., Barlow, E. and Sequeira, L. (1992) Extracellular polysaccharide is required for wild-type virulence of *Pseudomonas solanacearum*. *J. Bacteriol.*, **174** 1068-71.

Kazmierczak, B.I., Mielke, D.L., Russel, M. and Model, P. (1994) pIV, a filamentous phage protein that mediates phage export across the bacterial cell envelope, forms a multimer. *J. Mol. Biol.*, **238** 187-98.

Kearney, B. and Staskawicz, B.J. (1990) Widespread distribution and fitness contribution of *Xanthomonas campestris* avirulence gene, *avrBs2*. *Nature*, **346** 385-86.

Klement, Z. (1982) Hypersensitivity, in *Phytopathogenic Prokaryotes* (eds. M.S. Mount and G.H. Lacy), Academic Press, New York, pp. 149-77.

Knoop, V., Staskawicz, B. and Bonas, U. (1991) Expression of the avirulence gene, *avrBs3*, from *Xanthomonas campestris* pv. *vesicatoria* is not under the control of *hrp* genes and is independent of plant factors. *J. Bacteriol.*, **173** 7142-50.

Knutton, S., Rosenshine, I., Pallen, M.J., Nisan, I., Neves, B.C., Bain, C., Wolff, C., Dougan, G. and Frankel, G. (1998) A novel EspA-associated surface organelle of enteropathogenic *Escherichia coli* involved in protein translocation into epithelial cells. *EMBO J.*, **17** 2166-76.

Kobe, B. and Deisenhofer, J. (1994) The leucine-rich repeat: a versatile binding motif. *Trends Biochem. Sci.*, **19** 415-21.

Koster, M., Bitter, W., de Cock, H., Allaoui, A., Cornelis, G.R. and Tommassen, J. (1997) The outer membrane component, YscC, of the Yop secretion machinery of *Yersinia enterocolitica* forms a ring-shaped multimeric complex. *Mol. Microbiol.*, **26** 789-97.

Kubori, T., Matsushima, Y., Nakamura, D., Uralil, J., Lara-Tejero, M., Sukhan, A., Galan, J.E. and Aizawa, S.I. (1998) Supramolecular structure of the *Salmonella typhimurium* type III protein secretion system. *Science*, **280** 602-605.

Lai, E.-M. and Kado, C.I. (1998) Processed VirB2 is the major subunit of the promiscuous pilus of *Agrobacterium tumefaciens*. *J. Bacteriol.*, **180** 2711-17.

Leach, J.E. and White, F.F. (1996) Bacterial avirulence genes. *Annu. Rev. Phytopathol.*, **34** 153-79.

Leister, R.T., Ausubel, F.M. and Katagiri, F. (1996) Molecular recognition of pathogen attack occurs inside of plant cells in plant disease resistance specified by the *Arabidopsis* genes, *RPS2* and *RPM1*. *Proc. Natl. Acad. Sci. USA*, **93** 15497-502.

Li, J., Ochman, H., Groisman, E.A., Boyd, E.F., Solomon, F., Nelson, K. and Selander, R.K. (1995) Relationship between evolutionary rate and cellular location among the Inv/Spa invasion proteins of *Salmonella enterica*. *Proc. Natl. Acad. Sci. USA*, **92** 7252-56.

Lindgren, P.B., Peet, R.C. and Panopoulos, N.J. (1986) Gene cluster of *Pseudomonas syringae* pv. '*phaseolicola*' controls pathogenicity of bean plants and hypersensitivity of non-host plants (published erratum appears in *J. Bacteriol.* (1987) **169** 928). *J. Bacteriol.*, **168** 512-22.

Lindgren, P.B., Panopoulos, N.J., Staskawicz, B.J. and Dahlbeck, D. (1988) Genes required for pathogenicity and hypersensitivity are conserved and interchangeable among pathovars of *Pseudomonas syringae*. *Mol. Gen. Genet.*, **211** 4499-506.

Lindgren, P.B., Frederick, R., Govindarajan, A.G., Panopoulos, N.J., Staskawicz, B.J. and Lindow, S.E. (1989) An ice nucleation reporter gene system: identification of inducible pathogenicity genes in *Pseudomonas syringae* pv. *phaseolicola* (published erratum appears in *EMBO J.*, (1989) **8** 2121). *EMBO J.*, **8** 1291-301.

Loh, Y.T. and Martin, G.B. (1995) The disease-resistance gene, *Pto*, and the fenthion-sensitivity gene, *fen*, encode closely-related functional protein kinases. *Proc. Natl. Acad. Sci. USA*, **92** 4181-84.

Lorang, J.M. and Keen, N.T. (1995) Characterization of *avrE* from *Pseudomonas syringae* pv. *tomato*: an *hrp*-linked avirulence locus consisting of at least two transcriptional units. *Mol. Plant-Microbe Interact.*, **8** 49-57.

Lorang, J.M., Shen, H., Kobayashi, D., Cooksey, D. and Keen, N.T. (1994) *avrA* and *avrE* in *Pseudomonas syringae* pv. *tomato* PT23 play a role in virulence on tomato plants. *Mol. Plant-Microbe Interact.*, **7** 508-15.

Macnab, R.M. (1996) Flagella and motility, in *Escherichia coli and Salmonella: Cellular and Molecular biology* (eds. F.C. Neidhardt, R. Curtiss III, J.L. Ingraham, E.C.C. Lin, K.B. Low, B. Magasanik, W.S. Reznikoff, M. Riley, M. Schaechter and H.E. Umbarger), American Society for Microbiology Press, Washington DC, pp. 123-45.

Mansfield, J., Brown, I.R. and Maroofi, A. (1994) Bacterial pathogenicity and the plant response: ultrastructural, biochemical and physiological perspectives, in *Biotechnology and Plant Protection* (eds. D.D. Bills and S.-D. Kung), World Scientific Publishing Co., Singapore.

Marenda, M., Brito, B., Callard, D., Genin, S., Barberis, P., Boucher, C. and Arlat, M. (1998) PrhA controls a novel regulatory pathway required for the specific induction of *Ralstonia solanacearum hrp* genes in the presence of plant cells. *Mol. Microbiol.*, **27** 437-53.

Martin, D.W., Holloway, B.W. and Deretic, V. (1993a) Characterization of a locus determining the mucoid status of *Pseudomonas aeruginosa*: AlgU shows sequence similarities with a *Bacillus* sigma factor. *J. Bacteriol.*, **175** 1153-54.

Martin, G.B., Brommonschenkel, S.H., Chunwongse, J., Frary, A., Ganal, M.W., Spivey, R., Wu, T., Earle, E.D. and Tanksley, S.D. (1993b) Map-based cloning of a protein kinase gene conferring disease resistance in tomato. *Science*, **262** 1432-36.

Meinhardt, L.W., Krishnan, H.B., Balatti, P.A. and Pueppke, S.G. (1993) Molecular cloning and characterization of a sym plasmid locus that regulates cultivar-specific nodulation of soybean by *Rhizobium fredii* USDA257. *Mol. Microbiol.*, **9** 17-29.

Menard, R., Sansonetti, P. and Parsot, C. (1994) The secretion of the *Shigella flexneri* Ipa invasins is activated by epithelial cells and controlled by IpaB and IpaD. *EMBO J.*, **13** 5293-302.

Menard, R., Dehio, C. and Sansonetti, P.J. (1996) Bacterial entry into epithelial cells: the paradigm of *Shigella*. *Trends Microbiol.*, **4** 220-26.

Michiels, T., Vanooteghem, J.-C., Lambert de Rouvroit, C., China, B., Gustin, A., Boudry, P. and Cornelis, G.R. (1991) Analysis of *virC*, an operon involved in the secretion of Yop proteins by *Yersinia enterocolitica*. *J. Bacteriol.*, **173** 4994-5009.

Mills, S.D., Boland, A., Sory, M.P., van der Smissen, P., Kerbourch, C., Finlay, B.B. and Cornelis, G.R. (1997) *Yersinia enterocolitica* induces apoptosis in macrophages by a process requiring functional type III secretion and translocation mechanisms and involving YopP, presumably acting as an effector protein. *Proc. Natl. Acad. Sci. USA*, **94** 12638-43.

Mittal, S. and Davis, K.R. (1995) Role of the phytotoxin coronatine in the infection of *Arabidopsis thaliana* by *Pseudomonas syringae* pv. *tomato*. *Mol. Plant-Microbe Interact.*, **8** 165-71.

Nasser, W., Condemine, G., Plentier, R., Anker, D. and Robert-Baudouy, J. (1991) Inducing properties of analogs of 2-keto-3-deoxygluconate on the expression of pectinase genes of *Erwinia chrysanthemi*. *FEMS Microbiol. Letts.*, **65** 73-78.

Nasser, W., Bouillant, M.-L., Salmond, G. and Reverchon, S. (1998) Characterization of the *Erwinia chrysanthemi expI-expR* locus directing the synthesis of two *N*-acyl-homoserine lactone signal molecules. *Mol. Microbiol.*, **29** 1391-405.

Nealson, K.H., Platt, T. and Hastings, J.W. (1970) Cellular control of the synthesis and activity of the bacterial luminescent system. *J. Bacteriol.*, **104** 313-22.

Palmer, D.A. and Bender, C.L. (1995) Ultrastructure of tomato leaf tissue treated with the pseudomonad phytotoxin, coronatine, and comparison with methyl jasmonate. *Mol. Plant-Microbe Interact.*, **8** 683-92.

Parsot, C. (1994) *Shigella flexneri*: genetics of entry and intercellular dissemination in epithelial cells. *Curr. Topics Microbiol. Immunol.*, **192** 217-41.

Parsot, C., Menard, R., Gounon, P. and Sansonetti, P.J. (1995) Enhanced secretion through the *Shigella flexneri* Mxi-Spa translocon leads to assembly of extracellular proteins into macromolecular structures. *Mol. Microbiol.*, **16** 291-300.

Persson, C., Nordfelth, R., Holmstrom, A., Hakansson, S., Rosqvist, R. and Wolf-Watz, H. (1995) Cell-surface-bound *Yersinia* translocate the protein tyrosine phosphatase, YopH, by a polarized mechanism into the target cell. *Mol. Microbiol.*, **18** 135-50.

Pettersson, J., Nordfelth, R., Dubinina, E., Bergman, T., Gustafsson, M., Magnusson, K.E. and Wolf-Watz, H. (1996) Modulation of virulence factor expression by pathogen target cell contact. *Science*, **273** 1231-33.

Piper, K.R., Beck von Bodman, S. and Farrand, S.K. (1993) Conjugation factor of *Agrobacterium tumefaciens* regulates Ti plasmid transfer by autoinduction. *Nature*, **362** 448-50.

Pirhonen, M., Flego, D., Heikinheimo, R. and Palva, E.T. (1993) A small diffusible signal molecule is responsible for the global control of virulence and exoenzyme production in the plant pathogen, *Erwinia carotovora. EMBO J.*, **12** 2467-76.

Pirhonen, M.U., Lidell, M.C., Rowley, D.L., Lee, S.W., Jin, S., Liang, Y., Silverstone, S., Keen, N.T. and Hutcheson, S.W. (1996) Phenotypic expression of *Pseudomonas syringae avr* genes in *E. coli* is linked to the activities of the *hrp*-encoded secretion system. *Mol. Plant-Microbe Interact.*, **9** 252-60.

Plano, G.V. and Straley, S.C. (1995) Mutations in *yscC, yscD* and *yscG* prevent high-level expression and secretion of V antigen and Yops in *Yersinia pestis. J. Bacteriol.*, **177** 3843-54.

Plano, G.V., Barve, S.S. and Straley, S.C. (1991) LcrD, a membrane-bound regulator of the *Yersinia pestis* low-calcium response. *J. Bacteriol.*, **173** 7293-303.

Preston, G., Huang, H.C., He, S.Y. and Collmer, A. (1995) The HrpZ proteins of *Pseudomonas syringae* pvs. *syringae, glycinea* and *tomato* are encoded by an operon containing *Yersinia ysc* homologs and elicit the hypersensitive response in tomato but not soybean. *Mol. Plant-Microbe Interact.*, **8** 717-32.

Pugsley, A.P. (1993) The complete general secretory pathway in Gram-negative bacteria. *Microbiol. Rev.*, **57** 50-108.

Rahme, L.G., Mindrinos, M.N. and Panopoulos, N.J. (1991) Genetic and transcriptional organization of the *hrp* cluster of *Pseudomonas syringae* pv. *phaseolicola* (published erratum appears in *J. Bacteriol.* (1992) **174** 3840). *J. Bacteriol.*, **173** 575-86.

Rahme, L.G., Mindrinos, M.N. and Panopoulos, N.J. (1992) Plant and environmental sensory signals control the expression of *hrp* genes in *Pseudomonas syringae* pv. *phaseolicola. J. Bacteriol.*, **174** 3499-507.

Reverchon, S., Bouillant, M.-L., Salmond, G. and Nasser, W. (1998) Integration of the quorum-sensing system in the regulatory networks controlling virulence factor synthesis in *Erwinia chrysanthemi. Mol. Microbiol.*, **29** 1407-18.

Ried, J.L. and Collmer, A. (1986) Comparison of pectic enzymes produced by *Erwinia chrysanthemi, Erwinia carotovora* subsp. *carotovora* and *Erwinia carotovora* subsp. *atroseptica. Appl. Environ. Microbiol.*, **52** 305-10.

Ried, J.L. and Collmer, A. (1988) Construction and characterization of an *Erwinia chrysanthemi* mutant with directed deletion in all of the pectate lyase structural genes. *Mol. Plant-Microbe Interact.*, **1** 32-38.

Ritter, C. and Dangl, J.L. (1995) The *avrRpm1* gene of *Pseudomonas syringae* pv. *maculicola* is required for virulence on *Arabidopsis. Mol. Plant-Microbe Interact.*, **8** 444-53.

Roine, E., Wei, W., Yuan, J., Nurmiaho-Lassila, E.L., Kalkkinen, N., Romantschuk, M. and He, S.Y. (1997) Hrp pilus: an *hrp*-dependent bacterial surface appendage produced by *Pseudomonas syringae* pv. *tomato* DC3000. *Proc. Natl. Acad. Sci. USA*, **94** 3459-64.

Rommens, C.M., Salmeron, J.M., Baulcombe, D.C. and Staskawicz, B.J. (1995) Use of a gene expression system based on potato virus X to rapidly identify and characterize a tomato Pto homolog that controls fenthion sensitivity. *Plant Cell*, **7** 249-57.

Rosqvist, R., Forsberg, A., Rimpilainen, M., Bergman, T. and Wolf-Watz, H. (1990) The cytotoxic protein, YopE, of *Yersinia* obstructs the primary host defence. *Mol. Microbiol.*, **4** 657-67.

Rosqvist, R., Forsberg, A. and Wolf-Watz, H. (1991) Intracellular targeting of the *Yersinia* YopE cytotoxin in mammalian cells induces actin microfilament disruption. *Infect. Immun.*, **59** 4562-69.

Rosqvist, R., Magnusson, K.E. and Wolf-Watz, H. (1994) Target cell contact triggers expression and polarized transfer of *Yersinia* YopE cytotoxin into mammalian cells. *EMBO J.*, **1390** 964-72.

Russel, M. (1994) Phage assembly: a paradigm for bacterial virulence factor export? *Science*, **265** 612-14.

Ryan, C.L. and Farmer, E.E. (1991) Oligosaccharide signals in plants: a current assessment. *Annu. Rev. Plant Physiol. Plant Mol. Biol.*, **42** 651-74.

Sakai, T., Sasakawa, C. and Yoshikawa, M. (1988) Expression of four virulence antigens of *Shigella flexneri* is positively regulated at the transcriptional level by the 30-kDa VirF protein. *Mol. Microbiol.*, **2** 589-97.

Salmeron, J.M. and Staskawicz, B.J. (1993) Molecular characterization and *hrp* dependence of the avirulence gene, *avrPto*, from *Pseudomonas syringae* pv. *tomato*. *Mol. Gen. Genet.*, **239** 6-16.

Salmond, G.P.C. (1994) Secretion of extracellular virulence factors by plant pathogenic bacteria. *Annu. Rev. Phytopathol.*, **32** 181-200.

Sasakawa, C., Komatsu, K., Tobe, T., Suzuki, T. and Yoshikawa, M. (1993) Eight genes in region 5 that form an operon are essential for invasion of epithelial cells by *Shigella flexneri* 2a. *J. Bacteriol.*, **175** 2334-46.

Schulte, R. and Bonas, U. (1992) A *Xanthomonas* pathogenicity locus is induced by sucrose and sulfur-containing amino acids. *Plant Cell*, **4** 79-86.

Scofield, S.R., Tobias, C.M., Rathjen, J.P., Chang, J.H., Lavelle, D.T., Michelmore, R.W. and Staskawicz, B.J. (1996) Molecular basis of gene-for-gene specificity in bacterial speck disease of tomato. *Science*, **274** 2063-65.

Shen, H. and Keen, N.T. (1993) Characterization of the promoter of avirulence gene D from *Pseudomonas syringae* pv. *tomato*. *J. Bacteriol.*, **175** 5916-24.

Sheng, J. and Citovsky, V. (1996) *Agrobacterium*-plant cell DNA transport: have virulence proteins, will travel. *Plant Cell*, **8** 1699-710.

Shevchik, V.E., Robert-Baudouy, J. and Condemine, G. (1997) Specific interaction between OutD, an *Erwinia chrysanthemi* outer membrane protein of the general secretory pathway, and secreted proteins. *EMBO J.*, **16** 3007-16.

Shirasu, K., Koukolikova-Nicola, Z., Hohn, B. and Kado, C.I. (1994) An inner-membrane-associated virulence protein essential for T-DNA transfer from *Agrobacterium tumefaciens* to plants exhibits ATPase activity and similarities to conjugative transfer genes. *Mol. Microbiol.*, **11** 581-88.

Shurvinton, C.E., Hodges, L. and Ream, W. (1992) A nuclear localization signal and the C-terminal omega sequence in the *Agrobacterium tumefaciens* VirD2 endonuclease are important for tumor formation. *Proc. Natl. Acad. Sci. USA*, **89** 11837-41.

Sory, M.P. and Cornelis, G.R. (1994) Translocation of a hybrid YopE-adenylate cyclase from *Yersinia enterocolitica* into HeLa cells. *Mol. Microbiol.*, **14** 583-94.

Sory, M.P., Boland, A., Lambermont, I. and Cornelis, G.R. (1995) Identification of the YopE and YopH domains required for secretion and internalization into the cytosol of macrophages, using the *cyaA* gene fusion approach. *Proc. Natl. Acad. Sci. USA*, **92** 11998-2002.

Stachel, S.E., Timmerman, B. and Zambryski, P. (1986) Generation of single-stranded T-DNA molecules during the initial stages of T-DNA transfer for *Agrobacterium tumefaciens* to plant cells. *Nature*, **322** 706-12.

Steinberger, E.M. and Beer, S.V. (1988) Creation and complementation of pathogenicity mutants of *Erwinia amylovora*. *Mol. Plant-Microbe Interact.*, **1** 135-44.

Stevens, C., Bennett, M.A., Athanassopoulos, E., Tsiamis, G., Taylor, J.D. and Mansfield, J.W. (1998) Sequence variations in alleles of the avirulence gene, *avrPphE.R2*, from *Pseudomonas syringae* pv. *phaseolicola* lead to loss of recognition of the AvrPphE protein within bean cells and gain in cultivar specific virulence. *Mol. Microbiol.*, **29** 165-77.

Swarup, S., Yang, Y., Kingsley, M.T. and Gabriel, D.W. (1992) An *Xanthomonas citri* pathogenicity gene, *pthA*, pleiotropically encodes gratuitous avirulence on non-hosts. *Mol. Plant-Microbe Interact.*, **5** 204-13.

Swords, K.M.M., Dahlbeck, D., Kearney, B., Roy, M. and Staskawicz, B.J. (1996) Spontaneous and induced mutations in a single open reading frame alter both virulence and avirulence in *Xanthomonas campestris* pv. *vesicatoria avrBs2*. *J. Bacteriol.*, **178** 4661-69.

Tang, X., Frederick, R.D., Zhou, J., Halterman, D.A., Jia, Y. and Martin, G.B. (1996) Initiation of plant disease resistance by physical interaction of AvrPto and Pto kinase. *Science*, **274** 2060-63.

Thurn, K.K. and Chatterjee, A.K. (1985) Single-site chromosomal Tn*5* insertions affect the export of pectolytic and cellulolytic enzymes in *Erwinia chrysanthemi* EC16. *Appl. Environ. Microbiol.*, **50** 894-98.

van den Ackerveken, G., Marois, E. and Bonas, U. (1996) Recognition of the bacterial avirulence protein, AvrBs3, occurs inside the host plant cell. *Cell*, **87** 1307-16.

van Gijsegem, F., Gough, C., Zischek, C., Niqueux, E., Arlat, M., Genin, S., Barberis, P., German, S., Castello, P. and Boucher, C. (1995) The *hrp* gene locus of *Pseudomonas solanacearum*, which controls the production of a type III secretion system, encodes eight proteins related to components of the bacterial flagellar biogenesis complex. *Mol. Microbiol.*, **15** 1095-114.

Walters, K., Maroofi, A., Hitchin, E. and Mansfield, J. (1990) Gene for pathogenicity and ability to cause hypersensitive reaction cloned from *Erwinia amylovora*. *Physiol. Mol. Plant Pathol.*, **36** 509-21.

Walker, J.E., Saraste, M., Runswick, M.J. and Gay, N.J. (1982) Distantly-related sequences in the α- and β-subunits of ATP synthase, myosin, kinases and other ATP-requiring enzymes and a common nucleotide binding fold. *EMBO J.*, **1** 945-51.

Walton, J.D. (1996) Host-selective toxins: agents of compatibility. *Plant Cell*, **8** 1723-33.

Wandersman, C. (1996) Secretion across the bacterial outer membrane, in *Escherichia coli and Salmonella: Cellular and Molecular Biology* (eds. F.C. Neidhardt, R. Curtiss III, J.L. Ingraham, E.C.C. Lin, K.B. Low, B. Magasanik, W.S. Reznikoff, M. Riley, M. Schaechter and H.E. Umbarger), American Society for Microbiology Press, Washington DC, pp. 955-65.

Watarai, M., Tobe, T., Yoshikawa, M. and Sasakawa, C. (1995) Contact of *Shigella* with host cells triggers release of Ipa invasins and is an essential function of invasiveness. *EMBO J.*, **14** 2461-70.

Wei, Z.M., Laby, R.J., Zumoff, C.H., Bauer, D.W., He, S.Y., Collmer, A. and Beer, S.V. (1992a) Harpin, elicitor of the hypersensitive response produced by the plant pathogen, *Erwinia amylovora*. *Science*, **257** 85-88.

Wei, Z.M., Sneath, B.J. and Beer, S.V. (1992b) Expression of *Erwinia amylovora hrp* genes in response to environmental stimuli. *J. Bacteriol.*, **174** 1875-82.

Weiler, E.W., Kutchan, T.M., Gorba, T., Brodschelm, W., Niesel, U. and Bublitz, F. (1994) The *Pseudomonas* phytotoxin, coronatine, mimics octadecanoid signalling molecules of higher plants (published erratum appears in *FEBS Lett* (1994) **349** 317). *FEBS Lett.*, **345** 9-13.

Weiss, A.A., Johnson, F.D. and Burns, D.L. (1993) Molecular characterization of an operon required for *Pertussis* toxin secretion. *Proc. Natl. Acad. Sci. USA*, **90** 2970-74.

Wengelnik, K. and Bonas, U. (1996) HrpXv, an AraC-type regulator, activates expression of five out of six loci in the *hrp* cluster of *Xanthomonas campestris* pv. *vesicatoria*. *J. Bacteriol.*, **178** 3462-69.

Wengelnik, K., Marie, C., Russel, M. and Bonas, U. (1996a) Expression and localization of HrpA1, a protein of *Xanthomonas campestris* pv. *vesicatoria* essential for pathogenicity and induction of the hypersensitive reaction. *J. Bacteriol.*, **178** 1061-69.

Wengelnik, K., van den Ackerveken, G. and Bonas, U. (1996b) HrpG, a key *hrp* regulatory protein of *Xanthomonas campestris* pv. *vesicatoria* is homologous to two-component response regulators. *Mol. Plant-Microbe Interact.*, **9** 704-12.

Willis, D.K., Barta, T.M. and Kinscherf, T.G. (1991) Genetics of toxin production and resistance in phytopathogenic bacteria. *Experientia*, **47** 765-71.

Winans, S.C., Mantis, N.J., Chen, C.Y., Chang, C.H. and Han, D.C. (1994) Host recognition by the VirA, VirG two-component regulatory proteins of *Agrobacterium tumefaciens. Res. Microbiol.*, **145** 461-73.

Woestyn, S., Allaoui, A., Wattiau, P. and Cornelis, G.R. (1994) YscN, the putative energizer of the *Yersinia* Yop secretion machinery. *J. Bacteriol.*, **176** 1561-69.

Wolff, C., Nisan, I., Hanski, E., Frankel, G. and Rosenshine, I. (1998) Protein translocation into host epithelial cells by infecting enteropathogenic *Escherichia coli. Mol. Microbiol.*, **28** 143-55.

Wood, M.W., Rosqvist, R., Mullan, P.B., Edwards, M.H. and Galyov, E.E. (1996) SopE, a secreted protein of *Salmonella dublin*, is translocated into the target eukaryotic cell via a *sip*-dependent mechanism and promotes bacterial entry. *Mol. Microbiol.*, **22** 327-38.

Xiao, Y., Lu, Y., Heu, S. and Hutcheson, S.W. (1992) Organization and environmental regulation of the *Pseudomonas syringae* pv. *syringae* 61 *hrp* cluster. *J. Bacteriol.*, **174** 1734-41.

Xiao, Y., Heu, S., Yi, J., Lu, Y. and Hutcheson, S.W. (1994) Identification of a putative alternate sigma factor and characterization of a multicomponent regulatory cascade controlling the expression of *Pseudomonas syringae* pv. *syringae* Pss61 *hrp* and *hrmA* genes. *J. Bacteriol.*, **176** 1025-36.

Young, S.A., White, F.F., Hopkins, C.M. and Leach, J.E. (1994) AvrXa10 protein is in the cytoplasm of *Xanthomonas oryzae* pv. *oryzae. Mol. Plant-Microbe Interact.*, **7** 799-804.

Zhou, J., Tang, X. and Martin, G.B. (1997) The Pto kinase conferring resistance to tomato bacterial speck disease interacts with proteins that bind a *cis*-element of pathogenesis-related genes. *EMBO J.*, **16** 3207-18.

Zierler, M.K. and Galan, J.E. (1995) Contact with cultured epithelial cells stimulates secretion of *Salmonella typhimurium* invasion protein, InvJ. *Infect. Immun.*, **63** 4024-48.

3 Viral pathogenicity

Sarah Nettleship and Gary D. Foster

3.1 Introduction

Plant viruses can be defined as microscopic transmissible parasites, with small nucleic acid genomes, that require host cell components/metabolism to complete their infection cycle. The small nucleic acid genome equates to a relatively low number of encoded proteins, typically between four and ten. It was probably the simplicity of viruses that initially attracted many research workers to the science. However, recent research has begun to reveal that they may well have been misled.

Certainly, at first glance, and as described in many major virology/pathology texts (e.g. Matthews, 1991), viruses are simple parasites, much simpler in structure and genome organisation than cellular microbial pathogens. If you were to sit down in your armchair or at a desk and set about designing a plant virus, you could easily draw up a list of genes that your designer plant virus would require. The template for your viral genome would probably be single-stranded (ss) positive (+)-sense RNA, as this occurs in approximately 75% of all plant viruses. Assuming that all plant viruses are molecular parasites, deriving many of their requirements from their host, there would still be a few basic genes that the virus would have to encode for itself. As plants do not have specific RNA polymerases that copy viral RNA into RNA, you would require your own RNA-dependent RNA polymerase (or replicase) to copy your genome, and thus multiply. Once you have multiple copies of your RNA genome, you would require a coat protein (or capsid protein) to 'wrap-up' (encapsidate) your RNA and protect it from breakage, degradation, etc. Once you have entered one cell, produced multiple copies of yourself and encapsidated your genomes into virus particles, it would then be time to move on to the next cell and repeat the whole process again. The simplest way for a virus to move from cell-to-cell would be through the plants intracellular connections, the plasmodesmata. Unfortunately, the dimensions of viral RNA and/or viral particles are too large to pass through the plasmodesmata passively. In designing your simple virus, you would, therefore, want to encode a protein that could somehow alter the plasmodesmata to allow virus movement from cell-to-cell. Once cell-to-cell movement has taken the virus as far as the vascular tissue, it can be passively moved around the plant through the phloem, along with the flow

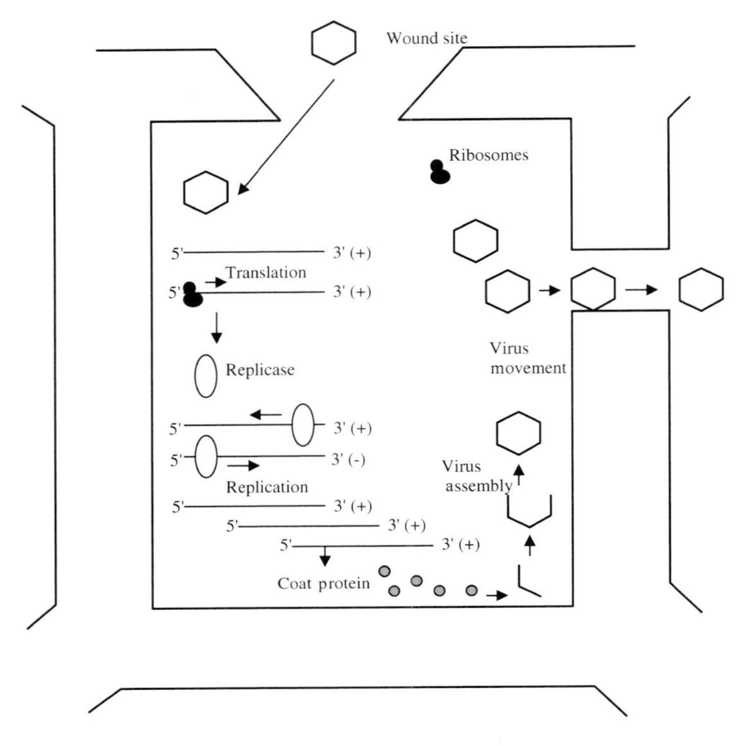

Figure 3.1 Simple infection cycle of a hypothetical plant virus. Major events are depicted, such as entry through the wound site, uncoating, translation, replication, encapsidation and virus movement from cell-to-cell.

of photoassimilates. Therefore, assuming that this basic virus can acquire all other needs from the plant, it needs to encode three basic proteins: a replicase to copy/multiply its genome; a coat protein to wrap-up/protect this genome; and a movement protein to allow passage from cell-to-cell.

To summarise the infection cycle of a basic virus (see Figure 3.1), it would enter through wounds by mechanical transmission. Once inside the cell, it would uncoat, express its genome, replicate, and move from cell-to-cell through the plasmodesmata. Once it had reached the vascular tissue, it would be rapidly transported through the phloem, allowing infection throughout the plant. Does such an ideal, simple virus exist? The answer certainly appeared to be yes a few years ago, in the form of tobacco mosaic virus (TMV). Scan any virology textbook or review article on plant viruses and it appeared that TMV could be described as one of the simplest viruses (see Figure 3.2). TMV was the first true virus to be described, in a publication by Beijerinck (1898), and has been studied intensively ever since (for a review of 100 years of TMV research, see

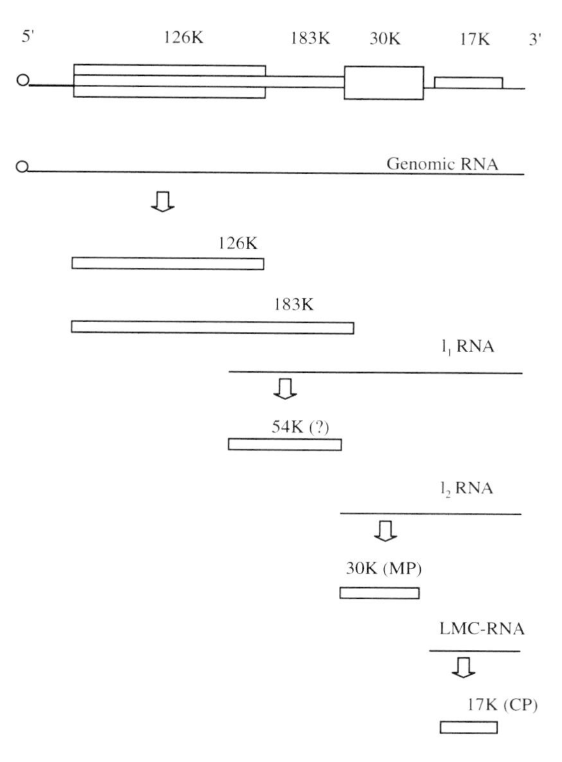

Figure 3.2 Genomic organisation of genomic and subgenomic RNAs of tobacco mosaic virus (TMV). Abbreviations: CP, coat protein; MP, movement protein.

Harrison and Wilson, 1999). However, in recent years, as scientists have probed the functions of the individual TMV proteins, the story that has emerged is that the simple model for TMV infection and the proposed roles of viral encoded proteins are no longer true. We now have an amazing insight into viral proteins with multiple roles, interacting with viral and host processes and resistance mechanisms in multiple ways.

3.2 Viral gene functions (then, now and in the future)

3.2.1 TMV as a very basic example

TMV is the type member of the tobamovirus genus, which produce rigid rod-shaped particles of ca. 300 × 18 nm, and contain a ss (+)-sense RNA, 6395 nucleotides long. As discussed in section 3.1, information on TMV has been collated for many years, and as a result simple models for viral replication and gene expression have been developed. Such a model for TMV is illustrated in Figure 3.2 and outlined below:

1) Viral particles enter the cell and the removal of the coat protein from the 5′ end of the TMV particles facilitates the translation of two proteins involved in the RNA-dependent RNA polymerase (replicase), the 126K and 183K proteins. The larger, 183K, protein is produced by readthrough of a leaky amber stop codon at the end of the 126K open reading frame (ORF).
2) The replicase proteins recognise sequences at the 3′ end of the genomic (+)-sense RNA and generate a complementary copy of negative (−)-sense RNA.
3) The replicase recognises sequences at the 3′ end of the (−)-sense RNA and generates genomic (+)-sense RNA.
4) The replicase proteins are also able to recognise internal subgenomic promoters within the (−)-sense RNA, allowing the generation of 3′ co-terminal (+)-sense subgenomic messenger ribonucleic acids (mRNAs).
5) A protein of 30K is translated from the 5′ end of subgenomic I_2. This 30K protein facilitates cell-to-cell movement of TMV.
6) The 17K coat protein is translated from the subgenomic RNA, designated LMC, and interacts with a specific sequence on the genomic RNA, designated the OAS (origin of assembly), to encapsidate the RNA within the viral particles.

It should be noted that a putative 54K protein might be translated from the 5′ end of subgenomic I_1, although such a protein has yet to be detected within infected tissues. The allocation of function to each of the viral proteins has been a major achievement over recent years (Harrison and Wilson, 1999). It is now clear that the 126K and 183K proteins are the main proteins involved in replication, the 30K movement protein (MP) facilitates cell-to-cell movement, and the 17K protein represents the viral coat protein (CP). However, continued research has revealed that each of these proteins may have additional roles; for example, it is clear that the CP protects the RNA genome by encapsidation into the virus particles. However, the CP has also been implicated in long-distance movement of TMV, with mutants unable to express the CP being unable to infect non-inoculated leaves (Takamatsu *et al.*, 1987; Dawson *et al.*, 1988). Transgenic *Nicotiana tabacum* plants expressing coat protein have been shown to compensate for coat-protein-defective mutants by increasing the systemic spread of such TMV mutants (Holt and Beachy, 1991).

Whilst it may be seen as a simple conclusion that the 30K protein facilitates cell-to-cell movement, it is obvious that such a protein must be multifunctional, considering what it has to achieve. This is a protein that is translated somewhere in the cytoplasm, that needs to make its way to the plant cell wall, find a plasmodesmata, alter the size exclusion and, at the same time, escort viral RNA and/or particles to and through the plasmodesmata. From recent research, it has become clear that in addition to signalling the enlargement of the plasmodesmata, the 30K MP has a repertoire of functions, of which at least six are currently recognised:

1) The enlargement of the plasmodesmata (Wolf *et al.*, 1989).
2) Binding of ssRNA in a non-specific manner, forming unfolded virus-specific ribonucleotide particles (vRNPs) that can penetrate the plasmodesmatal channel (Citovsky *et al.*, 1990, 1992).
3) Interacting with cytoskeletal elements, which may possibly help escort the vRNPs to the plasmodesmata (Heinlein *et al.*, 1995; McLean *et al.*, 1995).
4) By becoming phosphorylated, to regulate interactions with the plasmodesmata (Watanabe *et al.*, 1992; Citovsky *et al.*, 1993; Haley *et al.*, 1995; Kawakami *et al.*, 1999).
5) Binding to cell-wall-associated proteins, such as pectin methylesterase (PME) (reviewed in Rhee *et al.*, 2000).
6) MP has now been implicated in the control of translation of viral RNAs (Karporva *et al.*, 1997, 1999).

The identified functions of the TMV MP continue to expand. Indeed, the recent discovery that MP can be found associated with replication bodies, further implies a crucial central role for this protein in viral RNA and protein synthesis, and formation of movement complexes (Heinlein *et al.*, 1998).

The replicase proteins of TMV are made up of the 126K and 183K proteins, which are clearly complex with several conserved functional motifs within their sequence that can be attributed to the process of replication (reviewed in Buck, 1999). However, there is growing evidence that replicase proteins of TMV (Deom *et al.*, 1997) and other viruses (Gal-On *et al.*, 1994; Weiland and Edwards, 1996) may be involved in virus movement as well as replication. A link between virus replication and movement has also been suggested through mutations located within the 126K and 183K proteins of the masked strain of TMV that can affect levels of virus accumulation within the phloem (Derrick *et al.*, 1997).

Whilst multiple functions are now being assigned to the other TMV proteins, the role of the 54K protein, which may be encoded from the I_1 subgenomic mRNA, still remains a mystery. A subgenomic exists for the 54K, which can direct the translation of this protein *in vitro*, and the RNA can be found associated with polyribosomes *in planta* (Sulzinski *et al.*, 1985). However, no 54K protein has yet been detected within infected plant material, protoplasts or purified replicase complexes. The role and interaction of viral protein in the infection process of TMV may be further complicated when the role (if any) of the 54K protein is finally elucidated. Indeed, the story may not end there, as Morozov *et al.* (1993) have revealed the existence of an additional potential small ORF, beginning in the MP gene and extending into the CP gene. When cloned, this ORF directed the translation of a 4K protein *in vitro*, which had unusual properties within the *in vitro* translation system (Fedorkin *et al.*, 1995), although the putative protein encoded by this mRNA has yet to be detected *in vivo*.

3.2.2 Potyviruses as a more complicated example (and getting more complicated)

If TMV is a simple virus, encoding three basic classes of proteins (replicase, MP and CP), that is becoming increasingly complicated/intriguing, then what about those viruses encoding substantially more proteins, such as the potyviruses?

The potyviruses are the single largest genus of plant viruses and are probably the most important from an economic/agricultural standpoint (for a major review see Shukla *et al.*, 1994). Typically, potyviruses are aphid transmitted, with filamentous particles of ca. 12×900 nm, containing a single molecule of (+)-sense RNA ca. 8,500–10,000 nucleotides in length. The viral RNA is translated into a single large polyprotein that is autoproteolytically cleaved to provide at least nine mature viral proteins (see Figure 3.3 and Table 3.1). The assignment of primary functions to each of these proteins has been a major achievement over the years. However, research is now revealing, as with TMV, multiple roles for each viral protein, with ever increasing complexity of interactions between the viral proteins, and between the viral proteins and the host.

Reflecting the early simplistic models of viral infection processes, viral genes were originally ascribed just one function, which was determined by comparing viral nucleotide sequence/amino acid sequence with genes/proteins of known function or by studying the infection process of naturally-occurring or engineered mutant viruses. Taking potyviruses as an example, the functions that were assigned to the viral gene products in the comprehensive review by Dougherty and Carrington (1988) were (in order from the N-terminus of the potyviral polyprotein, with names given indicating the real or expected molecular weights of the proteins for tobacco etch virus [TEV]): 31K, cell-to-cell movement; 56K, insect transmission (helper component); 50K, polyprotein processing (proteinase); 70K (CI), replication; 6K (VPg), replication, binding viral genomic RNA; 49K, polyprotein processing (proteinase); 58K, replication (RNA-dependent RNA-polymerase); 30K, encapsidation of viral particles, insect transmission (capsid/coat protein) (summarised in Table 3.1). The 70K protein was known to be responsible for the formation of pinwheel-like structures in the cytoplasm of infected cells, and was therefore named the cytoplasmic (or cylindrical) inclusion (CI) protein. The 49K and 58K proteins became jointly known as the nuclear inclusion (NI) proteins, as in some

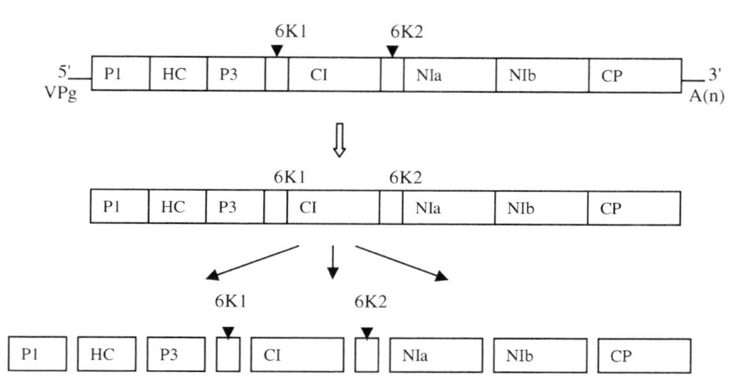

Figure 3.3 Genomic organisation of a typical potyvirus. Abbreviations: VPg, viral genome-linked protein; P1, P1 proteinase; HC, helper-component proteinase; P3, P3 protein; 6K1, 6K protein 1; CI, cytoplasmic/cylindrical inclusion protein; 6K2, 6K protein 2; NIa, nuclear inclusion protein a; NIb, nuclear inclusion protein b; CP, coat protein.

Table 3.1 Potyviral gene product functions

Gene	Original function(s)	Current function(s)
P1	Cell-to-cell movement	Proteinase, cell-to-cell movement (?)
HC-Pro	Insect transmission	Proteinase, insect transmission, replication, long-distance and cell-to-cell movement, host defence suppression
P3	Proteinase	Replication (?), symptom severity
6K1	(Not recognised)	Unknown
CI	Replication	RNA helicase, cell-to-cell movement, long-distance movement
6K2	(VPg) replication	Targeting of NIa to the endoplasmic reticulum
NIa-VPg	(Not recognised)	Binds genomic RNA, translation of polyprotein, replication, long-distance movement, cell-to-cell movement
NIa-Pro	Proteinase	Major proteinase
NIb	Replication	RNA-dependent RNA polymerase (replicase)
CP	Encapsidation, insect transmission	Encapsidation, insect transmission, cell-to-cell movement, long-distance movement, RNA involved in replication

potyviruses they aggregate in the nuclei of infected cells to form morphologically variable structures.

The current understanding of potyviral gene functions is somewhat more complicated. Although most of the proteins now have names, which makes comparison between viruses simpler, almost all the genes have now been ascribed multiple functions (see Figure 3.3 and Table 3.1). The 31K protein is now known as P1. Its cell-to-cell movement functions are still unconfirmed but it now has definite proteinase functions. More specifically, it has been shown to autocatalytically cleave itself out of the polyprotein at the junction with the next protein (HC-Pro). The 56K TEV HC-Pro also has proteolytic activity, being responsible for the remainder of the processing of the N-terminal third of the polyprotein. It is necessary for insect transmission of the virus *in vivo* (Dougherty and Carrington, 1988), and purified HC-Pro protein can act *in trans* to restore aphid-transmissibility to mutant virus isolates (Lopez-Moya and Pirone, 1998). The central region of HC-Pro has been shown to have important functions in the replication and long-distance movement of TEV (Cronin *et al.*, 1995; Kasschau *et al.*, 1997), and it has also been found to have cell-to-cell movement properties (Rojas *et al.*, 1997). More recently, a region of the TEV polyprotein encompassing P1, HC-Pro and the N-terminal of P3 (Anandalakshmi *et al.*, 1998; Kasschau and Carrington, 1998), and the PVY HC-Pro (Brigneti *et al.*, 1998) has been shown to act as a suppressor of host defences, a function that is discussed in more detail in section 3.3.1.

Based on comparisons with a proteinase amino acid sequence from poliovirus, it has been suggested that the predicted product of the P3 potyviral gene (50K) may be involved in polyprotein processing (Dougherty and Carrington, 1988), but evidence for this is limited. All known cleavage sites in the potyviral polyprotein have had other proteinases ascribed to them, and not all potyviral P3s show sequence similarity to known proteinases. As the potyviral ORF is translated in full to form the polyprotein, which is subsequently cleaved to produce mature proteins, it is not possible that the P3 protein is never present in infected cells, but until recently the status of the potyviral P3 protein was uncertain. Subsequently, it has been shown by immunocytology that P3 protein is present in cylindrical inclusions (Rodriguez-Cerezo *et al.*, 1993) and nuclear inclusion bodies (Langenberg and Zhang, 1997). Engineered mutations in the P3 gene can eliminate viral viability (Klein *et al.*, 1994), suggesting a role for P3 in viral replication, although P3 protein was shown not to interact with RNA *in vitro* (Merits *et al.*, 1998). A role for P3 in determining symptom severity is also possible (Chu *et al.*, 1997).

Following publication of Dougherty and Carrington's 1988 review of potyviral gene functions, a possible extra gene product has been identified from the presence of conserved proteinase recognition sequences at the 3′ end of the P3 cistron (Riechmann *et al.*, 1992). This putative 6K protein (6K1) has never been identified in virus-infected cells, and it has been suggested that in plum pox potyvirus (PPV) it represents the result of a regulatory mechanism that functions via cleavage of the P3 protein (Riechmann *et al.*, 1995).

The CI protein (70K for TEV) has been proven by sequence identity and functionality to have RNA helicase and ATPase activity (Lain *et al.*, 1990), and preferentially binds ssRNA in a non-sequence-specific manner (Merits *et al.*, 1998). An important role in genome replication is implied by the helicase activity of CI, an assumption supported by the finding that some amino acid substitutions in the TEV CI can abolish viral replication in protoplasts (Carrington and Whitham, 1998). In the same study, it was shown that other mutations could affect cell-to-cell movement without altering replication efficiency, and some possibly affected long-distance movement of the virus without detriment to the other functions of CI. These two findings supported evidence from ultrastructural studies for other potyviruses that cylindrical inclusion bodies are often associated with cell membranes and plasmodesmata, and appear to channel virions or aggregated CP/RNA into neighbouring cells (Rodriguez-Cerezo *et al.*, 1997; Roberts *et al.*, 1998).

Downstream of the CI gene is another putative small gene, 6K2, which has been the focus of more studies than the 6K1 cistron. The 6K2 protein exists as a mixed population of free protein and partially processed polyprotein in combination with its C-terminal neighbour, NIa, which directs this processing autoproteolytically. It has been proved conclusively through reporter-fusion studies that the TEV 6K2 protein targets and integrates with endoplasmic reticulum (ER)-derived membranes, and can also direct proteins to which it is fused to the ER (Schaad *et al.*, 1997a). As the NIa-VPg has nuclear targeting signals near its N-terminus, which direct its movement into the nucleus where it aggregates with NIb to form nuclear inclusions, the pre-processed 6K2-NIa is subject to competition between these two localisation signals. It has been shown that the presence of the TEV 6K2 in the polyprotein out-competes the nuclear localisation signal of NIa, and that NIa controls the targeting of its own subpopulations to the cytoplasm or nucleus via autoproteolysis of the 6K2 protein (Restrepo-Hartwig and Carrington, 1992). This processing of the 6K2 protein appears to be essential for replication. When the 6K2-NIa cleavage site is disrupted by mutagenesis, the resulting viruses are non-viable. Insertion of amino acids at some positions within the 6K2 protein has also been shown to reduce viability (Restrepo-Hartwig and Carrington, 1994).

NIa (49K for TEV) is a protein with two functionally distinct domains. The C-terminus possesses proteolytic activity and is the major proteinase of potyviruses responsible for processing the C-terminal two-thirds of the polyprotein. NIa controls its own autocatalytic excision from the polyprotein at the CI-6K2 and NIa-NIb junctions, followed by differential processing at the 6K2-NIa junction, as outlined above. NIa also effects an internal cleavage at a suboptimal recognition sequence to yield a subpopulation of the N- and C-terminal halves of the protein, (NIa-)VPg and NIa-Pro, respectively. VPg (viral genome-linked protein) binds the genomic RNA at its 5′ end, and is thought to be functionally comparable to the 5′-cap structures of mRNA; a view supported by the finding that in a yeast two-hybrid system, turnip mosaic virus (TuMV) VPg interacts with the *Arabidopsis thaliana* protein, eIF(iso)4E, a eukaryotic initiation factor with an essential role in initiation of translation from capped mRNAs (Wittmann *et al.*, 1997). VPg may also act as a primer for replication of the viral genome, although its presence does not seem to be essential for the viability of all viruses that encode it (Shukla *et al.*, 1994). In the case of TEV, free VPg is essential for viral viability, as a virus engineered to remove the VPg-proteinase cleavage site was unable to replicate in protoplasts (Carrington *et al.*, 1993). Mutational analysis has been used to prove that the TEV VPg has a long-distance movement function (Schaad *et al.*, 1997b). Chimaeric viruses were produced using two isolates of the virus, both capable of replication and cell-to-cell movement in a given host, but only one capable of long-distance movement. By swapping regions of the fully viable isolate into the long-distance movement defective isolate, the phenotype determinant was mapped to a region of the VPg. VPg may also play a role in cell-to-cell movement. The virulence determinant for the *va* resistance gene of *N. tabacum* c.v. TN86, which functions via impaired cell-to-cell movement, has been isolated to the VPg of tobacco vein mottling virus (TVMV) (Nicolas *et al.*, 1997).

The other nuclear inclusion protein, NIb (58K for TEV), has been shown to be an RNA-dependent RNA polymerase, or replicase, which would clearly be required for genome amplification (Hong and Hunt, 1996). NIb has been shown *in vivo* and *in vitro* to interact with NIa, although more recent studies have not resulted in a unifying theory as to which domain of NIa interacts with NIb. In the case of TVMV, Hong *et al.* (1995) used the yeast two-hybrid system to show that VPg, but not NIa-Pro, interacted with NIb, and Fellers *et al.* (1998) found that the TVMV VPg interacted with and stimulated the polymerase activity of NIb *in vitro*. Regarding TEV, Daros *et al.* (1999) showed that the NIa-Pro/NIb interaction is important for viral genome amplification in a yeast expression system. Potyviral replication is known to occur in association with cytoplasmic membranes and the proven relationships between NIb, NIa and 6K2 indicate that it will not be long before an integrated theory of potyviral replication complexes is completed.

NIb has also been shown to interact with potyviral capsid or coat proteins in the yeast two-hybrid system (Hong *et al.*, 1995). CP is the only true structural potyviral protein, although VPg also occurs in virions. The CP has an obvious role in the physical protection of the viral particles, and since the N- and C-terminal domains are exposed on the outside of virions, these would be expected to interact with non-viral proteins, an assumption supported by the high sequence variability of the CP N-terminus (Shukla *et al.*, 1994). CP was shown to be necessary for aphid transmission of TVMV, in a study that proved that a single amino acid difference could eliminate aphid transmissibility (Atreya *et al.*, 1990). It has also been shown that a specific interaction between CP and HC-Pro amino acid residues is required for aphid transmissibility, accumulation and movement of potyviruses (Blanc *et al.*, 1997; Anderjeva *et al.*, 1999).

The TEV CP has been shown to have movement properties in genetic studies. Substitution or deletion of CP amino acids results in impairment or elimination of movement out of initially infected cells. This impairment has been complemented *in trans* by plants expressing wild-type CP transgenes. In a closer analysis, long-distance transport functions have been mapped to the N- and C-terminal domains of the protein, whereas the core of the protein appears to be necessary for cell-to-cell movement (Dolja *et al.*, 1994, 1995). In TEV and TVMV, the maintenance of a net charge at the N-terminus appears to be required for the long-distance movement function of the CP (Lopez-Moya and Pirone, 1998). In pea seed-borne mosaic potyvirus (PSbMV), it has been possible to map the long-distance movement determinant to one amino acid, although this does not appear to provide a full explanation (Andersen and Johansen, 1998). In microinjection experiments, CP (and HC-Pro) have been demonstrated to act as potyviral movement proteins, increasing the size exclusion limit for plasmodesmata and facilitating the passage of viral RNA (Rojas *et al.*, 1997).

Counter-intuitively, it has been demonstrated that the CP is not a prerequisite for genome amplification in TEV. Using frameshift-stop mutants and 3′ deletion mutants at comparable nucleotides, it has been shown that genome amplification requires translation to somewhere between codons 138 and 189, but the protein product is unnecessary, and there is also a requirement for a *cis*-acting element within the 3′ end of the CP RNA (Mahajan *et al.*, 1996). Further work has shown that the secondary structure of the CP RNA and the 3′ non-translated region (NTR) are the functional elements of this *cis*-active region (Haldeman-Cahill *et al.*, 1998). The potyviral non-translated regions have also been shown to have distinct functions. The nucleotide sequence of the 3′-NTR of TVMV can affect symptom severity (Rodriguez-Cerezo *et al.*, 1991), whilst the 5′-NTR has been shown to be capable of initiating translation, even when located at an internal position (Levis and Astiermanifacier, 1993), and synergistically regulates translation in conjunction with the poly(A) tail, much as 5′ caps do in mRNAs (Gallie *et al.*, 1995).

3.3 Virus-host interactions

As can be seen from the examples described above, the interactions between viral proteins and their roles within the infection process are complex. Has there been a similar expansion in our knowledge of how viral proteins influence host responses to virus infection?

Many studies have been carried out on the analyses of infected tissue, revealing a range of gross changes, including changes in levels of respiration, metabolic pathways, host mRNA and protein accumulation (Goodman *et al.*, 1986; Tesci *et al.*, 1994; Geri *et al.*, 1999). However, in many of the earlier studies it was unclear whether these changes were caused directly by virus replication or were a consequence of symptom expression. A picture is now emerging of very specific and precise interaction between viruses and hosts. Elegant experiments by Wang and Maule (1995) have revealed that virus replication is a highly regulated event in terms of its location relevant to the infection front. Their work has clearly demonstrated that replication in PSbMV is restricted to a small area behind the infection front (up to ca. 12 cells behind). More surprising has been their discovery that there is a well-defined zone in which there is a marked shut-off of host gene transcription, extending ca. 8 cells behind the infection front. Once the replication front has moved on, the cells recover and levels of host gene transcription return to normal. It is unclear which viral proteins (if any) are involved in such precise control of the host transcription, but it is fascinating to observe such effects on host gene transcription (a nuclear event) caused by a virus that carries out its infection cycle within the cytoplasm.

It is now emerging from other studies that viral proteins can have very precise effects on the host response to viral infection; examples that illustrate this are outlined below. Such viral genes may be regarded as suppressors of host defences, gene silencers or enhanced susceptibility genes.

3.3.1 Potyvirus proteins involved in suppression of host defence

One of the best examples of how a virus can interact with its host has been revealed in a study directed at how two different viruses interact during a double infection of the same host. When two or more viruses co-infect the same plant, they can act synergistically, producing symptoms greater than the expected sum of the two individual infections. In the well-studied example of potato virus X (PVX)/potyviral synergism in tobacco, PVY, TVMV, TEV and pepper mottle virus (PMV) are all capable of interacting with PVX to increase symptom severity and PVX accumulation. Using transgenic plants expressing subsets of the TVMV or TEV genomes, it has been shown that expression of the 5′-proximal sequence of the potyviral genome, corresponding to the 5′-NTR, P1, HC-Pro and about one quarter of the P3 gene (termed P1/HC-Pro), is sufficient to produce this synergism, indicating that replication of the potyviral genome is not necessary for this interaction (Vance *et al.*, 1995). The increase in accumulation of PVX conferred by the potyvirus is due, at least in part, to a change in regulation of PVX RNA replication. PVX replication in single cells is normally shut off after 12–24 h, whereas potyviruses normally replicate at a slower rate, for longer. In PVX/potyviral synergism, PVX replication occurs on a timescale similar to that for potyviruses, suggesting that the presence of the potyvirus is countering a host defence mechanism that would otherwise inhibit PVX replication (Pruss *et al.*, 1997; Carrington and Whitham, 1998).

Expression of the TEV P1/HC-Pro sequence also enhances the pathogenicity and accumulation of TMV and cucumber mosaic virus (CMV) (Pruss *et al.*, 1997). By expressing subsets of the TEV P1/HC-Pro region from a PVX vector, it has been shown that expression of just the HC-Pro gene is sufficient to enhance symptom severity, and that the HC-Pro RNA alone is insufficient to produce this effect. The potyviral 5′-NTR, P1 and P3 regions appear to be dispensable with respect to symptom expression in PVX/potyviral synergistic disease, but enhanced PVX negative strand RNA accumulation in protoplasts is reliant on the presence of the entire P1/HC-Pro sequence (Pruss *et al.*, 1997).

One possible mechanism for PVX/potyviral synergistic disease is potyvirus-driven transactivation of PVX replication. The expression of synergistically-active TEV HC-Pro has been shown to be capable of enhancing the expression of a (non-viral) reporter gene driven by a PVX subgenomic promoter, to levels 100-fold higher than when a mutated HC-Pro gene with no synergism capability was expressed (Pruss *et al.*, 1997). Three independent studies have subsequently shown that suppression of post-transcriptional gene silencing (PTGS) operates in PVX/potyviral synergistic disease. Anandalakshmi *et al.* (1998) used a reversal of silencing assay to show that TEV P1/HC-Pro expression was able to reverse PTGS of a transgenically expressed *uidA*, encoding beta-glucuronidase (GUS), reporter gene. In the same study, TEV P1/HC-Pro polyprotein or just HC-Pro protein, expressed from a PVX vector, was also shown to reduce the efficiency of virus-induced gene silencing (VIGS) of a green fluorescent protein (GFP) transgene. Kasschau and Carrington (1998) showed that a TEV P1/HC-Pro transgene was able to suppress PTGS of a GUS transgene. They also showed, in a more detailed analysis, that the suppression of silencing acted at the post-transcriptional level, rather than by reducing the GUS transgene transcription to a level below the PTGS initiation threshold, as GUS transcription levels were the same or even higher in plants also expressing the P1/HC-Pro transgene. Infecting *Nicotiana benthamiana* transgenically expressing GFP with a PVX vector carrying PVY genes, Brigneti *et al.* (1998) narrowed the region proposed to be the PTGS suppressor in PVY to the HC-Pro gene. They also showed that suppression of silencing by PVY probably operates by blocking the maintenance of previously-set PTGS, as leaves that had been completely silenced were able to express GFP following PVY infection.

3.3.2 Cucumovirus proteins involved in suppression of host defence

In a concurrent study with cucumber mosaic virus (Beclin *et al.*, 1998), infection of transgenic tobacco plants at an early stage of development (before silencing had been established) with CMV resulted in the prevention of initiation of co-suppression of a transgene and homologous host gene (nitrate reductase *Nia*), and prevented or reduced the extent of PTGS of a *uidA* transgene. Infection with CMV after *Nia* co-suppression or *uidA* PTGS had become established resulted in prevention of propagation of silencing into new leaves. This effect was also demonstrated in *Arabidopsis* plants expressing *uidA*. As it was shown that silencing was not active in meristems, the mechanism of suppression of silencing was concluded to occur in newly developing leaves, either by inhibiting the initiation of PTGS or by blocking the propagation of a systemic silencing signal. In the same study, it was demonstrated that tomato black ring nepovirus (TBRV) did not suppress silencing at any stage, which corresponds with the observation that plants are able to recover from TBRV infection via a PTGS-like mechanism, although this has never been reported for CMV.

In the same study in which Brigneti *et al.* (1998) proved that mature PVY HC-Pro protein was necessary and sufficient for suppression of PTGS, it was established by expressing individual CMV genes from a PVX vector that the component of CMV responsible for the suppression of PTGS is the 2b gene product, which is otherwise required for long-distance movement of CMV (Ding *et al.*, 1995). The proposed mechanism of PTGS suppression by CMV in this *N. benthamiana* study concurred with the findings of Beclin *et al.* (1998) in tobacco and *Arabidopsis*, that suppression of silencing only occurred in leaves that emerged after virus infection. The fact that PVY HC-Pro- and Cmv2b-mediated PTGS suppressions have been shown to operate at different places in the plant suggests that they may operate via different mechanisms, a hypothesis supported by the finding that TEV P1/HC-Pro (functionally analogous to the PVY protein) can enhance CMV infection (Pruss *et al.*, 1997), which would not be expected if HC-Pro and Cmv2b had identical functions.

Tomato aspermy cucumovirus (TAV), closely related to CMV, expresses an analogous 2b protein (Tav2b), which when expressed from a TMV vector does not suppress PTGS in tobacco (*N. tabacum* c.v. Samsun nn). In this host, TMV-expressed Tav2b was shown to elicit strong hypersensitive response (HR)-mediated resistance, which was also directed against the TMV vector. The incompatible resistance response was abolished by substituting two Tav2b amino acids with those found at the same positions in Cmv2b. When wild-type TAV, or CMV carrying the Tav2b gene, infects this host, there is no such resistance response and no significant induction of PR genes. Therefore, the plant response to Tav2b expression is dependent on the background from which it is expressed. However, in the closely-related plant species, *N. benthamiana*, the TMV-Tav2b vector does not elicit an HR-mediated defence response and is able to replicate fully. In this host, it acts as a suppressor of PTGS, as it prevents the establishment of silencing of a GFP transgene in new leaves by a mechanism that appears to be the same as that for Cmv2b, although the Tav2b-mediated PTGS suppression occurs several days in advance of the Cmv2b effect. Taken as a whole, these results suggest that the induction of the HR resistance response to Tav2b in *N. tabacum* may be an example of a counter-defensive strategy to virus-induced suppression of PTGS, and that the fact that this defence response is not elicited against Tav2b expressed from a cucumovirus background may be an example of a 'counter-counter strategy' by the virus (Li *et al.*, 1999).

Clearly, the ability to effect host responses to viral infection will have a major impact on how we view susceptible interactions, and has a bearing on the interaction of resistance genes with their target virus. Indeed, as outlined in the next section, recent investigations have revealed that many of the virus/resistance gene interactions are also not as simple as originally proposed.

3.4 Virus interactions with resistance genes

The *Rx* gene of potato confers extreme resistance (immunity) to PVX. It is a single dominant gene and in potato it occurs in addition to two hypersensitive PVX resistance genes, *Nb* and *Nx*. Until 1980, *Rx* was capable of conferring resistance to all known isolates of PVX; at which point, Moreira *et al.* (1980) found an *Rx* resistance-breaking strain of PVX, PVX_{HB}, which was also capable of infecting *Nb* and *Nx* plants. No evidence for selection to produce *Rx* resistance-breaking isolates of PVX has been found, as PVX isolates of pathogenicity group 4, which are able to infect *Nx*/*Nb* plants, did not mutate to produce *Rx*-virulent isolates when grafted onto *Rx*-containing *Solanum tuberosum* cv. Cara stock (Jones, 1985). Furthermore, no other naturally-occurring *Rx* resistance-breaking strains of PVX have ever been isolated.

In 1986, *Rx* was shown to operate in isolated protoplasts (Adams *et al.*, 1986). Cara protoplasts were not able to support full replication of the PVX_{DX} isolate, which was able to replicate to ten times the Cara level in protoplasts of cultivars that did not contain *Rx*. PVX_{HB} was able to replicate to a high titre in Cara protoplasts, indicating that these protoplasts were capable of supporting virus replication. As the HR is thought to be a tissue-related phenomenon that is not expressed at the protoplast level, the demonstration of *Rx* resistance in protoplasts indicates that it does not require HR for its expression. The mechanism of *Rx* resistance was postulated to affect viral replication, as uncoating and particle assembly were both demonstrated in the resistant protoplasts. However, since the Cara protoplasts support some replication of PVX_{DX}, whereas whole plants are completely immune, another resistance mechanism is probably active at the whole plant level.

Publications in quick succession elucidated the molecular determinants of *Rx* resistance. By producing hybrid complementary deoxyribonucleic acid (cDNA) clones of PVX_{HB} and PVX_{UK3} (an *Rx*-sensitive isolate), Kavanagh *et al.* (1992) demonstrated that the CP gene of PVX_{HB} was the resistance-breaking determinant. Köhm *et al.* (1993) introduced frame-shift mutations into the CP genes of clones of *Rx*-virulent and *Rx*-avirulent isolates. They demonstrated that *Rx* resistance depends on a feature of the CP of the avirulent isolates, as resistance to the avirulent isolate was abolished by the frame-shift, whereas *Rx* resistance-breaking is a passive phenomenon, as the frame-shift in the virulent isolate did not change its phenotype. In the same study, they showed that the resistance induced by an avirulent isolate was able to affect a virulent PVX isolate, or an unrelated virus (CMV) co-inoculated with it. They suggested that *Rx* resistance involves separate recognition and response phases, culminating in a generalised antiviral state in the plant cells. Goulden *et al.* (1993) focused the virulence determinant of the CP to a threonine residue at amino acid position 121, which was required for induction of the *Rx* resistance response in potato. They also found this residue to be important for the hypersensitive resistance response to PVX in *Gomphrena globosa* (Goulden and Baulcombe, 1993), suggesting that these species have homologous recognition mechanisms leading to distinct resistance functions.

In a later paper, Querci *et al.* (1995) demonstrated that the possession of the threonine residue at position 121 in PVX_{HB} was sufficient to induce a resistance response in four other *Solanum* species. A fifth species could not be infected with PVX_{HB}, although it could be infected with mutant strains possessing the same threonine residue, implying that there may be other determinants of virulence. The amino acid at position 121 had previously been demonstrated to be one of two residues unique to PVX_{HB} that could possibly alter the secondary structure of the predicted protein (Querci *et al.*, 1993). In a further paper, Bendahmane *et al.* (1995) proved the resistance elicitation activity of the CP, supporting the evidence for a generalised antiviral response by demonstrating that TMV accumulation was also suppressed in protoplasts infected with an avirulent strain of PVX. By expressing PVX CP from a TMV vector, they also demonstrated that no other PVX genes were required for elicitation of *Rx* resistance.

More recent papers have concentrated on elucidating the mechanism of action of *Rx*. Gilbert *et al.* (1998) used actinomycin D to inhibit transcription of nuclear genes in Cara protoplasts inoculated with an avirulent strain of PVX. They found no evidence of any additional PVX replication in the presence of the transcription inhibitor than without it, proving that the *Rx* resistance response is not reliant on host gene induction post-infection, and that it must therefore utilise host components found in protoplasts prior to infection. In the same study, they attempted to produce *Rx* plants transgenically expressing an avirulent PVX CP gene. Two methods were employed. Firstly, *Agrobacterium tumefaciens* mediated transformation with a CaMV 35S-driven CP gene, which resulted in no viable progeny being produced from 421 explants, despite successful transformation with a virulent CP gene. In order to rule out the possibility of inefficient transformation as an explanation for the lack of progeny, crosses of the *Rx* plant to *rx* plants carrying the avirulent CP transgene were conducted. Although segregants expressing both *Rx* and the CP gene germinated, no mature plants were obtained expressing both transcripts to normal levels, suggesting that the *Rx*/avirulent CP genotype is lethal. One explanation for this outcome is that *Rx* is capable of producing a hypersensitive response, which contradicts the previous findings in protoplasts (Adams *et al.*, 1986) but is supported by the fact that necrotic reactions do occur in *Rx*/PVX interactions in some systems (Kavanagh *et al.*, 1992; Goulden and Baulcombe, 1993).

More recently, Bendahmane *et al.* (1999) have cloned and sequenced the *rx* gene from the Cara cultivar of potato. By studying the predicted amino acid sequence, they have concluded that it bears some similarities to the NBS-LRR class of resistance genes (see chapter 5 in this volume), which function via the hypersensitive response. By transgenically expressing the cloned *Rx* gene in *Rx* potato, *N. benthamiana* and *N. tabacum*, they have been able to demonstrate that this gene alone is able to confer extreme resistance to PVX, without any induction of HR. In order to reconcile these two contradictory findings, and those of Gilbert *et al.* (1998), constitutively expressed CP constructs have been transformed into mature, *Rx*-expressing *N. benthamiana* leaves by agroinfiltration. An HR leading to death of the infiltrated area was observed when the CP construct was from an avirulent PVX strain. When a resistance-breaking CP construct was used, there was no such cell death. This indicates that *Rx* is capable of initiating a necrotic response when the elicitor of resistance is overexpressed, and explains why Gilbert *et al.* (1998) were not able to produce whole *Rx* plants transgenically expressing avirulent PVX CP.

In order to characterise the relationship between extreme resistance and HR for *Rx* more fully, a series of experiments on transgenic tobacco have been conducted. These have involved expression of PVX CP (*Rx* elicitor) from a TMV vector (elicitor of the HR via the tobacco *N* gene). When high titres of hybrid virus, carrying either the avirulent or resistance-breaking PVX CP, were inoculated onto plants containing just the *N* gene, an HR was induced, presumably by the *N* gene. When hybrid virus containing the *Rx* resistance-breaking PVX CP was inoculated onto plants with an *N*/*Rx* genotype, there was also an HR, directed by the *N* gene. However, when hybrid virus carrying the avirulent PVX CP was inoculated onto *N*/*Rx* plants, there was no HR. These findings indicate that the extreme resistance and HR responses are independent, and that extreme resistance is epistatic to HR.

3.5 Conclusions

Like the proverbial onion, it would appear that as one layer of answers is peeled back, another layer of questions and complexities is revealed. Whilst we have just celebrated 100 years of TMV research (Harrison and Wilson, 1999) and may reflect on some success in allocating major functions to each of the viral gene products, it appears that there are still major challenges. These challenges include developing a fuller understanding of how all the viral proteins interact in virus-specific roles, and also how these viral proteins interact with the host.

In the introduction to this chapter, it was suggested that perhaps those attracted to plant virology because of the simplicity of their 'pets' of interest had been deceived. Perhaps they have been 'saved'. The authors firmly believe that viruses, for the most part, mimic what occurs in the world around them. Therefore, if plant viral proteins can have multiple functions and interact with their host at a myriad of levels, then so can the proteins encoded by other more 'complex' (!) microbial pathogens. So to those fungal and bacterial pathologists, celebrating the fact that they have finally assigned a function to their chosen gene, virologists say look and look again. Then, once you have determined all possible functions for your gene, start to look at how each of those functions may interact with the host. To have that multiplied by the number of genes encoded by typical cellular microbial pathogens is a daunting prospect. Perhaps plant virologists were not misled after all, and do have some very simple 'pets' to study. Those non-virologists, working on their cellular microbial pathogens, face difficult, though interesting, times ahead.

Acknowledgements

The authors would like to thank the Editors of this volume for their help and patience during the writing of this chapter. Many thanks to Andy Bailey and Diana Foster for their proof-reading and useful suggestions.

References

Adams, S.E., Jones, R.A.C. and Coutts, R.H.A. (1986) Expression of potato virus X resistance gene *Rx* in potato leaf protoplasts. *J. Gen. Virol.*, **67** 2341-45.

Anandalakshmi, R., Pruss, G.J., Ge, X., Marathe, R., Mallory, A.C., Smith, T.H. and Vance, V.B. (1998) A viral suppressor of gene silencing in plants. *Proc. Natl. Acad. Sci. USA*, **95** 13079-84.

Andersen, K. and Johansen, I.E. (1998) A single conserved amino acid in the coat protein gene of pea seed-borne mosaic potyvirus modulates the ability of the virus to move systemically in *Chenopodium quinoa*. *Virology*, **241** 304-11.

Andrejeva, J., Puurand, U., Merits, A., Rabenstein, F., Jarvekulg, L. and Valkonen, J.P.T. (1999) Potyvirus helper component proteinase and coat protein (CP) have coordinated functions in virus-host interactions and the same CP motif affects virus transmission and accumulation. *J. Gen. Virol.*, **80** 1133-39.

Atreya, C.D., Raccah, B. and Pirone, T.P. (1990) A point mutation in the coat protein abolishes aphid transmissibility of a potyvirus. *Virology*, **178** 161-65.

Beclin, C., Berthome, R., Palauqui, J.C., Tepfer, M. and Vaucheret, H. (1998) Infection of tobacco or *Arabidopsis* plants by CMV counteracts systemic post-transcriptional silencing of nonviral trans(genes). *Virology*, **252** 313-17.

Beijerinck, M.J. (1898) Concerning a contagium vivum fluidum as cause of the spot disease of tobacco leaves. Verhandelingen der Koninkyke akademie Wettenschappen te Amsterdam, **65** 3-21. Translation published in English as Phytopathological Classics Number 7 (1942). American Phytophathological Society Press, St. Paul, Minnesota.

Bendahmane, A., Köhm, B.A., Dedi, C. and Baulcombe, D.C. (1995) The coat protein of potato virus X is a strain-specific elicitor of *Rx1*-mediated virus resistance in potato. *Plant J.*, **8** 933-41.

Bendahmane, A., Kanyuka, K. and Baulcombe, D.C. (1999) The *Rx* gene from potato controls separate virus resistance and cell death responses. *Plant Cell*, **11** 781-91.

Blanc, S., Lopez-Moya, J.J., Wang, R.Y., Garcia-Lampasona, S., Thornbury, D.W. and Pirone, T.P. (1997) A specific interaction between coat protein and helper component correlates with aphid transmission of a potyvirus. *Virology*, **231** 141-47.

Brigneti, G., Voinnet, O., Li, W.X., Ji, L.H., Ding, S.W. and Baulcombe, D.C. (1998) Viral pathogenicity determinants are suppressors of transgene silencing in *Nicotiana benthamiana. EMBO J.*, **17** 6739-46.

Buck, K.W. (1999) Replication of tobacco mosaic virus RNA. *Philos. Trans. R. Soc. Lond. B*, **354** 613-27.

Carrington, J.C. and Whitham, S.A. (1998) Viral invasion and host defense: strategies and counter-strategies. *Curr. Opin. Plant Biol.*, **1** 336-41.

Carrington, J.C., Haldeman, R., Dolja, V.V. and Restrepo-Hartwig, M.A. (1993) Internal cleavage and trans-proteolytic activities of the VPg-proteinase (NIa) of tobacco etch potyvirus *in vivo. J. Virol.*, **67** 6995-7000.

Carrington, J.C., Jensen, P.E. and Schaad, M.C. (1998) Genetic evidence for an essential role for potyvirus CI protein in cell-to-cell movement. *Plant J.*, **14** 393-400.

Chu, M.H., Lopez-Moya, J.J., Llave-Correas, C. and Pirone, T.P. (1997) Two separate regions in the genome of the tobacco etch virus contain determinants of the wilting response of tabasco pepper. *Mol. Plant-Microbe Interact.*, **10** 472-80.

Citovsky, V., Knorr, D., Schuster, G. and Zambryski, P. (1990) The P30 movement protein of tobacco mosaic virus is a single-stranded nucleic acid binding protein. *Cell*, **60** 637-47.

Citovsky, V., Wong, M.L., Shaw, A., Prasad, B.V.V. and Zambryski, P. (1992) Visualization and characterization of tobacco mosaic virus movement protein binding to single-stranded nucleic acids. *Plant Cell*, **4** 397-411.

Citovsky, V., McLean, B.G., Zupan, J. and Zambryski, P. (1993) Phosphorylation of tobacco mosaic virus cell-to-cell movement protein by a developmentally-regulated plant cell-wall-associated protein kinase. *Genes Dev.*, **7** 904-10

Cronin, S., Verchot, J., Haldeman-Cahill, R., Schaad, M.C. and Carrington, J.C. (1995) Long-distance movement factor—a transport function of the potyvirus helper component-proteinase. *Plant Cell*, **7** 549-59.

Daros, J.A., Schaad, M.C. and Carrington, J.C. (1999) Functional analysis of the interaction between VPg-proteinase (NIa) and RNA polymerase (NIb) of tobacco etch potyvirus, using conditional and suppressor mutants. *J. Virol.*, **73** 8732-40.

Dawson, W.O., Bubrick, P. and Grantham, G.L. (1988) Modifications of the tobacco mosaic virus coat protein gene affecting replication, movement and symptomology. *Phytopathology*, **78** 783-89.

Deom, C.M., Quan, S. and He, X.Z. (1997) Replicase proteins as determinants of phloem-dependent long-distance movement of tobamoviruses in tobacco. *Protoplasma*, **199** 1-8.

Derrick, P.M., Carter, S.A. and Nelson, R.S. (1997) Mutations of the tobacco mosaic tobamovirus 126 and 183 kDa proteins: effects on phloem-dependent virus accumulation and synthesis of viral proteins. *Mol. Plant-Microbe Interact.*, **10** 589-96.

Ding, S.W., Li, W.X. and Symons, R.H. (1995) A novel naturally-occurring hybrid gene encoded by a plant RNA virus facilitates long-distance virus movement. *EMBO J.*, **14** 5762-72.

Dolja, V.V., Haldeman, R., Robertson, N.L., Dougherty, W.G. and Carrington, J.C. (1994) Distinct functions of capsid protein in assembly and movement of tobacco etch potyvirus in plants. *EMBO J.*, **13** 1482-91.

Dolja, V.V., Haldeman-Cahill, R., Montgomery, A.E., Vandenbosch, K.A. and Carrington, J.C. (1995) Protein determinants involved in cell-to-cell and long-distance movement of tobacco etch potyvirus. *Virology*, **206** 1007-16.

Dougherty, W.G. and Carrington, J.C. (1988) Expression and function of potyviral gene products. *Annu. Rev. Phytopathol.*, **26** 123-43.

Fedorkin, O.N., Denisenko, O.N., Sitkov, A.S., Zelenina, D.A., Lukashova, L.I., Morozov, S.Y. and Atabekov, J.G. (1995) The tomato mosaic virus small gene product forms stable complex with translation elongation-factor, EF-1-alpha. *Doklady Akademii. Nauk SSSR*, **343** 703-704.

Fellers, J., Wan, J.R., Hong, Y.L., Collins, G.B. and Hunt, A.G. (1998) *In vitro* interactions between a potyvirus-encoded genome-linked protein and RNA-dependent RNA polymerase. *J. Gen. Virol.*, **79** 2043-49.

Gallie, D.R., Tanguay, R.L. and Leathers, V. (1995) The tobacco etch viral 5′-leader and poly(A) tail are functionally synergistic regulators of translation. *Gene*, **165** 233-38.

Gal-On, A., Kaplan, I., Rossinck, M.J. and Palukaitis, P. (1994) The kinetics of infection of zucchini squash by cucumber mosaic virus indicate a function for RNA1 in virus movement. *Virology*, **205** 280-89.

Geri, C., Cecchini, E., Giannakou, M.E., Covey, S.N. and Milner, J.J. (1999) Altered patterns of gene expression in *Arabidopsis* elicited by cauliflower mosaic virus (CaMV) infection and by CaMV gene VI transgene. *Mol. Plant-Microbe Interact.*, **12** 377-84.

Gilbert, J., Spillane, C., Kavanagh, T.A. and Baulcombe, D.C. (1998) Elicitation of *Rx*-mediated resistance to PVX in potato does not require new RNA synthesis and may involve a latent hypersensitive response. *Mol. Plant-Microbe Interact.*, **11** 833-35.

Goodman, R.N., Kiraly, Z. and Wood, K.R. (1986) *The Biochemistry and Physiology of Plant Disease.* University of Missouri Press, Columbia.

Goulden, R.N. and Baulcombe, D.C. (1993) Functionally homologous host components recognize potato virus X in *Gomphrena globosa* and potato. *Plant Cell*, **5** 921-30.

Goulden, M.G., Köhm, B.A., Santa-Cruz, S., Kavanagh, T.A. and Baulcombe, D.C. (1993) A feature of the coat protein of potato virus X affects both induced virus resistance in potato and viral fitness. *Virology*, **197** 293-302.

Haldeman-Cahill, R., Daros, J.A. and Carrington, J.C. (1998) Secondary structures in the capsid protein coding sequence and 3′ nontranslated region involved in amplification of the tobacco etch virus genome. *J. Virol.*, **72** 4072-79.

Haley, A., Hunter, T., Kiberstis, P. and Zimmern, D. (1995) Multiple serine phosphorylation sites on the 30 kDa TMV cell-to-cell movement protein synthesised in tobacco protoplasts. *Plant J.*, **8** 715-24.

Harrison, B.D. and Wilson, T.M.A. (1999) Tobacco mosaic virus: pioneering research for a century. *Philos. Trans. R. Soc. Lond. B*, **354** 517-685.

Heinlein, M., Epel, B.L., Padgett, H.S. and Beachy, R.N. (1995) Interaction of tobamovirus movement proteins with the plant cytoskeleton. *Science*, **270** 1983-85.

Heinlein, M., Padgett, H.S., Gens, J.S., Pickard, B.G., Casper, S.J., Epel, B.L. and Beachy, R.N. (1998) Changing patterns of localization of the tobacco mosaic virus movement protein and replicase to the endoplasmic reticulum and microtubules during infection. *Plant Cell*, **10** 1107-20.

Holt, C.A. and Beachy, R.N. (1991) *In vivo* complementation of infectious transcripts from mutant tobacco mosaic virus cDNAs in transgenic plants. *Virology*, **181** 109-17.

Hong, Y.L. and Hunt, A.G. (1996) RNA polymerase activity catalyzed by a potyvirus-encoded RNA-dependent RNA polymerase. *Virology*, **226** 146-51.

Hong, Y.L., Levay, K., Murphy, J.F., Klein, P.G., Shaw, J.G. and Hunt, A.G. (1995) A potyvirus polymerase interacts with the viral coat protein and VPg in yeast cells. *Virology*, **214** 159-66.

Jones, R.A.C. (1985) Further studies on resistance-breaking strains of potato virus X. *Plant Pathol.*, **34** 182-89.

Karpova, O.V., Ivanov, K.I., Rodionova, N.P., Dorokhov, Y.L. and Atabekov, J.G. (1997) Nontranslatability and dissimilar behavior in plants and protoplasts of viral RNA and movement protein complexes formed *in vitro*. *Virology*, **230** 11-21.

Karpova, O.V., Rodionova, N.P., Ivanov, K.I., Kozlovsky, S.V., Dorokhov Y.L. and Atabekov, J.G. (1999) Phosphorylation of tobacco mosaic virus movement protein abolishes its translation repressing ability. *Virology*, **261** 20-24.

Kasschau, K.D. and Carrington, J.C. (1998) A counter-defensive strategy of plant viruses: suppression of post-transcriptional gene silencing. *Cell*, **95** 461-70.

Kasschau, K.D., Cronin, S. and Carrington, J.C. (1997) Genome amplification and long-distance movement functions associated with the central domain of tobacco etch potyvirus helper component proteinase. *Virology*, **228** 251-62.

Kavanagh, T., Goulden, M., Santa Cruz, S., Chapman, S., Barker, I. and Baulcombe, D.C. (1992) Molecular analysis of a resistance-breaking strain of potato virus X. *Virology*, **189** 609-17.

Kawakami, S., Padgett, H.S., Hosokawa, D., Okada, Y., Beachy, R.N. and Watanabe, Y. (1999) Phosphorylation and/or presence of serine 37 in the movement protein of tomato mosaic virus is essential for intracellular localization and stability *in vivo*. *J. Virol.*, **73** 6831-40.

Klein, P.G., Klein, R.R., Rodriguez-Cerezo, E., Hunt, A.G. and Shaw J.G. (1994) Mutational analysis of the tobacco vein mottling virus genome. *Virology*, **204** 759-69.

Köhm, B.A., Goulden, M.G., Gilbert, J.E., Kavanagh, T.A. and Baulcombe, D.C. (1993) A potato virus X resistance gene mediates an induced, non-specific resistance in protoplasts. *Plant Cell*, **5** 913-20.

Lain, S., Riechmann, J.L. and Garcia, J.A. (1990) RNA helicase: a novel catalytic activity associated with protein encoded by a positive strand RNA virus. *Nucl. Acids Res.*, **18** 7003-7006.

Langenberg, W.G. and Zhang, L.Y. (1997) Immunocytology shows the presence of tobacco etch virus P3 protein in nuclear inclusions. *J. Struct. Biol.*, **118** 243-47.

Levis, C. and Astiermanifacier, S. (1993) The 5′ untranslated region of PVY RNA, even located in an internal position, enables initiation of translation. *Virus Genes*, **7** 367-79.

Li, H.W., Lucy, A.P., Guo, H.S., Li, W.X., Ji, L.H., Wong, S.M. and Ding, S.W. (1999) Strong host resistance targeted against a viral suppressor of the plant gene silencing defence mechanism. *EMBO J.*, **18** 2683-91.

Lopez-Moya, J.J. and Pirone, T.P. (1998) Charge changes near the N-terminus of the coat protein of two potyviruses affect virus movement. *J. Gen. Virol.*, **79** 161-65.

Mahajan, S., Dolja, V.V. and Carrington, J.C. (1996) Roles of the sequence encoding tobacco etch virus capsid protein in genome amplification: requirements for the translation process and a *cis*-active element. *J. Virol.*, **70** 4370-79.

Matthews, R.E.F. (1991) *Plant Virology*, 3rd edition, Academic Press Inc., San Diego, California, USA.

McLean, B.G., Zupan, J. and Zambryski, P. (1995) Tobacco mosaic virus movement protein associates with the cytoskeleton in tobacco cells. *Plant Cell*, **7** 2101-14.

Merits, A., Guo, D.Y. and Saarma, M. (1998) Vpg, coat protein and five non-structural proteins of potato A potyvirus bind RNA in a sequence-unspecific manner. *J. Gen. Virol.*, **79** 3123-27.

Moreira, A., Jones, R.A.C. and Fribourg, C.E. (1980) Properties of a resistance-breaking strain of potato virus X. *Ann. Appl. Biol.*, **95** 93-103.

Morozov, S.Y., Denisenko, O.N., Zelenina, D.A., Fedorkin, O.N., Solovyev, A.G., Maiss, E., Casper, R. and Atabekov, J.G. (1993) A novel open reading frame in tobacco mosaic virus genome coding for a putative small, positively-charged protein. *Biochemie*, **75** 659-65.

Nicolas, O., Dunnington, S.W., Gotow, L.F., Pirone, T.P. and Hellmann, G.M. (1997) Variations in the VPg protein allow a potyvirus to overcome *va* gene resistance in tobacco. *Virology*, **237** 452-59.

Pruss, G., Ge, X., Shi, X.M., Carrington, J.C. and Vance, V.B. (1997) Plant viral synergism: the potyviral genome encodes a broad-range pathogenicity enhancer that transactivates replication of heterologous viruses. *Plant Cell*, **9** 859-68.

Querci, M., van der Vlugt, R., Goldbach, R. and Salazar, L.F. (1993) RNA sequence of potato virus X strain HB. *J. Gen. Virol.*, **74**, 2251-55.

Querci, M., Baulcombe, D.C., Goldbach, R.W. and Salazar, L.F. (1995) Analysis of the resistance-breaking determinants of potato virus X (PVX) strain HB on different potato genotypes expressing extreme resistance to PVX. *Phytopathology*, **85** 1003-10.

Restrepo-Hartwig, M.A. and Carrington, J.C. (1992) Regulation of nuclear transport of a plant potyvirus protein by autoproteolysis. *J. Virol.*, **66** 5662-66.

Restrepo-Hartwig, M.A. and Carrington, J.C. (1994) The tobacco etch potyvirus 6-kilodalton protein is membrane-associated and involved in viral replication. *J. Virol.*, **68** 2388-97.

Rhee, Y., Tzfira, T., Chen, M-H., Waigmann, E. and Citovsky, V. (2000) Cell-to-cell movement of tobacco mosaic virus: enigmas and explanations. *Mol. Plant Pathol.*, **1** 33-40.

Riechmann, J.L., Lain, S. and Garcia, J.A. (1992) Highlights and prospects of potyvirus molecular biology. *J. Gen. Virol.*, **73** 1-16.

Riechmann, J.L., Cervera, M.T. and Garcia, J.A. (1995) Processing of the plum pox virus polyprotein at the P3-6K(1) junction is not required for virus viability. *J. Gen. Virol.*, **76** 951-56.

Roberts, I.M., Wang, D., Findlay, K. and Maule, A.J. (1998). Ultrastructural and temporal observations of the potyvirus cylindrical inclusions (CIs) shows that the CI protein acts transiently in aiding virus movement. *Virology*, **245** 173-81.

Rodriguez-Cerezo, E., Klein, P.G. and Shaw, J.G. (1991) A determinant of disease severity is located in the 3-terminal noncoding region of the RNA of a plant virus. *Proc. Natl. Acad. Sci. USA*, **88** 9863-67.

Rodriguez-Cerezo, E., Ammar, E.D., Pironi, T.P. and Shaw, J.G. (1993) Association of the nonstructural P3 viral protein with cylindrical inclusions in potyvirus-infected cells. *J. Gen. Virol.*, **74** 1945-49.

Rodriguez-Cerezo, E., Findlay, K., Shaw, J.G., Lomonossoff, G.P., Qui, S.G., Linstead, P., Shanks, M. and Risco, C. (1997) The coat and cylindrical inclusion proteins of a potyvirus are associated with connections between plant cells. *Virology*, **236** 296-306.

Rojas, M.R., Zerbini, F.M., Allison, R.F., Gilbertson, R.L. and Lucas, W.J. (1997) Capsid protein and helper component proteinase function as potyvirus cell-to-cell movement proteins. *Virology*, **237** 283-95.

Schaad, M.C., Jensen, P.E. and Carrington, J.C. (1997a) Formation of plant RNA virus replication complexes on membranes: role of an endoplasmic reticulum-targeted viral protein. *EMBO J.*, **16** 4049-59.

Schaad, M.C., Lellis, A.D. and Carrington, J.C. (1997b) VPg of tobacco etch potyvirus is a host genotype-specific determinant for long-distance movement. *J. Virol.*, **71** 8624-31.

Shukla, D.D., Ward, C.W. and Brunt, A.A. (1994) *The Potyviridae*. CAB International, Wallingford, UK.

Sulzinski, M.A., Gabard, K.A., Palukaitis, P. and Zaitlin, M. (1985) Replication of tobacco mosaic virus. VIII. Characterisation of a third subgenomic TMV RNA. *Virology*, **145** 132-40.

Takamatsu, N., Ishikawa, M., Meshi, T. and Okada, Y. (1987) Expression of bacterial chloramphenicol acetyltransferase gene in tobacco plants mediated by TMV-RNA. *EMBO J.*, **6** 307-11.

Tecsi, L.I., Maule, A.J., Smith, A.M. and Leegood, R.C. (1994) Metabolic alterations in cotyledons of *Cucurbita pepo* infected with cucumber mosaic virus. *J. Exp. Bot.*, **45**, 1541-51.

Vance, V.B., Berger, P.H., Carrington, J.C., Hunt, A.G. and Shi, X.M. (1995) 5′-proximal potyviral sequences mediate potato virus X potyviral synergistic disease in transgenic tobacco. *Virology*, **206** 583-90.

Wang, D. and Maule, A.J. (1995) Inhibition of host gene expression associated with plant virus replication. *Science*, **267** 229-31.

Watanabe, Y., Ogawa, T. and Okada, Y. (1992) *In vivo* phosphorylation of the 30-kDa protein of tobacco mosaic virus. *FEBS Letts.*, **313** 181-84.

Weiland, J.J. and Edwards, M.C. (1996) A single nucleotide substitution in the alpha a gene confers oat pathogenicity to barley stripe mosaic virus strain ND18. *Mol. Plant-Microbe Interact.*, **9** 62-67.
Wittmann, S., Chatel, H., Fortin, M.G. and Laliberte, J.F. (1997) Interaction of the viral protein genome linked of turnip mosaic potyvirus with the translational eukaryotic initiation factor (iso) 4E of *Arabidopsis thaliana* using the yeast two-hybrid system. *Virology*, **234** 84-92.
Wolf, S., Deom, C.M., Beachy, R.N. and Lucas, W.J. (1989) Movement protein of tobacco mosaic virus modified plasmodesmatal size exclusion limit. *Science*, **246** 377-79.

4 Genetic analysis and evolution of plant disease resistance genes

Peter N. Dodds, Gregory J. Lawrence,
A. Pryor and Jeffrey G. Ellis

4.1 Introduction

Plants contain a complex array of genes capable of recognising invading pathogens and initiating active defence mechanisms. These genes typically operate as single dominant resistance genes and each resistance gene generally controls resistance to a single species of pathogen, and very frequently to only some isolates of that pathogenic species. Genetic analysis of host-pathogen interactions has shown that there is a 'gene-for-gene' interaction between each resistance allele and a corresponding avirulence gene in the pathogen that leads to the recognition and resistance response (Flor, 1971). This relationship suggests that resistance genes may encode receptors that recognise the direct or indirect products of pathogen avirulence genes (van den Ackerveken *et al.*, 1993; Matthieu *et al.*, 1994; Yucel *et al.*, 1994).

The present review discusses the impact of recent findings on two of the main questions facing researchers in this field. Firstly, how is the specificity of resistance genes for particular avirulence genes determined? Secondly, how do new resistance specificities evolve? Over the last few years, the isolation of a significant number of disease resistance genes and the ability to analyse the effects of changes within these sequences is beginning to provide direct evidence of the roles of structural features in determining specificity. Likewise, comparison of resistance gene family sequences and molecular analysis of genetic events at resistance gene loci are providing insights into the evolutionary processes acting on resistance genes.

4.2 Features of cloned resistance genes

A number of plant resistance genes have been cloned and fall into several classes of related sequences (Table 4.1, and associated references; see also review by Ellis and Jones, 1998). The majority of resistance genes encode proteins classified as NBS-LRR proteins because they contain a nucleotide binding site (NBS) domain and a leucine–rich repeat (LRR) domain. This class of resistance proteins can be further separated into two subgroups. The first group consists

Table 4.1 Classes of plant disease resistance gene sequences

Class	Gene	Species	Pathogen	Reference
NBS-LRR				
	Prf	Tomato	*Pseudomonas syringae*	Salmeron *et al.*, 1996
	I2C	Tomato	*Fusarium oxysporum*	Ori *et al.*, 1997
	Mi	Tomato	Root knot nematode and potato aphid	Milligan *et al.*, 1998 and Rossi *et al.*, 1998
	Rpm1	*Arabidopsis*	*Pseudomonas syringae*	Grant *et al.*, 1995
	Rps2	*Arabidopsis*	*Pseudomonas syringae*	Mindrinos *et al.*, 1994
	Rps5	*Arabidopsis*	*Pseudomonas syringae*	Warren *et al.*, 1998
	Rpp8	*Arabidopsis*	*Peronospora parasitica*	McDowell *et al.*, 1998
	Xa1	Rice	*Xanthomonas oryzae*	Yoshimura *et al.*, 1998
	Dm3	Lettuce	*Bremia lactucae*	Meyers *et al.*, 1998a,b
	Rp1	Maize	*Puccinia sorghi*	Collins *et al.*, 1999
	Cre3	Wheat	Cereal cyst nematode	Lagudah *et al.*, 1997
TIR-NBS-LRR				
	Rpp5	*Arabidopsis*	*Peronospora parasitica*	Parker *et al.*, 1997
	Rpp1 A,B,C	*Arabidopsis*	*Peronospora parasitica*	Botella *et al.*, 1998
	N	Tobacco	Tobacco mosaic virus	Whitham *et al.*, 1994
	L	Flax	*Melampsora lini*	Lawrence *et al.*, 1995 and Ellis *et al.*, 1999
	M	Flax	*Melampsora lini*	Anderson *et al.*, 1997
Extracellular LRR				
	Cf-2, 4,5,9	Tomato	*Cladosporium fulvum*	Jones *et al.*, 1994; Dixon *et al.*, 1996, 1998; and Parniske *et al.*, 1997
Kinase				
	Pto	Tomato	*Pseudomonas syringae*	Martin *et al.*, 1993
Extracellular LRR/ Intracellular kinase				
	Xa21	Rice	*Xanthomonas oryzae*	Song *et al.*, 1995

Abbreviations: NBS, nucleotide-binding site; LRR, leucine-rich repeat; TIR, domain with homology to the cytoplasmic domain of the Toll receptor of *Drosophila* and the interleukin-1 receptor of mammals.

of proteins that have a distinct N-terminal region (TIR domain) that resembles the cytoplasmic domains of the *Drosophila* Toll protein and the mammalian interleukin-1 receptor protein (IL-1R). Proteins in the second group lack this region, although several of these may have a leucine zipper (LZ) or coiled-coil (CC) domain in the N-terminal region (see chapter 5 in this volume). In addition to this large group of genes, several other resistance genes encode proteins with different structures. The products of the tomato *Cf* genes consist almost entirely of an LRR domain predicted to be extracellular and attached to a short transmembrane anchor at the C-terminal. The *Pto* gene, also from tomato, encodes a cytoplasmic ser/thr protein kinase, while the *Xa21* gene from rice is a combination of these two types, encoding a transmembrane protein with an extracellular LRR and an intracellular protein kinase. The $Hs1^{pro-1}$ nematode resistance gene from sugar beet (not shown in Table 4.1) (Cai *et al.*, 1997) probably represents a novel resistance gene class (Ellis and Jones, 1998).

It is likely that some regions of related resistance proteins have a common function, perhaps in signal transduction, while other regions are involved in specific recognition of factors produced by different pathogens. Although little is yet known about how resistance proteins work, the similarity of some of these regions to known proteins provides some clues as to their possible roles. For example, the presence of sequence motifs known to form nucleotide binding sites in adenosine triphosphate/guanosine triphosphate (ATP/GTP) binding proteins (Traut, 1994) in the products of many resistance genes suggests that hydrolysis of ATP/GTP may be involved in their function. It was recently noted (Chinnaiyan *et al.*, 1997; van der Biezen and Jones, 1998) that the NBS domains of resistance proteins are similar to the corresponding domains of the animal cell apoptosis regulators, CED-4 and Apaf-1. A number of conserved motifs not present in other nucleotide binding proteins are also shared between this region of resistance proteins and Apaf-1/CED-4. Given the common, although not universal, involvement of hypersensitive cell death in resistance responses, this raises the intriguing possibility that signalling by resistance proteins may bear some similarity to the processes that lead to apoptosis in animal cells (see chapter 7 in this volume).

The TIR domains of some resistance genes may have similar functions to the homologous regions of the *Drosophila* Toll and mammalian IL-1R proteins. These proteins are membrane-bound receptors with functions in innate cellular resistance responses in animals (Lemaitre *et al.*, 1996; Medzhitov *et al.*, 1997). When the extracellular domains of Toll and IL-1R have bound their respective ligands, their intracellular domains interact with components of the signalling pathway to induce expression of antimicrobial genes (Volpe *et al.*, 1997; Yang and Steward, 1997). The similarity of TIR domains of resistance proteins to the signalling domains of Toll and IL-1R suggests that these regions may be involved in signalling in disease resistance responses, although there is now some evidence that they also influence recognition (Luck, 1998; Ellis *et al.*, 1999; see also section 4.3).

The signal transduction pathways utilised by different classes of resistance proteins may be different (see chapter 6 in this volume). Two *Arabidopsis* mutants distinguish signalling pathways used by TIR-NBS-LRR and NBS-LRR resistance proteins. The *eds1* mutation abolishes resistance to a number of isolates of *Pseudomonas syringae* and *Peronospora parasitica* in *Arabidopsis* (Parker *et al.*, 1996; Falk *et al.*, 1999), while the *ndr1* mutation affects a different subset of resistance genes to these pathogens (Century *et al.*, 1995, 1997). Cloned resistance genes affected by *eds1* are all of the TIR-NBS-LRR type, while *ndr1*-dependent resistance genes that have been cloned are of the NBS-LRR type, although *Rpp8* (NBS-LRR class) is not affected by either mutation (Aarts *et al.*, 1998). This result also suggests that the N-terminal regions of these proteins (the TIR domain and potential LZ/CC, respectively) determine which signalling pathway they utilise. However, there are other sequence features, within the NBS and LRR domains, that distinguish these classes of resistance protein, and they may also be involved in signalling processes.

4.3 Control of resistance gene specificity

A perusal of the resistance genes listed in Table 4.1 and the host pathogen systems in which they operate reveals some interesting observations. Firstly, genes of the most common class, encoding cytoplasmic proteins with NBS and LRR domains, recognise widely diverse pathogens, including viruses, bacteria, fungi, nematodes and aphids. The pathogens recognised by these resistance genes all either replicate entirely within plant cells (tobacco mosaic virus), directly secrete proteins into the plant cell (bacterial pathogens), make close contact with the plant cell membrane through the formation of haustoria (rust and mildew fungi) or physically penetrate plant cells (nematodes and aphids). Thus, there is the opportunity for intracellular resistance proteins to interact with molecules produced by these pathogens. In contrast, *Cladosporium fulvum* grows entirely within the extracellular spaces in tomato leaves, without penetration of plant cell walls. The *Cf* genes in tomato for resistance to *C. fulvum* are the only example, so far, of resistance genes encoding extracellular membrane-bound LRR proteins, and no other class of resistance gene has been found conferring resistance to this pathogen. This observation may be related to the different mode of infection employed by *C. fulvum* compared to other pathogens listed in Table 4.1. More resistance genes of this type may be found to other pathogens that have a similar growth style to *C. fulvum*. Indeed, Bevan *et al.* (1998) found a gene related to *Cf-2* in *Arabidopsis*, which is not known to be infected by *C. fulvum*.

It is also apparent that pathogen species can be recognised by different types of resistance genes. For instance, the rice resistance gene, *Xa21*, encodes a membrane-bound receptor kinase, while *Xa1* encodes an NBS-LRR, but both are active against strains of the same pathogen, *Xanthomonas oryzae*. Furthermore, in *Arabidopsis*, genes of both the NBS-LRR and TIR-NBS-LRR subgroups confer resistance to the fungal pathogen, *P. parasitica*, and also to the bacterial pathogen, *P. syringae*.

Members of the NBS-LRR family of resistance genes are highly versatile in their ability to determine resistance specificity against an extreme range of pathogens. Their structural features suggest a model where the NBS domain (and perhaps the TIR or LZ/CC domains) functions in signalling, while the LRR may be involved in recognition. Similarly, the Xa21 resistance protein from rice resembles receptor-kinases with an extracellular ligand binding domain (LRR) and an intracellular signalling domain (protein kinase). LRRs are found in many eukaryotic proteins and mediate specific protein-protein interactions (Kobe and Diesenhofer, 1994), including a direct involvement in ligand binding of several mammalian hormone receptors (Braun *et al.*, 1991). Thus, these domains of resistance genes are likely to be involved in recognition of avirulence determinants. Recent experiments involving domain swaps and sequence comparisons between alleles of the *L* locus for rust resistance in flax generally confirm this prediction, but also point to further refinements (Ellis *et al.*, 1999). For instance, when the LRR encoding region from *L2* was combined with the TIR-NBS encoding region of *L10* or *L6*, the resulting chimeric gene had identical rust resistance specificity to *L2*, indicating that this region contains all of the information necessary to distinguish these specificities. Also, the *L6* and *L11* sequences only differ in the LRR encoding region, so again this region must determine the differences between their specificities.

However, other evidence suggests that the TIR and/or NBS domains also influence specificity. For instance, the combination of the *L10* LRR encoding region with the *L2* TIR-NBS encoding region resulted in a chimeric resistance gene with a novel specificity. This *L2/L10* hybrid gene provided resistance to some but not all of the rust races recognised by *L10*, but gave no resistance to rust races not recognised by *L10*. It is not yet clear whether the avirulence gene detected by this chimeric L protein is related to the *A-L10* avirulence gene. Furthermore, the L6 and L7 proteins differ by only 11 amino acids and these are all located in the TIR domain, again suggesting that alterations in this region affect recognition specificity. Further analysis of chimeric genes with exchanges in this region showed that just three of these amino acid changes are sufficient to generate this specificity difference (Luck, 1998). Thus, although it is likely that the LRR region plays the primary role in formation of a ligand binding surface, it appears that recognition specificity may be determined by complex interactions within resistance proteins.

LRRs are also involved in interactions between proteins in signal transduction pathways (e.g. the interaction between yeast adenylate cyclase and the RAS protein; Suzuki *et al.*, 1990). Recently, Warren *et al.* (1998) found that an amino acid substitution in the N-terminal region of the LRR of the RPS5 protein not only abolished *Rps5* resistance but also led to partial suppression of other resistance genes in *Arabidopsis* to both *P. syringae* and *P. parasitica*. This region of the RPS5 protein may be involved in a signalling process and the pleiotropic effect of this mutation on other resistance genes may be due to its effect on a mutual signalling pathway, such as by titration of a common signal transduction factor. Consistent with this suggestion is the observation that this part of the LRR is relatively well conserved among NBS-LRR genes, while the more C-terminal region is variable.

The *Cf* resistance genes of tomato consist almost entirely of LRRs (Dixon *et al.*, 1996, 1998; Parniske *et al.*, 1997), and so it is likely that both specificity and signalling interactions are mediated by at least part of the LRR region. In fact, the C-terminal half of the LRR in these proteins is highly conserved, while the N-terminal region is highly variable in both sequence and LRR repeat unit copy number (Parniske *et al.*, 1997; Dixon *et al.*, 1998). Thus, the C-terminal LRR region may be involved in signalling, possibly by binding to downstream proteins or perhaps by mediating dimerisation, while the N-terminal region is likely to be responsible for ligand recognition.

4.4 Do resistance proteins interact directly with avirulence determinants?

It is principally the 'gene-for-gene' relationship that provides the strongest evidence for resistance proteins being specific receptors for pathogen-derived ligands. However, with one exception discussed later in this section, it has not yet been possible to demonstrate any direct interaction between resistance proteins and their corresponding avirulence proteins, where these are identified. This is beginning to lead to the notion that resistance proteins may recognise protein complexes formed between avirulence proteins and other plant proteins (Cheng *et al.*, 1998). Indeed, the product of the *Avr-9* gene from *C. fulvum* induces a resistance response only in tomato lines containing *Cf-9*, but the Avr-9 peptide binds to high affinity sites on tomato cell membranes that are present in both resistant and susceptible lines (Kooman-Gersmann *et al.*, 1996). Expression of the Cf-9 protein in *Arabidopsis* did not lead to the expression of an Avr-9 binding site in this species (de Wit, 1997).

This hypothesis has a certain theoretical sense to it. Many 'avirulence' proteins are thought to actually have functions in promoting pathogen growth (i.e. a pathogenicity function), possibly by manipulating aspects of the plant's metabolism or response mechanisms (Collmer, 1998; Lauge and de Wit, 1998). For instance, the 2b protein of tomato aspermy virus acts as a virulence factor during infection of *Nicotiana benthamiana* by repressing post-transcriptional gene silencing, but activates a hypersensitive resistance response when expressed in tobacco (Li *et al.*, 1999). The pathogenicity function of pathogen 'avirulence' products necessarily existed before plant resistance genes evolved the ability to recognise those avirulence gene products. Thus, it is logical to expect that the interaction between an avirulence protein and its cellular target in the plant would be highly developed. Rather than try to pre-empt this high affinity interaction, newly evolving resistance proteins, perhaps with a relatively low affinity for their target, would be more likely to interact with the already formed avirulence protein/host protein complex.

In this context, it is interesting to note that *Pto*, the only resistance protein shown to interact directly with its corresponding avirulence protein (Scofield *et al.*, 1996; Tang *et al.*, 1996), requires a second gene, *Prf*, to confer resistance. *Pto* is unique among resistance genes in encoding an intracellular protein kinase with no associated LRR, while *Prf* is a member of the common NBS-LRR class of resistance gene. The Pto kinase has also been shown to interact with other plant proteins, including DNA-binding proteins that bind to sequences found in the promoters of pathogenesis-related genes (Zhou *et al.*, 1997), and also a second protein kinase that may have a role in induction of the hypersensitive response (Zhou *et al.*, 1995). This may be a reflection of the role of Pto in AvrPto-induced resistance, but, if it reflects a more general role in defence mechanisms, then *Pto* would make an attractive target for manipulation by the pathogen. It is possible that *Pto* is the cellular target of the AvrPto protein, whose function may be to inhibit defence responses during infection, and that *Prf* recognises the complex formed between these proteins.

Prf is also required for the expression of fenthion sensitivity in tomato, which results from an interaction between this insecticide molecule and Fen, a protein kinase encoded by a gene family member closely related to *Pto* (Martin *et al.*, 1994; Salmeron *et al.*, 1994). This suggests that the Prf protein may be capable of detecting perturbations in the structure of individual proteins of the Pto family caused by interactions with other molecules. It is possible that some resistance proteins may respond solely to alterations in host proteins caused by avirulence products, without any direct contact with the avirulence product. A possibly related phenomenon is that the Rpm1 resistance protein in *Arabidopsis* recognises two apparently unrelated avirulence products from *P. syringae* (Bisgrove *et al.*, 1994). This may be because these avirulence products have the same (or at least a highly similar) target in the host plant. On a similar note, the nematode resistance gene, *Mi*, in tomato is also active in resistance to aphids (Rossi *et al.*, 1998). Rather than these unrelated organisms producing an identical avirulence determinant, it may be that different virulence products from aphids and nematodes modify the same host protein, which then serves as a trigger of *Mi*-mediated resistance. These examples provide exceptions to the generalised gene-for-gene hypothesis and it will be interesting to see whether further characterisation of these resistance genes and their ligands can reconcile this discrepancy.

Another resistance locus whose function deviates from the strict gene-for-gene hypothesis of disease resistance is the *Rp8* rust resistance locus in maize. Resistance is conferred only when the plant is heterozygous for the *Rp8A* and *Rp8B* alleles, but not when either allele is homozygous or when the *Rp8C* allele is present (Delaney *et al.*, 1998). In this case, recognition may require a heterodimeric resistance protein complex, possibly involving the formation of a ligand-binding surface incorporating regions of both proteins.

4.5 Organisation of resistance gene loci

Many resistance genes occur in complex loci that contain multiple copies of closely-related gene sequences. For instance, the *Dm3* locus in lettuce contains at least 22 sequences related to the *Dm3* resistance gene, spanning over 3.5 Mbp

(Meyers *et al.*, 1998a), while the *M* rust resistance gene in flax occurs in a complex locus of about 15 related genes, spanning 300–1000 kbp (Anderson *et al.*, 1997). The structure of a complex locus can be highly polymorphic within a species and variants of the same locus may contain different numbers of gene copies. Even when similar gene numbers are present at variants of a complex locus, it may not be possible to distinguish precise allelic relationships between individual genes within the cluster.

The concept of haplotype is used to describe the precise complement of related genes occurring in a particular variant of a complex locus. In some cases, mutational analysis suggests that only one of the related sequences within a particular resistance haplotype expresses a detectable resistance specificity (for instance at the *M* locus in flax; Anderson *et al.*, 1997), and where extensive sequence data are available, many of the copies present at a locus are found to be pseudogenes. For example, all but two of the seven related sequences at the *Xa21* locus in rice are interrupted by transposable elements or stop codons (Song *et al.*, 1997) and, of nine fully-sequenced genes at the *Dm3* locus in lettuce, three contained nonsense or frame-shift mutations (Meyers *et al.*, 1998b). Similarly, at the *Rpp5*-homologous locus in the Columbia ecotype of *Arabidopsis*, which has been completely sequenced as part of the *Arabidopsis* genome sequencing project, only two of eight *Rpp5* homologues are intact (Bevan *et al.*, 1998). In other cases, such as the *Cf-4/9* locus in tomato (Parniske *et al.*, 1997), most of the genes at the locus contain uninterrupted open reading frames and are transcribed. In this case, transformation experiments and unequal crossing-over events that removed the *Cf-4* and *Cf-9* genes revealed that additional resistance specificities, independent of the *Avr-4* and *Avr-9* avirulence genes, are encoded by other members of the *Cf-4/9* gene family (Parniske *et al.*, 1997). In addition, the *Cf-2* locus in tomato contains two identical copies of the gene sequence that encodes *Cf-2* resistance (Dixon *et al.*, 1996). Interestingly, the *Rpp1* locus in *Arabidopsis* contains four TIR-NBS-LRR sequences and three of these genes have distinct recognisable disease resistance specificities (Botella *et al.*, 1998). This indicates that the entire haplotype of complex resistance loci may contribute to the observed resistance specificities that map to that locus. This ability to encode multiple resistance specificities at a single locus may be a significant selective force favouring gene duplication and the evolution of complex loci.

Some resistance genes do occur in simple loci. For instance, the *L* locus of flax encodes an allelic series of multiple resistance specificities with 13 resistance alleles. It is interesting that this simple locus encodes such a large number of specificities, more than most complex loci. There are also some simple loci that encode only a single known resistance allele (e.g. *Rps2, Rpm1* and *Xa1*). Grant *et al.* (1998) have analysed the *Rpm1* locus in a variety of lines of *Arabidopsis*, and in all those lines that lack *Rpm1* resistance, the *Rpm1* gene is completely absent, being replaced by a 98 base-pair (bp) sequence of unknown origin. A similar situation was observed for the *Rpm1* homologous locus in the related species, *Brassica napus*; that is, some lines contain a *Rpm1* homologue at this locus while others, although containing syntenous flanking markers, have no detectable nucleotide similarity to *Rpm1* (Grant *et al.*, 1998).

Large numbers of functional resistance gene loci have been described in plant genomes that are active against a variety of pathogens. In addition, it is now clear that most plant genomes contain hundreds of sequences related to known resistance genes. Botella *et al.* (1997) identified *Arabidopsis* expressed sequence tags (ESTs) with homology to resistance genes (mostly to just the LRR domains) and mapped these to 47 loci, many of which contained repeated sequences. With the ability to identify new genes of the NBS-LRR subclass by degenerate polymerase chain reaction (PCR), large numbers of resistance gene-like sequences are now being identified in diverse species as potential candidates to encode useful or interesting resistance genes. For example, Collins *et al.* (1998) used this approach to identify resistance gene candidates in maize, and this led to the isolation of the *Rp1-D* resistance gene (Collins *et al.*, 1999). This approach was also used to isolate genes at the *Dm3* locus in lettuce (Shen *et al.*, 1998). Large numbers of resistance gene homologues have also been identified in soybean (Kanazin *et al.*, 1996; Yu *et al.*, 1996), potato (Leister *et al.*, 1996), wheat (Seah *et al.*, 1998), rice and barley (Leister *et al.*, 1998, 1999). Many of these resistance gene homologues map close to known resistance gene loci or clusters, but many map to regions with no known functional disease resistance loci. Perhaps some resistance gene homologues function against pathogen isolates found in natural populations that have not been tested on the host plants in laboratory or agricultural situations. Many resistance-like genes may be non-polymorphic, having become fixed in the population, and such fixed genes may be involved in determining some cases of non-host resistance. It is also possible that some resistance gene homologues perform different functions in the plant, unrelated to disease resistance.

Unrelated resistance loci are often clustered together in the genome. For instance, the major resistance gene clusters of *Arabidopsis* contain multiple linked loci for resistance to different pathogens (Kunkel, 1996), and many *Arabidopsis* resistance gene homologues also map to locations within these clusters (Botella *et al.*, 1997). Similarly, resistance-gene-related sequences are also clustered in soybean (Kanazin *et al.*, 1996; Yu *et al.*, 1996), potato (Leister *et al.*, 1996), wheat (Seah *et al.*, 1998), rice and barley (Leister *et al.*, 1999). Three rust resistance loci closely linked to the *Rp1* locus in maize do not contain genes that are closely related to *Rp1* (Collins *et al.*, 1999). Perhaps the best illustration of this phenomenon comes from the sequence of a 1.9 Mbp region of chromosome 4 of *Arabidopsis* (Bevan *et al.*, 1998). This region contains the Columbia ecotype version of the *Rpp5* resistance gene locus, which includes eight gene sequences closely related to *Rpp5* within 90 kbp. Two of these contain full length open reading frames (the others are interrupted by retro-elements or frame-shifts) and could encode the *Rpp4* resistance gene, which maps to this locus in the Columbia ecotype. There are also three other resistance gene-like sequences within the 1.9 Mbp region: one related to each of the TIR-NBS-LRR, NBS-LRR and extracellular LRR classes of resistance genes. Interestingly, the *Prf* gene, which is not part of a multigene family, occurs within the region comprising the *Pto* complex locus in tomato (Salmeron *et al.*, 1996).

4.6 Evolution of resistance genes by divergent selection

Wild plant/pathogen populations are typically highly polymorphic both in resistance and avirulence phenotypes (Frank, 1992). Bevan *et al.* (1993) analysed natural populations of groundsel (*Senecio vulgaris*) infected by powdery mildew (*Erisyphe fischeri*) and found high levels of variation for both resistance phenotypes in the host plants and virulence phenotypes in the pathogen isolates. Interestingly, no plants in these populations were resistant to all of the mildew isolates, and nor were any mildew isolates virulent on all of the plants. The pathogen isolates with the widest host plant range were the rarest class among the mildew population, and conversely the widest resistance phenotypes were the least common among the groundsel population. Although the selective forces driving the co-evolution of plants and their pathogens are complex and difficult to quantify (Frank, 1992; Rausher, 1996), the maintenance of a high level of genetic diversity at resistance gene loci is clearly favoured.

Selection for diversity has left its mark in the DNA sequences of resistance genes. Several authors have recently noted evidence for diversifying selection acting on resistance gene loci (Parniske *et al.*, 1997; Botella *et al.*, 1998; McDowell *et al.*, 1998; Meyers *et al.*, 1998b). This has been detected by comparing the DNA sequences of individual genes within a family and determining the frequency of synonymous nucleotide substitutions (Ks), which do not alter the amino acids encoded by the DNA sequence, to the frequency of nonsynonymous substitutions (Ka), which lead to altered amino acid sequences. In most genes, selection favours amino acid conservation because most amino acid changes are deleterious to the function of the protein product, and thus $Ks > Ka$. Under neutral selection (for instance in pseudogenes), Ks and Ka are approximately equal (Hughes, 1995). In cases where amino acid variation is favoured, Ka may be greater than Ks. This situation was first described for the major histocompatibility complex (MHC) class I genes of the vertebrate immune system, which are involved in antigen presentation to T-cells (Hughes and Nei, 1988). The antigen recognition sites of these proteins are variable within populations and between species, and exhibit higher Ka than Ks values, whereas the remainder of the gene is subject to conservative selection ($Ka < Ks$).

The ability of this test to identify the sites in the MHC protein that are involved in antigen recognition makes it a useful tool to identify similar sites in other genes subject to diversifying selection. For instance, Ishimizu *et al.* (1998) analysed stylar self-incompatibility (S-RNase) gene sequences from plants in the family Rosaceae and found four regions with Ka > Ks. The amino acids encoded by these sequences were predicted to fold into two epitopes on the surface of the protein that may be involved in interactions with pollen encoded self-incompatibility determinants.

Evidence for diversifying selection has now been described for several plant resistance gene families (Table 4.2 and accompanying references). In these cases, analysis has focused on a particular region of the protein, the xxLxLxx motif within each repeat unit of the LRR. This attention is based on the comparison with the porcine ribonuclease inhibitor (PRI) protein, whose known structure is used as the model for interpreting the likely structures of other LRR proteins. In PRI, the xxLxLxx motifs in adjacent repeat units form a surface of parallel β-sheet/β-turn structures, which contains many of the residues that form direct contacts with the ligand protein, ribonuclease A (Kobe and Diesenhofer, 1995). Thus, the corresponding region in LRR-containing resistance proteins may be involved in binding to their ligands. When

Table 4.2 Relative rates of non-synonymous (Ka) and synonymous (Ks) nucleotide substitution in resistance gene families

Gene family	Ka/Ks		
	Whole gene	LRR[a]	
		xxLxLxx	Unframed
L-alleles[b]	1.3	3.8	1.13
Cf-4/9[c]		2.1	0.59
Dm3[d]		2.1	0.63
Xa21[e]		2.1	0.52
Rpp1[f]	0.91	2.7	0.65
Rpp8[g]		2.4	0.8

[a]The leucine-rich repeat (LRR) encoding regions of resistance gene families were divided into two regions, one encoding the 'x' amino acid positions in the β-strand motif, xxLxLxx (excluding the conserved leucine residues), and one encoding the amino acids outside this motif (unframed). [b]The coding sequences of 11 *L*-alleles (Ellis *et al.*, 1999) were aligned and Ka/Ks ratios were determined using the diverge program in the GCG suite. The average Ka/Ks ratio for the 55 pairwise comparisons is shown. [c]Data for the 5′ LRR region (Parniske *et al.*, 1997). [d]Data for the 3′ LRR region (Meyers *et al.*, 1998b). [e]Data for the B, D and F members of the *Xa21* gene family (Meyers *et al.*, 1998b). [f]Botella *et al.*, 1998. [g]Nucleotide substitutions in the larger motif, xxLxLxxxx, were analysed (McDowell *et al.*, 1998).

nucleotide substitution rates have been analysed in resistance gene families, elevated Ka/Ks ratios have been observed in this region. This is consistent with a role in ligand binding, as selection for amino acid variation within this motif would favour the evolution of new and advantageous recognitional specificities.

A similar analysis has been performed with alleles of the *L* locus for rust resistance in flax (Table 4.2). In this gene family, Ka exceeds Ks for the gene as a whole, possibly because *L* locus alleles are more closely related to each other (at least 95% nucleotide identity for all gene pairs) than members of the other resistance gene families, and so there is less dilution of positively selected variation by background non-selected variation. However, as in other resistance gene families, this high overall Ka/Ks ratio reflects a concentration of non-synonymous substitutions in the β-strand motif of the LRR. The Ka/Ks ratios for the TIR and NBS domains as a whole are less than one; however, when window analysis of substitution rates was performed, two peaks of $Ka/Ks > 1$ were found within the TIR domain (G. Huttley and P. Dodds, unpublished results). This is interesting, as there is now evidence that changes in the TIR domain due to interallelic recombination can result in changes in recognition specificity (Ellis *et al.*, 1999; Luck, 1998). This may be because the TIR domain can influence the tertiary structure of the protein so as to alter the binding surface presented by the LRR, or may reflect a direct role in ligand binding for the TIR domain.

It is important to note that such analyses are limited, in that codons are analysed in groups so the contribution of individual codons cannot be resolved. Thus, not all amino acids within the xxLxLxx motif are necessarily involved in ligand binding, and conversely residues outside this motif may also contribute to recognitional specificity. Indeed, in the PRI-ribonuclease A complex, 17 out of 28 contact points involve amino acids in the xxLxLxx motif and 11 contact points involve amino acids in the loop region adjacent to the carboxy end of the β-sheet/β-turn (Kobe and Diesenhofer, 1995). In a similar complex formed between the human ribonuclease inhibitor (hRI) and the ribonuclease angiogenin, 19 amino acids within the xxLxLxx motif of hRI make strong contacts with angiogenin, and 6 contact points occur outside this motif (Papageorgiou *et al.*, 1997).

Diversifying selection is unusual among gene families. Endo *et al.* (1996) analysed 3595 gene families (represented by 20,000 gene sequences present in the genbank database) and found only 17 families (0.5%) with higher rates of non-synonymous substitution than synonymous across the entire gene, as was observed for *L* alleles (Table 4.2). Most of these gene families encoded surface antigen proteins from human pathogens, which evolve new variations at a rapid rate. With a more sensitive test using window analysis, it was estimated that about 5% of the gene families contained regions subject to positive selection in which Ka/Ks was greater than one (Endo *et al.*, 1996).

4.7 Evolution of resistance genes by recombination

Much of the variation amongst resistance gene families appears to result from the shuffling of polymorphic sites by recombination between individual genes in those families. This is true both between allelic family members, as in the *L* alleles of flax, and between members of gene families that occur in complex loci, such as *Cf-2/5* (Dixon *et al.*, 1998) and *Cf-4/9* (Parniske *et al.*, 1997). This process is inferred from the observation of patchworks of sequence similarities shared between gene family members. Direct observations of sequence exchanges are rare, although numerous potential recombination events have been detected among alleles of the *L* locus in flax (reviewed by Islam and Shepherd, 1991). These events were relatively frequent, about 1/1000 gametes, and resulted in either complete loss of both parental specificities or, in some cases, the expression of a new specificity.

Recently, some of these variants have been analysed at a molecular level and it has been found that they have indeed resulted from intragenic recombination within the coding region of the *L* gene (Ellis *et al.*, 1997, and unpublished results; Luck, 1998). These recombinants encode novel combinations of polymorphic residues in the new proteins. For example, a recombination event between *L2* and *L10* led to a loss of detectable resistance. This recombinant allele encoded a chimeric protein with a TIR domain derived from *L2* and an NBS-LRR region from *L10*, which only differed from the L10 protein by five amino acids. A subsequent recombination between this new allele and the *L9* allele led to the expression of a novel resistance specificity by a recombinant protein that contained the L10 LRR region and the TIR-NBS region of L9. Interestingly, this specificity is indistinguishable from that expressed by the chimeric *L* allele produced *in vitro* containing the L2 TIR/NBS and L10 LRR described above (Ellis *et al.*, 1999). Similarly, a recombinant allele derived from the *L2* and *L6* alleles encoded a protein with only three amino acid changes from the L6 protein, all within the TIR domain, that expressed a specificity identical to the *L7* allele (Luck, 1998). These results demonstrate that recombination can result in novel variation both in resistance gene sequence and specificity. The high frequency of such events, and the evidence for past recombination shown by the sequences of *L* alleles, suggests that recombination makes a significant contribution to the generation of diversity and evolution of this resistance locus.

Recombination at the complex *Rp1* rust resistance locus of maize has also been studied extensively. Recombination at *Rp1*, inferred by the exchange of flanking restriction fragment length polymorphism (RFLP) markers, is associated with loss of resistance (Sudupak *et al.*, 1993), combining of multiple resistance specificities onto one chromosome (Hu and Hulbert, 1996), the generation of new specificities (Richter *et al.*, 1995) and disease lesion mimic phenotypes (Hu *et al.*, 1996). Pairing between repeated sequences at the *Rp1* locus can be variable, as recombinants with altered phenotypes derived from heterozygous parents have been observed with both parental combinations of flanking markers. In addition, 27 *Rp1-D* mutants isolated from approximately 200,000 progeny of *Rp1-D* homozygotes showed at least nine different deletions of various *Rp1* gene family members, presumably reflecting a range of chromosome-mispairing configurations adopted during unequal crossing-over events at this complex locus (Collins *et al.*, 1999). Although some of the *Rp1-D* recombinations may have occurred in intergenic regions, preliminary analysis of two recombinants indicates that both occurred within the reading frames of *Rp1* homologous genes and one of these is associated with a novel phenotype, presumably due to the expression of a recombinant protein (Pryor and Hulbert, unpublished data). Selection for the loss of *Rp1-D* resistance in this experiment places constraints on the types of recombination events that can be detected, as the cross-overs must alter or remove *Rp1-D* on one of the recombinant chromosomes. The likely position of the *Rp1-D* gene at one end of the gene cluster (N. Collins, personal communication) severely restricts the potential pairing alignments that can give rise to such an event, and thus the frequency of recombination within the *Rp1* complex is likely to be higher than suggested by the number of *Rp1-D* mutants in this screen.

Dixon *et al.* (1998) analysed 12,000 F_2 tomato plants resulting from a cross between *Cf-2* and *Cf-5* homozygous lines and found one susceptible recombinant that had lost both resistance genes by a recombination event that occurred between the *Cf-2.2* gene and one of the *Cf-2/5* homologous genes in the *Cf-5* haplotype, *Hcr2-5B*. This event produced a chimeric gene with 5′ sequences from *Cf-2.2* and 3′ sequences from *Hcr2-5B*, but no resistance phenotype was associated with this recombinant gene. Parniske *et al.* (1997) also observed unequal crossing-over between the *Cf-4* and *Cf-9* haplotypes in tomato, leading to the loss of both *Cf-4* and *Cf-9* specificities. However, these events all occurred in the intergenic regions and so did not result in the production of novel gene variants.

Sequence exchanges can also occur between repeated regions within resistance genes. For example, Parker *et al.* (1997) isolated a variant of the *Rpp5* gene that contained an in-frame deletion of four LRR repeat units (although this amazingly did not affect resistance function). Similarly, Anderson *et al.* (1997) identified three loss-of-function mutants of the *M* gene in flax with the same internal deletion of a repeated sequence in the LRR. Several alleles of the *L* locus in flax contain internal deletions or expansions that may have resulted from unequal crossing-over in repeated regions of the genes (Ellis *et al.*, 1999). This type of variation is also evident among *Cf-2/5* and *Cf-4/9* homologues that differ in the number of LRR units they contain (Thomas *et al.*, 1997; Dixon *et al.*, 1998).

The weight of evidence from resistance gene sequence comparisons and analysis of recombination at resistance gene loci suggests that both intragenic (between allelic genes) and intergenic (between paralogous genes at complex loci) recombination events are important in generating diversity at resistance gene loci. Again, parallels with the vertebrate immune system abound. Class II MHC loci show evidence of recombination between allelic gene pairs and also between paralogous genes in complex loci (Andersson and Mikko, 1995). Similarly, Hughes (1995) suggested that sequence exchange occurs between paralogous genes at MHC class I loci, even between active genes and pseudogenes. Therefore, even non-expressed/able copies at complex loci can serve as a reservoir of variability and contribute to the evolution of new specificities by recombining with functional genes. In fact, at the *L* locus of flax, recombination between *L6* and the *L* allele present in the flax line Hoshangabad, which has no detectable rust resistance specificity, caused an alteration in specificity similar to that described above for recombination between *L2* and *L6* (i.e. the recombinant gene expressed the *L7* specificity; Luck, 1998). Given the preexistence of nucleotide variation, recombination can potentially generate new alleles at a substantially greater rate than point substitution. For instance, during the evolutionary lifetime of human populations, only four new class I MHC alleles have been generated by novel point mutation, compared to over 80 new alleles resulting from recombination between preexisting polymorphic sites (Parham and Ohta, 1996).

4.8 Concluding remarks

Recently, great advances have been made in understanding the basis and evolution of disease resistance specificity in plants. There is now direct evidence to support a model of specificity determination by resistance genes based on interactions between leucine rich repeats and their putative ligands. Likewise, a picture of resistance gene evolution has emerged based, firstly, on the accumulation of point mutations and, subsequently, on the shuffling of polymorphisms between members of gene families both by intragenic and intergenic recombination. Given the wealth of raw material now available in resistance gene sequences, significant progress should continue to be made as further molecular analysis of resistance gene structure proceeds.

References

Aarts, N., Metz, M., Holub, E., Staskawicz, B.J., Daniels, M.J. and Parker, J.E. (1998) Different requirements for *EDS1* and *NDR1* by disease resistance genes define at least two *R*-gene-mediated pathways in *Arabidopsis*. *Proc. Natl. Acad. Sci. USA*, **95** 10306-11.

Andersson, L. and Mikko, S. (1995) Generation of MHC class II diversity by intra- and intergenic recombination. *Immunol. Rev.*, **143** 6-12.

Anderson, P.A., Lawrence, G.J., Morrish, B.C., Ayliffe, M.A., Finnegan, E.J. and Ellis, J.G. (1997) Inactivation of the flax rust resistance gene, *M*, associated with loss of a repeated unit within the leucine-rich repeat coding region. *Plant Cell*, **9** 641-51.

Bevan, J.R., Crute, I.R. and Clarke, D.D. (1993) Variation for virulence in *Erysiphe fischeri* from *Senecio vulgaris*. *Plant Pathol.*, **42** 622-35.

Bevan, M., Bancroft, I., Bent, E., Love, K., Goodman, H., Dean, C., *et al.* (1998) Analysis of 1.9 Mb of contiguous sequence from chromosome 4 of *Arabidopsis thaliana*. *Nature*, **391** 485-88.

Bisgrove, S.R., Simonich, M.T., Smith, N.M., Sattler, A. and Innes, R.W. (1994) A disease resistance gene in *Arabidopsis* with specificity for two different pathogen avirulence genes. *Plant Cell*, **6** 927-33.

Botella, M.A., Coleman, M.J., Hughes, D.E., Nishimura, M.T., Jones, J.D.G. and Somerville, S.C. (1997) Map positions of 47 *Arabidopsis* sequences with sequence similarity to disease resistance genes. *Plant J.*, **12** 1197-211.

Botella, M.A., Parker, J.E., Frost, L.N., Bittner-Eddy, P.D., Beynon, J.L., Daniels, M.J., Holub, E.B. and Jones, J.D.G. (1998) Three genes of the *Arabidopsis Rpp1* complex resistance locus recognise distinct *Peronospora parasitica* avirulence determinants. *Plant Cell*, **10** 1847-60.

Braun, T., Schofield, P.R. and Sprengel, R. (1991) Amino-terminal leucine-rich repeats in gonadotropin receptors determine hormone selectivity. *EMBO J.*, **10** 1885-90.

Cai, D., Kleine, M., Kifle, S., Harloff, H.-J., Sandal, N.N., Marcker, K.A., Klein-Lankhorst, R.M., Salentjin, E.M.J., Lange, W., Stiekema, W.J., *et al.* (1997) Positional cloning of a gene for nematode resistance in sugar beet. *Science*, **275** 832-34.

Century, K.S., Holub, E.B. and Staskawicz, B.J. (1995) *NDR1*, a locus of *Arabidopsis thaliana* that is required for disease resistance to both a bacterial and a fungal pathogen. *Proc. Natl. Acad. Sci. USA*, **92** 6597-601.

Century, K.S., Shapiro, A.D., Repetti, P.P., Dahlbeck, D., Holub, E. and Staskawicz, B.J. (1997) *NDR1*, a pathogen-induced component required for *Arabidopsis* disease resistance. *Science*, **278** 1963-65.

Cheng, J., Smith-Becker, J. and Keen, N.T. (1998) Genetics of plant-pathogen interactions. *Curr. Opin. Plant Biotechnol.*, **9** 202-207.

Chinnaiyan, A.M., Chaudhary, D., O'Rourke, K., Koonin, E.V. and Dixit, V.M. (1997) Role of CED-4 in the activation of CED-3. *Nature*, **338** 728-29.

Collins, N.C., Webb, C.A., Seah, S., Ellis, J.G., Hulbert, S.H. and Pryor, A. (1998) The isolation and mapping of disease resistance gene analogs in maize. *Mol. Plant-Microbe Interact.*, **11** 968-78.

Collins, N., Drake, J., Ayliffe, M., Sun, Q., Ellis, J., Hulbert, S. and Pryor, T. (1999) Molecular characterisation of the maize *Rp1-D* rust resistance haplotype and its mutants. *Plant Cell*, **11** 1365-76.

Collmer, A. (1998) Determinants of pathogenicity and avirulence in plant pathogenic bacteria. *Curr. Opin. Plant Biol.*, **1** 329-35.

Delaney, D.E., Webb, C.A. and Hulbert, S.H. (1998) A novel rust resistance gene showing overdominance. *Mol. Plant-Microbe Interact.*, **11** 242-45.

De Wit, P.J.G.M. (1997) Pathogen avirulence and plant resistance: a key role for recognition. *Trends Plant Sci.*, **2** 452-58.

Dixon, M.S., Jones, D.A., Keddie, J.S., Thomas, C.M., Harrison, K. and Jones, J.D.G. (1996) The tomato *Cf-2* disease resistance locus comprises two functional genes encoding leucine-rich repeat proteins. *Cell*, **84** 451-59.

Dixon, M.S., Hatzixanthis, K., Jones, D.A., Harrison, K. and Jones, J.D.G. (1998) The tomato *Cf-5* disease resistance gene and six homologues show pronounced allelic variation in leucine-rich repeat copy number. *Plant Cell*, **10** 1915-25.

Ellis, J. and Jones, D. (1998) Structure and function of proteins controlling strain-specific pathogen resistance in plants. *Curr. Opin. Plant Biol.*, **1** 288-93.

Ellis, J.G., Lawrence, G., Ayliffe, M., Anderson, P., Collins, N., Finnegan, J., Frost, D., Luck, J. and Pryor, T. (1997) Advances in the molecular genetic analysis of the flax-flax rust interaction. *Annu. Rev. Phytopathol.*, **35** 271-91.

Ellis, J.G., Lawrence, G.J., Luck, J.E. and Dodds, P.N. (1999) Identification of regions in alleles of the flax rust resistance gene, *L*, that determine differences in gene-for-gene specificity. *Plant Cell*, **11** 495-506.

Endo, T., Ikeo, K. and Gojobori, T. (1996) Large scale search for genes on which positive selection may operate. *Mol. Biol. Evolut.*, **13** 685-90.

Falk, A., Feys, B.J., Frost, L.N., Jones, J.D.G., Daniels, M.J. and Parker, J.E. (1999) *EDS1*, an essential component of *R* gene-mediated disease resistance in *Arabidopsis* has homology to eukaryotic lipases. *Proc. Natl. Acad. Sci. USA*, **96** 3292-97.

Flor, H.H. (1971) Current status of the gene-for-gene concept. *Annu. Rev. Phytopathol.*, **9** 275-96.

Frank, S.A. (1992) Models of plant-pathogen co-evolution. *Trends Genet.*, **8** 213-19.

Grant, M.R., Godiard, L., Straube, E., Ashfield, T., Lewald, J., Sattler, A., Innes, R.W. and Dangl, J.L. (1995) Structure of the *Arabidopsis RPM1* gene enabling dual specificity disease resistance. *Science*, **269** 843-46.

Grant, M.R., McDowell, J.M., Sharpe, A., de Torres Zabala, M., Lydiates, D.J. and Dangl, J.E. (1998) Independent deletions of a pathogen-resistance gene in *Brassica* and *Arabidopsis*. *Proc. Natl. Acad. Sci. USA*, **95** 15843-48.

Hu, G. and Hulbert, S. (1996) Construction of 'compound' rust resistance genes in maize. *Euphytica*, **87** 47-51.

Hu, G., Richter, T.E., Hulbert, S. and Pryor, T. (1996) Disease lesion mimicry caused by mutations in the rust resistance gene, *Rp1*. *Plant Cell*, **8** 1367-76.

Hughes, A.L. (1995) Origin and evolution of HLA class I pseudogenes. *Mol. Biol. Evolut.*, **12** 247-58.

Hughes, A.L. and Nei, M. (1988) Pattern of nucleotide substitution at major histocompatibility complex class I loci reveals overdominant selection. *Nature*, **335** 167-70.

Ishimizu, T., Endo, T., Yamaguchi-Kabata, Y., Nakamura, K.T., Sakiyama, F. and Norioka, S. (1998) Identification of regions in which positive selection may act in S-RNase of Rosaceae: implication for S-allele-specific recognition sites in S-RNases. *FEBS Lett.*, **440** 337-42.

Islam, M.R. and Shepherd, K.W. (1991) Present status of genetics of rust resistance in flax. *Euphytica*, **55** 255-67.

Jones, D.A., Thomas, C.M., Hammond-Kossack, K.E., Balint-Kurti, P.J. and Jones, J.D.G. (1994) Isolation of the tomato *Cf-9* gene for resistance to *Cladosporium fulvum* by transposon tagging. *Science*, **266** 789-93.

Kanazin, V., Marek, L.F. and Shoemaker, R.C. (1996) Resistance gene analogs are conserved and clustered in soybean. *Proc. Natl. Acad. Sci. USA*, **93** 11746-50.

Kobe, B. and Diesenhofer, J. (1994) The leucine-rich repeat: a versatile binding motif. *Trends Biol. Sci.*, **19** 415-21.

Kobe, B. and Diesenhofer, J. (1995) A structural basis of the interactions between leucine-rich repeats and protein ligands. *Nature*, **374** 183-86.

Kooman-Gersmann, M., Honee, G., Bonnema, G. and de Wit, P.J.G.M. (1996) A high-affinity binding site for the AVR9 peptide elicitor of *Cladosporium fulvum* is present on plasma membranes of tomato and other solanaceous plants. *Plant Cell*, **8** 929-38.

Kunkel, B.N. (1996) A useful weed put to work: genetic analysis of disease resistance in *Arabidopsis thaliana*. *Trends Genet.*, **12** 63-69.

Lagudah, E.S., Moullet, O. and Appels, R. (1997) Map-based cloning of a gene sequence encoding a nucleotide-binding domain and leucine-rich region at the *Cre3* nematode resistance locus of wheat. *Genome*, **40** 659-65.

Lauge, R. and de Wit, P.J.G.M. (1998) Fungal avirulence genes—structure and possible functions. *Fungal Genet. Biol.*, **24** 285-97.

Lawrence, G.J., Finnegan, E.J., Ayliffe, M.A. and Ellis J.G. (1995) The *L6* gene for flax resistance is related to the *Arabidopsis* bacterial resistance gene, *Rps2*, and the tobacco viral resistance gene, *N*. *Plant Cell*, **7** 1195-206.

Leister, D., Ballvora, A., Salamini, F. and Gebhardt, C. (1996) A PCR-based approach for isolating pathogen resistance genes from potato with potential for wide application in plants. *Nature Genet.*, **14** 421-29.

Leister, D., Kurth, J., Laurie, D.A., Yano, D.A., Sasaki, T., Devos, K., Graner, A. and Schulze-Lefert, P. (1998) Rapid reorganisation of resistance gene homologues in cereal genomes. *Proc. Natl. Acad. Sci. USA*, **95** 370-75.

Leister, D., Kurth, J., Laurie, D.A., Yano, D.A., Sasaki, T., Graner, A. and Schulze-Lefert, P. (1999) RFLP and physical mapping of resistance gene homologues in rice (*O. sativa*) and barley (*H. vulgare*). *Theor. Appl. Genet.*, **98** 509-20.

Lemaitre, B., Nicolas, E., Michaut, L., Reichhart, J.-M. and Hoffmann, J.A. (1996) The dorsoventral regulatory cassette spatzle/Toll/cactus controls the potent antifungal response in *Drosophila* adults. *Cell*, **86** 973-83.

Li, H.-W., Lucy, A.P., Guo, H.-S., Li, W.-X., Ji, L.-H., Wong, S.-M. and Ding, S.-W. (1999) Strong host resistance targeted against a viral suppressor of the plant gene silencing defence mechanism. *EMBO J.*, **18** 2683-91.

Luck, J.E. (1998) Transposition and recombination events in flax rust resistance genes. PhD thesis, Australian National University, Australia.

Martin, G.B., Brommonschenkel, S.H., Chunwongse, J., Frary, A., Gana, M.W., Spivey, R., Wu, T., Earle, E.D. and Tanksley, S.D. (1993) Map-based cloning of a protein kinase gene conferring disease resistance in tomato. *Science*, **262** 1432-36.

Martin, G.B., Frary, A., Wu, T., Brommonschenkel, S., Chunwongse, J., Earle, E.D. and Tanksley, S.D. (1994) A member of the *Pto* gene family confers sensitivity to fenthion resulting in rapid cell death. *Plant Cell*, **6** 1543-52.

Matthieu, H.A.J.J., Cozijnsen, T.J. and de Wit, P.J.G.M. (1994) Host resistance to a fungal tomato pathogen lost by a single base-pair change in an avirulence gene. *Nature*, **367** 384-86.

McDowell, J.M., Dhandaydham, M., Long, T.A., Aarts, M.G.M., Goff, S., Holub, E.B. and Dangl, J.E. (1998) Intragenic recombination and diversifying selection contribute to the evolution of downy mildew resistance at the *Rpp8* locus of *Arabidopsis*. *Plant Cell*, **10** 1861-74.

Medzhitov, R., Preston-Hurlburt, P. and Janeway, C.A. (1997) A human homologue of the *Drosophila* Toll protein signals activation of adaptive immunity. *Nature*, **338** 394-97.

Meyers, B.C., Chin, D.B., Shen, K.A., Sivaramakrishnan, S., Lavelle, D.O., Zhang, Z. and Michelmore, R.W. (1998a) The major resistance gene cluster in lettuce is highly duplicated and spans several megabases. *Plant Cell*, **10** 1817-32.

Meyers, B.C., Shen, K.A., Rohani, P., Gaut, B.S. and Michelmore, R.W. (1998b) Receptor-like genes in the major resistance locus of lettuce are subject to divergent selection. *Plant Cell*, **10** 1833-46.

Milligan, S.B., Bodeau, J., Yaghoobi, J., Kaloshian, I., Zabel, P. and Williamson, V.M. (1998) The root knot nematode resistance gene, *Mi*, from tomato is a member of the leucine zipper, nucleotide-binding, leucine-rich repeat family of plant genes. *Plant Cell*, **10** 1307-19.

Mindrinos, M., Katagiri, F., Yu, G.L. and Ausubel, F. (1994) The *A. thaliana* disease resistance gene, *RPS2*, encodes a protein containing a nucleotide-binding site and leucine-rich repeats. *Cell*, **78** 1089-99.

Ori, N., Eshed, Y., Paran, I., Presting, G., Aviv, D., Tanksley, S., Zamir, D. and Fluhr, R. (1997) The *I2C* family from the wilt disease resistance locus *I2* belongs to the nucleotide-binding, leucine-rich repeat superfamily of plant resistance genes. *Plant Cell*, **9** 521-32.

Papageorgiou, A.C., Shapiro, R. and Acharya, K.R. (1997) Molecular recognition of human angiogenin by placental ribonuclease inhibitor—an X-ray crystallographic study at 2.0 Å resolution. *EMBO J.*, **16** 5162-77.

Parham, P. and Ohta, T. (1996) Population biology of antigen presentation by MHC class I molecules. *Science*, **272** 67-74.

Parker, J.E., Holub, E.B., Frost, L.N., Falk, A., Gunn, N.D. and Daniels, M.J. (1996) Characterisation of *eds1*, a mutation in *Arabidopsis* suppressing resistance to *Peronospora parasitica* specified by several different *RPP* genes. *Plant Cell*, **8** 2033-46.

Parker, J.E., Coleman, M.J., Szabo, V., Frost, L.N., Schmidt, R., van der Biezen, E.A., Moores, T., Dean, C., Daniels, M.J. and Jones, J.D.G. (1997) The *Arabidopsis* downy mildew resistance gene, *Rpp5*, shares similarity to the Toll and interleukin-1 receptors with *N* and *L6*. *Plant Cell*, **9** 879-94.

Parniske, M., Hammond-Kosack, K.E., Golstein, C., Thomas, C.M., Jones, D.A., Harrison, K., Wulff, B.B.H. and Jones, J.D.G. (1997) Novel disease resistance specificities result from sequence exchange between tandemly repeated genes at the *Cf-4/9* locus of tomato. *Cell*, **91** 821-32.

Rausher, M.D. (1996) Genetic analysis of co-evolution between plants and their natural enemies. *Trends Genet.*, **12** 212-17.

Richter, T.E., Pryor, T.J., Bennetzen, J.L. and Hulbert, S.H. (1995) New rust resistance specificities associated with recombination in the *Rp1* complex in maize. *Genetics*, **141** 373-81.

Rossi, M., Goggin, F.L., Milligan, S.B., Kaloshian, I., Ullman, D.E. and Williamson, V.M. (1998) The nematode resistance gene, *Mi*, of tomato confers resistance against the potato aphid. *Proc. Natl. Acad. Sci. USA*, **95** 9750-54.

Salmeron, J.M., Barker, S.J., Carland, F.M., Mehta, A.Y. and Staskawicz, B.J. (1994) Tomato mutants altered in bacterial disease resistance provide evidence for a new locus controlling pathogen recognition. *Plant Cell*, **6** 511-20.

Salmeron, J.M., Oldroyd, G.E.D., Rommens, C.M.T., Scofield, S.R., Kim, H.-S., Lavelle, D.T., Dahlbeck, D. and Staskawicz, B.J. (1996) Tomato *Prf* is a member of the leucine-rich repeat class of plant disease resistance genes and lies embedded within the *Pto* kinase gene cluster. *Cell*, **86** 123-33.

Scofield, S.R., Tobias, C.M., Rathjen, J.P., Chang, J.H., Lavelle, D.T., Michelmore, R.W. and Staskawicz, B.J. (1996) Molecular basis of gene-for-gene specificity in bacterial speck disease of tomato. *Science*, **274** 2063-65.

Seah, S., Sivasithamparan, K., Karakousis, A. and Lagudah, E.S. (1998) Cloning and characterisation of a family of disease resistance gene analogs from wheat and barley. *Theor. Appl. Genet.*, **97** 937-45.

Shen, K.A., Meyers, B.C., Islam-Faridi, M.N., Chin, D.B., Stelly, M. and Michelmore R.W. (1998) Resistance gene candidates identified using PCR with degenerate oligonucleotide primers map to resistance gene clusters in lettuce. *Mol. Plant-Microbe Interact.*, **11** 815-23.

Song, W.-Y., Wang, G.-L., Chen, L.-L., Kim, H.-S., Pi, L.-Y., Holsten, T., Gardner, J., Wang, B., Zhai, W.-X., Zhu, L.-H., Fauquet, C. and Ronald, P. (1995) A receptor-like protein encoded by the rice disease resistance gene, *Xa21*. *Science*, **270** 1804-806.

Song, W.-Y., Pi, L.-Y., Wang, G.-L., Gardner, J., Holsten, T. and Ronald, P. (1997) Evolution of the rice *Xa21* disease resistance gene family. *Plant Cell*, **9** 1279-87.

Sudupak, M.A., Bennetzen, J.L. and Hulbert, S.H. (1993) Unequal exchange and meiotic instability of disease resistance genes in the *Rp1* region of maize. *Genetics*, **131** 119-25.

Suzuki, N., Choe, H.-R., Nishida, Y., Yamawaki-Kataoka, Y., Ohnishi, S., Tamaoki, T. and Katoka, T. (1990) Leucine-rich repeats and carboxyl terminus regions are required for interaction of yeast adenylate cyclase with RAS proteins. *Proc. Natl. Acad. Sci. USA*, **87** 8711-15.

Tang, X., Frederick, R.D., Zhou, J., Halterman, D.A., Jia, Y. and Martin, G.B. (1996) Physical interaction of avrPto and the Pto kinase defines a recognition event involved in plant disease resistance. *Science*, **274** 2060-63.

Thomas, C.M., Jones, D.A., Parniske, M., Harrison, K., Balint-Kurti, P., Hatzixanthis, K. and Jones, J.D.G. (1997) Characterisation of the tomato *Cf-4* gene for resistance to *Cladosporium fulvum* identifies sequences that determine recognitional specificity in Cf-4 and Cf-9. *Plant Cell*, **9** 2209-24.

Traut, T.W. (1994) The functions and consensus motifs of nine types of peptide segments that form different types of nucleotide-binding sites. *Eur. J. Biochem.*, **222** 9-19.

van den Ackerveken, G.F.J.M., Vossen, P. and de Wit, P.J.G.M. (1993) The avr9 race specific elicitor of *Cladosporium fulvum* is processed by endogenous and plant proteases. *Plant Physiol.*, **103** 91-96.

van der Biezen, E.A. and Jones, J.D.G. (1998) The NB-ARC domain: a novel signalling motif shared by plant resistance genes products and regulators of cell death in animals. *Curr. Biol.*, **8** R226-R227.

Volpe, F., Clatworthy, J., Kaptein, A., Maschera, B., Griffin, A.-M. and Ray, K. (1997) The IL-1 receptor accessory protein is responsible for the recruitment of the interleukin-1 receptor associated kinase to the IL-1/IL-1 receptor I complex. *FEBS Lett.*, **419** 41-44.

Warren, R.F., Henk, A., Mowery, P., Holub, E. and Innes, R.W. (1998) A mutation within the leucine-rich repeat domain of the *Arabidopsis* disease resistance gene, *Rps5*, partially suppresses multiple bacterial and downy mildew resistance genes. *Plant Cell*, **10** 1439-52.

Whitham, S., Dinesh-Kumar, S.P., Choi, D., Hehl, R., Corr, C. and Baker, B. (1994) The product of the tobacco mosaic virus resistance gene, *N*: similarity to Toll and the interleukin-1 receptor. *Cell*, **78** 1101-15.

Yang, J. and Steward, R. (1997) A multimeric complex and the nuclear targeting of the *Drosophila* rel protein dorsal. *Proc. Natl. Acad. Sci. USA*, **94** 14524-29.

Yoshimura, S., Yamanouchi, U., Katayose, Y., Toki, S., Wang, Z.-Z., Kono, I., Kurata, N., Yano, M., Iwata, N. and Sasaki, T. (1998) Expression of *Xa1*, a bacterial blight-resistance gene in rice, is induced by bacterial inoculation. *Proc. Natl. Acad. Sci. USA*, **95** 1663-68.

Yu, Y.G., Buss, G.R. and Maroof, M.A.S. (1996) Isolation of a superfamily of candidate disease resistance genes in soybean, based on a conserved nucleotide-binding site. *Proc. Natl. Acad. Sci. USA*, **93** 11751-56.

Yucel, I., Midland, S.L., Sims, J.J. and Keen, N.T. (1994) Class I and Class II *avrD* alleles direct the production of different products in Gram-negative bacteria. *Mol. Plant-Microbe Interact.*, **7** 148-50.

Zhou, J., Loh, Y.-T., Bressan, R.A. and Martin, G.B. (1995) The tomato gene, *Pti1*, encodes a serine/threonine kinase that is phosphorylated by *Pto* and is involved in the hypersensitive response. *Cell*, **83** 925-35.

Zhou, J., Tang, X. and Martin, G.B. (1997) The Pto kinase conferring resistance to tomato bacterial speck disease interacts with proteins that bind a *cis*-element of pathogenesis-related genes. *EMBO J.*, **16** 3207-18.

5 Resistance genes and resistance protein function

David A. Jones

5.1 Introduction

Despite the discovery of resistance genes (*R* genes) early in the twentieth century and of the host-pathogen gene-for-gene relationship before the middle of the twentieth century, most progress in our understanding of the nature and function of *R* genes has only recently been achieved. The first *R* gene involved in a gene-for-gene interaction to be cloned and characterised was the tomato *Pto* gene for resistance to bacterial speck disease caused by *Pseudomonas syringae* pv. *tomato* (Martin *et al.*, 1993). Since then, there has been a rapid accumulation of *R* genes cloned from a diverse range of plant species, including both monocotyledons and dicotyledons, and conferring resistance to a diverse range of pathogens, including viruses, bacteria, fungi, oomycetes, nematodes and insects (Table 5.1).

Based on DNA sequence information, these *R* genes fall into five classes (Table 5.2). The largest class encodes cytoplasmic proteins containing a carboxyl-terminal (C-terminal) leucine-rich repeat (LRR) domain and a central nucleotide binding site (NBS). The involvement of the LRR motif in protein-protein or receptor-ligand interaction has been well documented (Kobe and Deisenhofer, 1994), and its role in resistance proteins is presumed to be that of direct or indirect recognition of the pathogen avirulence gene (*Avr* gene) product. The NBS binds adenosine triphosphate (ATP) or guanosine triphosphate (GTP), or possibly deoxy-ATP or -GTP (dATP or dGTP), and this binding is presumed to be a regulatory switch. The NBS-LRR resistance proteins can be subdivided according to their amino-terminal (N-terminal) domains. Some contain homology to the cytoplasmic domains of the *Drosophila* Toll receptor and the mammalian and avian interleukin-1 (IL-1) receptor (TIR domain). This domain has been implicated in interaction with downstream components of the Toll/IL-1 signalling pathway. Others contain leucine zipper (LZ) domains. The LZ is also a protein interaction motif and is able to form an amphipathic alpha helix that interacts with similar LZ domains in other proteins to form a coiled-coil (CC) structure. Both these domains are presumed to be effector domains involved in downstream signalling. The remaining NBS-LRR proteins have not been reported to contain TIR or LZ domains.

Table 5.1 Cloned resistance genes: classified according to the kind of resistance they confer and their source

Resistance	Host	Pathogen*	*R* gene	Reference
Viral	Tobacco	Tobacco mosaic virus	*N*	Whitham *et al.*, 1994
	Potato	Potato virus X	*Rx*	Bendahmane *et al.*, 1999
Bacterial	Tomato	*Pseudomonas syringae*	*Pto*	Martin *et al.*, 1993
		pv. *tomato*	*Prf*	Salmeron *et al.*, 1996
	Arabidopsis	*P. s.* pv. *tomato*	*RPS2*	Bent *et al.*, 1994
				Mindrinos *et al.*, 1994
			RPS5	Warren *et al.*, 1998
			RPS4	Gassmann *et al.*, 1999
		P. s. pv. *maculicola*	*RPM1*	Grant *et al.*, 1995
	Rice	*Xanthomonas oryzae*	*Xa21*	Song *et al.*, 1995
		pv. *oryzae*	*Xa1*	Yoshimura *et al.*, 1998
Fungal	Tomato	*Cladosporium fulvum*	*Cf-9*	Jones *et al.*, 1994
			Cf-2	Dixon *et al.*, 1996
			Cf-4	Thomas *et al.*, 1997
			Cf-5	Dixon *et al.*, 1998
		Fusarium oxysporum	*I2*	Ori *et al.*, 1997
		pv. *lycopersici*		Simons *et al.*, 1998
	Flax	*Melampsora lini*	*L6*	Lawrence *et al.*, 1995
			M	Anderson *et al.*, 1997
	Rice	*Magnaporthe grisea*	*Pi-b*	Wang *et al.*, 1999
			Pi^{ta}	Valent, unpublished
	Maize	*Puccinia sorghi*	*Rp1-D*	Collins *et al.*, 1999
Oomycete	*Arabidopsis*	*Peronospora parasitica*	*RPP5*	Parker *et al.*, 1997
			RPP1	Botella *et al.*, 1998
			RPP10	
			RPP14	
			RPP8	McDowell *et al.*, 1998
	Lettuce	*Bremia lactuca*	*Dm3*	Meyers *et al.*, 1998a,b
Nematode	Wheat	*Heterodera avenae*	*Cre3*	Lagudah *et al.*, 1997
	Sugar beet	*Heterodera schachtii*	$Hs1^{pro-1}$	Cai *et al.*, 1997
	Tomato	*Melodogyne incognita*	*Mi***	Milligan *et al.*, 1998
				Vos *et al.*, 1998
Insect	Tomato	*Macrosiphon euphorbiae*	*Meu1***	Rossi *et al.*, 1998
				Vos *et al.*, 1998

*A number of the *Pseudomonas syringae* pv. *tomato* strains used in host/pathogen studies carry avirulence genes derived from other *Pseudomonas syringae* pathovars (pv.) (see Table 5.3).
***Mi* and *Meu1* are the same gene.

The remaining *R* genes fall into four monotypic groups. The tomato *Cf* genes for resistance to the leaf mould fungus, *Cladosporium fulvum*, encode membrane-anchored glycoproteins with a large extracytoplasmic N-terminal LRR domain, a single transmembrane (TM) domain and a small cytoplasmic C-terminal domain (Jones *et al.*, 1994; Dixon *et al.*, 1996, 1998; Thomas *et al.*, 1997). The extracytoplasmic domain consists almost entirely of LRRs. Although these proteins have features consistent with a receptor function, they

Table 5.2 Cloned resistance genes: classified according to the kind of protein they encode

Protein	Host	Pathogen	*R* gene
TIR-NBS-LRR	Tobacco	Tobacco mosaic virus	*N*
	Flax	*Melampsora lini*	*L6, M*
	Arabidopsis	*Pseudomonas syringae* pv. *tomato*	*RPS4*
		Peronospora parasitica	*RPP1, RPP5, RPP10, RPP14*
LZ-NBS-LRR	Tomato	*P. s.* pv. *tomato*	*Prf*
		Melodogyne incognita	*Mi*
		Macrosiphon euphorbiae	*Meu1*
	Potato	Potato virus X	*Rx*
	Arabidopsis	*P. s.* pv. *tomato*	*RPS2, RPS5*
		P. s. pv. *maculicola*	*RPM1*
		Peronospora parasitica	*RPP8*
NBS-LRR	Tomato	*Fusarium oxysporum* pv. *lycopersici*	*I2*
	Lettuce	*Bremia lactuca*	*Dm3*
	Wheat	*Heterodera avenae*	*Cre3*
	Rice	*Xanthomonas oryzae* pv. *oryzae*	*Xa1*
		Magnaporthe grisea	*Pi-b*, Pi^{ta}
	Maize	*Puccinia sorghi*	*Rp1-D*
LRR-TM	Tomato	*Cladosporium fulvum*	*Cf-2, Cf-4, Cf-5, Cf-9*
LRR-TM-PK	Rice	*Xanthomonas oryzae* pv. *oryzae*	*Xa21*
PK	Tomato	*P. s.* pv. *tomato*	*Pto*
Novel	Sugar beet	*Heterodera schachtii*	$Hs1^{pro\text{-}1}$

Abbreviations: TIR, Toll and interleukin-1 receptor homology; NBS, nucleotide binding site; LRR, leucine-rich repeat; TM, transmembrane domain; PK, serine/threonine protein kinase; LZ, leucine zipper.

lack any obvious signalling capacity. However, the LRR domain is thought to comprise a variable N-terminal region involved in ligand recognition and a conserved C-terminal region involved in interaction with a signalling partner (Dixon *et al.*, 1996; Jones and Jones, 1997; Thomas *et al.*, 1997).

The rice *Xa21* gene for resistance to the blight bacterium, *Xanthomonas oryzae* pv. *oryzae*, encodes a protein resembling the Cf proteins with the addition of a serine/threonine protein kinase (PK) domain at the C-terminus (Song *et al.*, 1995). Thus, Xa21 has obvious recognition and signalling capacity.

The tomato *Pto* gene encodes a PK (Martin *et al.*, 1993), which has obvious signalling capacity, but its role as a receptor for the corresponding AvrPto avirulence protein is less obvious. The *Prf* gene which encodes a LZ-NBS-LRR protein is required for Pto function (Salmeron *et al.*, 1996). The receptor and signalling roles of these two proteins are discussed in section 5.8.

Finally, the sugar beet *Hs1*$^{pro-1}$ gene for resistance to the cyst nematode, *Heterodera schachtii*, encodes a protein proposed to resemble the Cf proteins (Cai *et al.*, 1997), but is more likely to encode a novel cytoplasmic protein (Ellis and Jones, 1998). This review focuses primarily upon the NBS-LRR resistance proteins and also upon Pto, because an NBS-LRR protein is involved in its function.

5.2 The TIR domain

One subgroup of NBS-LRR resistance proteins, which includes proteins encoded by the tobacco *N* gene (Whitham *et al.*, 1994), flax *L6* (Lawrence *et al.*, 1995) and *M* (Anderson *et al.*, 1997) genes, and the *Arabidopsis RPP1, RPP10, RPP14* (Botella *et al.*, 1998), *RPP5* (Parker *et al.*, 1997) and *RPS4* (Gassmann *et al.*, 1999) genes, contains an N-terminal TIR domain with homology to the cytoplasmic domains of the Toll receptor of *Drosophila melanogaster* and the IL-1 receptor (IL-1R) of birds and mammals. The IL-1R triggers an innate immune response through the production of proinflammatory cytokines, immune helper factors and antimicrobial proteins, such as defensins (Singh *et al.*, 1998; Hoffman *et al.*, 1999). The Toll receptor was originally identified through its role in *Drosophila* embryonic development, but the same signalling pathway is used in later stages of development to trigger innate immunity to bacterial and fungal pathogens. A family of Toll receptors has been identified in both *Drosophila* (Toll and 18-wheeler) (Eldon *et al.*, 1994) and mammals (TLRs 1–5) (Medzhitov *et al.*, 1997; Chaudhary *et al.*, 1998b; Rock *et al.*, 1998).

Toll signalling is required for the production of the antifungal peptide, drosomycin (Lemaitre *et al.*, 1996), and the antifungal and antibacterial peptide, metchnikowin (Levashina *et al.*, 1998), and also plays a role in the production of the antibacterial peptides, cecropin, defensin and drosocin (Lemaitre *et al.*, 1996). Similarly, signalling through 18-wheeler appears to control the production of the antibacterial peptides, attacin and cecropin (Williams *et al.*, 1997). In mammals, TLR 2 and 4, in conjunction with CD14 (an extracytoplasmic LRR protein anchored to the membrane by a glycophosphoinositol linkage) (Ferrero and Goyert, 1988), and LBP (lipopolysaccharide binding protein) are involved in the recognition of bacterial lipopolysaccharides (Kirschning *et al.*, 1998; Yang *et al.*, 1998a; Chow *et al.*, 1999; Zhang *et al.*, 1999a). They trigger an inflammatory response and the production of antimicrobial peptides, such as defensins, via the IL-1R signalling pathway. Plants also produce antimicrobial peptides, including defensins, and although the signalling mechanisms underlying their production are not well understood, they may be induced as part of the primary resistance response and in systemic-acquired resistance (Terras *et al.*, 1995; Bowling *et al.*, 1997; Pennickx *et al.*, 1998).

There are striking similarities between the Toll and IL-1R signalling pathways, which involve a number of homologous signalling components and similar mechanisms based on homophilic oligomerisation (i.e. interaction between homologous domains on different proteins), adaptor proteins and phosphorylation cascades. In Toll signalling (Figure 5.1), the TIR domain is involved in the activation of the serine/threonine protein kinase, Pelle (homologous to Pto and the kinase domain of Xa21) (Shelton and Wasserman, 1993), and the accessory protein, Tube (Letsou *et al.*, 1991). Interestingly, Tube and Pelle

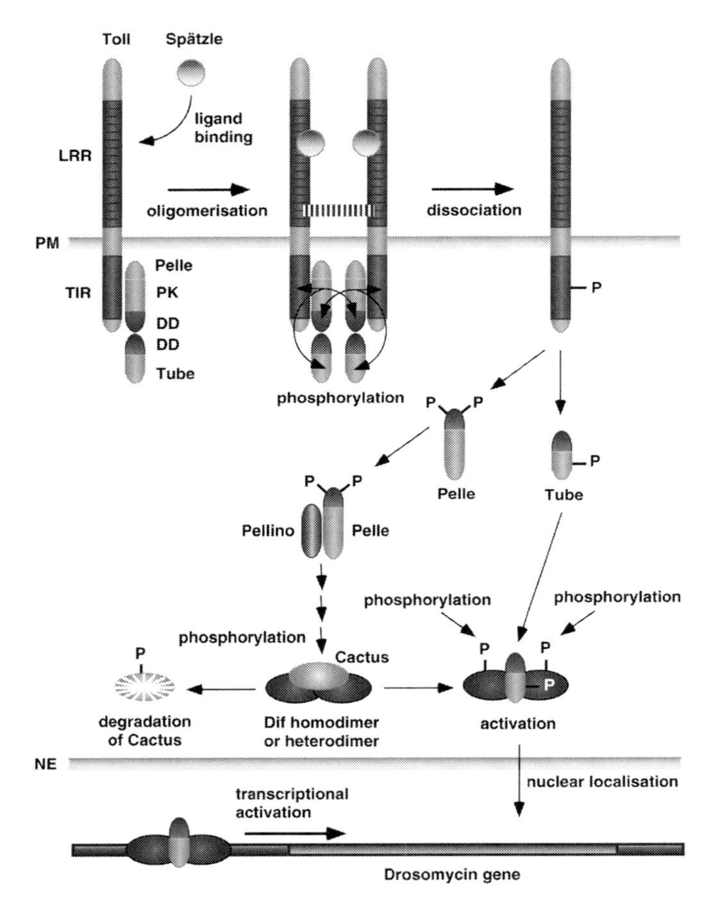

Figure 5.1 A model for the Toll signalling pathway leading to the activation of innate immunity in *Drosophila*, in this case to the production of the antifungal peptide, drosomycin (adapted from Shen and Manley, 1998). Abbreviations: LRR, leucine-rich repeat; PM, plasma membrane; TIR, Toll/interleukin-1 (IL-1) receptor domain; PK, serine/threonine protein kinase; DD, death domain; P, phosphorylated residue; NE, nuclear envelope.

share N-terminal death domains (so-called because of their presence in some proteins that regulate cell death; but absent from Pto and Xa21) (Feinstein *et al.*, 1995) and interact with one another via these domains (Edwards *et al.*, 1997; Shen and Manley, 1998). Pelle is capable of autophosphorylating its own death domain (Shen and Manley, 1998) and is also able to phosphorylate Tube (Großhans *et al.*, 1994). Pelle is also capable of binding and phosphorylating the TIR domain of Toll (Shen and Manley, 1998). Pelle has no TIR domain, so its interaction with the TIR domain of Toll is clearly heterophilic in nature (i.e. between non-homologous domains in each of the interacting proteins). Toll, Pelle and Tube are thought to form an inactive ternary complex in the absence of the Spätzle ligand, but after Spätzle binding, the ternary complexes are thought to oligomerise to activate Pelle (Shen and Manley, 1998; Großhans *et al.*, 1999). Pelle is then thought to phosphorylate all three components of the complex, causing their dissociation (Shen and Manley, 1998). Activated Pelle binds, but does not phosphorylate, a downstream protein, Pellino, via the kinase domain of Pelle (Großhans *et al.*, 1999). The role of Pellino is not yet known.

However, activation of Pelle eventually results directly or indirectly in the phosphorylation of an inhibitory ankyrin repeat protein, Cactus (Geisler *et al.*, 1992; Kidd, 1992), which is bound to a transcription factor, Dorsal (Steward, 1987). Phosphorylation dependent degradation of Cactus, allows the phosphorylation and nuclear localisation of Dorsal and the activation of genes responsible for formation of the dorsal-ventral axis in the *Drosophila* embryo (Belvin *et al.*, 1995). The connection between the Pelle/Pellino complex and the phosphorylation of Cactus is not yet known. The role of activated Tube is not fully understood, but it is thought to interact with and modulate the activity of Dorsal (Norris and Manley, 1995; Yang and Steward, 1997). Dorsal is a member of the Rel family of transcription factors, of which Dif (dorsal-like immunity factor) is another member closely related to Dorsal. Dif is activated by both Toll and 18-wheeler signalling (Williams *et al.*, 1997) and is required for the production of drosomycin and defensin in the adult fly (Meng *et al.*, 1999) and is also involved in the production of cecropin (Petersen *et al.*, 1995). A third member of the Rel family, Relish, is not inhibited by Cactus but carries its own ankyrin repeats, which presumably play a role in cytoplasmic localisation similar to that in Cactus (Dushay *et al.*, 1996). Relish is also thought to play a role in innate immunity in adult flies (Dushay *et al.*, 1996).

In IL-1R and TLR signalling in mammals (Figure 5.2), the TIR domain is involved in the activation of the serine/threonine protein kinases, IRAK and IRAK-2 (interleukin-1 receptor associated kinases; homologues of Pelle, Pto and the kinase domain of Xa21, that also carry N-terminal death domains like Pelle) (Muzio *et al.*, 1997), via the accessory proteins IL-1RAcP (interleukin -1 receptor accessory protein) (Greenfeder *et al.*, 1995; Volpe *et al.*, 1997) and MyD88 (myeloid differentiation antigen 88; no homology with Tube) (Muzio *et al.*, 1997; Wesche *et al.*, 1997b). IL-1 binds to IL-1R (Sims *et al.*, 1988) not

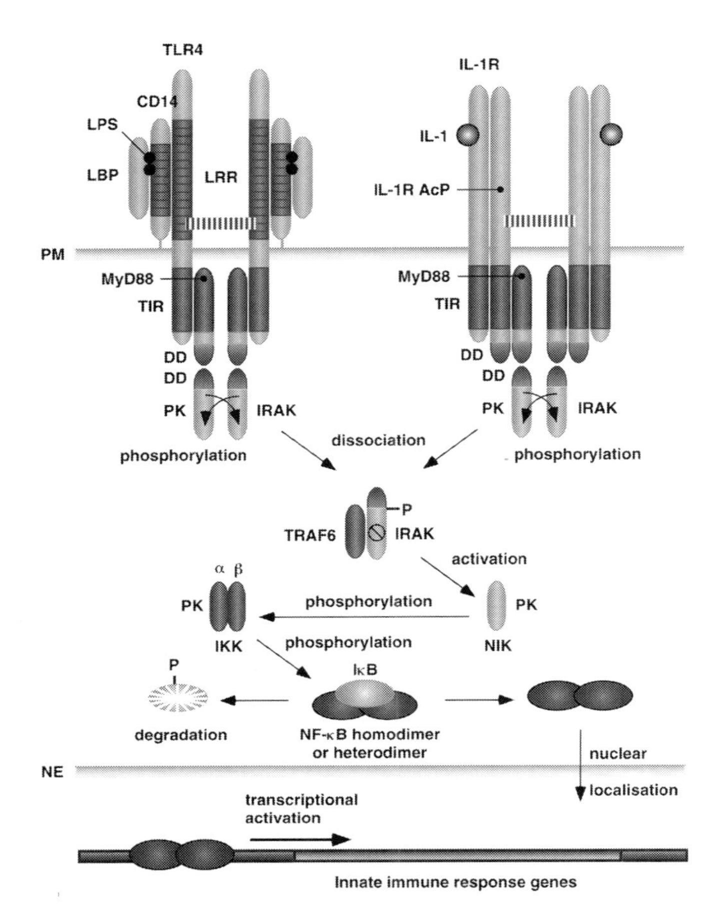

Figure 5.2 A model for the TLR and IL-1R signalling pathways leading to the activation of innate immunity in mammals. For TLR signalling, the response of TLR4 to bacterial LPS is illustrated (adapted from Hoffmann *et al.*, 1999). The model depicts the signalling events that occur after ligand-induced oligomerisation of receptor complexes. Abbreviations: TLR, Toll-like receptor; LPS, lipopolysaccharide; LBP, lipopolysaccharide-binding protein; LRR, leucine-rich repeat; PM, plasma membrane; IL-1R, interleukin-1 receptor; IL-1, interleukin-1; IL-1R AcP, interleukin-1 receptor accessory protein; TIR, Toll/IL-1 receptor domain; IRAK, interleukin-1 receptor associated kinase; PK, serine/threonine protein kinase; DD, death domain; P, phosphorylated residue; TRAF6, tumor necrosis factor receptor-associated factor 6; NIK, NF-κB inducing kinase; IKK, IκB kinase; IκB, inhibitor of NF-κB; NF-κB, nuclear factor κB; NE, nuclear envelope; MyD88, myeloid differentiation antigen 88.

IL-1RAcP (Greenfeder *et al.*, 1995), but IL-1RAcP is required to transmit the IL-1 binding signal (Huang *et al.*, 1997; Volpe *et al.*, 1997; Wesche *et al.*, 1997a) by recruiting MyD88, which in turn recruits IRAK (Wesche *et al.*, 1997b; Burns *et al.*, 1998). IL-1RAcP is a transmembrane protein similar to IL-1R and likewise carries a cytoplasmic TIR domain. MyD88 is an adaptor protein with an N-terminal death domain homologous to those of IRAK and IRAK-2, and a C-terminal TIR domain homologous to those of IL-1R, IL-1RAcP and the TLRs. MyD88 is recruited to the IL-1R/ IL-1RAcP complex or activated TLRs by homophilic oligomerisation of TIR domains (Muzio *et al.*, 1997, 1998; Medzhitov *et al.*, 1998); IRAK and IRAK-2 are then recruited to the IL-1R/ IL-1RAcP/MyD88 or TLR/MyD88 complex by homophilic oligomerisation of death domains (Wesche *et al.*, 1997b; Burns *et al.*, 1998; Medzhitov *et al.*, 1998), and IRAK becomes phosphorylated following oligomerisation of the receptor complexes (Cao *et al.*, 1996; Yamin and Miller, 1997). IRAK is recruited preferentially to IL-1AcP and TLR1, whereas IRAK-2 is recruited preferentially to IL- 1R (Cao *et al.*, 1996; Muzio *et al.*, 1997, 1998; Medzhitov *et al.*, 1998). Interestingly, IRAK-2 appears to be a defective kinase, but still plays a role in signalling (Vig *et al.*, 1999). Likewise, kinase-defective mutants of IRAK can also play a role in signalling, suggesting that kinase activity is not required after activation, but rather the protein is involved in the assembly of downstream signalling components (Knop and Martin, 1999; Vig *et al.*, 1999).

Phosphorylation of IRAK targets it for degradation (Yamin and Miller, 1997), so it is not clear to what extent phosphorylation is required for activation. However, phosphorylated IRAK dissociates from the IL-1R/IL-1RAcP/ MyD88 or TLR/MyD88 complex and the kinase domain of IRAK and then interacts with TRAF6 (tumour necrosis factor receptor associated factor 6; no homology with Pellino) (Cao *et al.*, 1996; Medzhitov *et al.*, 1998). The IRAK/TRAF6 complex directly or indirectly activates a kinase cascade mediated by NIK (NF-κB inducing kinase) (Malinin *et al.*, 1997; Song *et al.*, 1997) and IKK (IκB kinase) (Zandi *et al.*, 1997), resulting in the phosphorylation and degradation of IκB (inhibitor of NF-κB; a homologue of Cactus) (Brown *et al.*, 1995) and the release and nuclear translocation of the transcription factor NF-κB (nuclear factor-κB; a homologue of Dorsal) (Malek *et al.*, 1998). NF-κB stimulates the transcription of proinflammatory cytokines and immunostimulatory factors associated with resistance to bacterial and viral infection (Hoffmann *et al.*, 1999).

A possible structural homology between the TIR domain and the guanosine triphosphatase (GTPase) domain of ras (reported by Hopp, 1995) was noted previously (Jones and Jones, 1997). More recently, the TIR domain has been reported to have structural homology with the acidic pocket $(\beta/\alpha)_5$ domain of bacterial two-component response regulators, such as CheY (Rock *et al.*, 1998). Based on the way CheY functions, this domain may bind a divalent cation and undergo phosphorylation, thereby triggering a conformational change that allows interaction with downstream signalling components (Rock *et al.*, 1998). The possible phosphorylation of TIR domains has not yet been investigated for any TIR protein other than Toll, nor has the further possibility that the phosphorylated TIR domain of Toll has an additional role in signalling.

The multitude of Toll-like receptors, Pelle/IRAK- like kinases, Cactus/IκB α-like inhibitors and Dorsal/NF-κB-like transcription factors and their combinatorial possibilities underscore the complexity of the signalling underlying the activation of innate immunity in both *Drosophila* and humans, which somewhat parallels that emerging for plant disease resistance. Although plants may share homologous upstream components, few plant homologues to the downstream components of the Toll, TLR or IL-1R signalling have yet been reported. In *Arabidopsis*, the wild-type gene corresponding to the *nim1/npr1* (non-inducible immunity/non-expresser of PR genes) mutation was found to encode an ankyrin repeat protein (Cao *et al.*, 1997; Ryals *et al.*, 1997), thought to be a plant homologue of IκBα (Ryals *et al.*, 1997). This mutation results in a failure to activate PR gene expression in response to signals, such as salicylic acid, which would normally activate systemic-acquired resistance, and in response to *R*-gene-dependent activation of plant defences (Shah *et al.*, 1997). NIM1/NPR1 is, therefore, an activator of PR gene expression rather than an inhibitor, as would be expected by analogy with IκBα; and, rather than binding a plant homologue of NF-κB, NIM1/NPR1 interacts with plant bZIP (basic leucine zipper) transcription factors to activate transcription of PR genes (Zhang *et al.*, 1999b). NIM1/NPR1 may not, therefore, be an authentic IκB homologue, but instead is somewhat more reminiscent of Relish described above.

The absence of plant homologues to the downstream components of the Toll, TLR or IL-1R signalling pathways suggests that they remain to be found or that downstream signalling is mediated by different components. Likewise, EDS1 is a component required for TIR-NBS-LRR resistance protein signalling in *Arabidopsis* (Falk *et al.*, 1999), but no counterpart has yet been identified in animal TIR signalling systems, suggesting that the reciprocal is also true. Nevertheless, the various modes of signalling through TIR domains and serine/threonine kinases in these systems broaden the range of possibilities that should be considered for resistance gene signalling and will help to shape the direction of future experimentation. Homophilic oligomerisation is a recurrent theme that emerges, but the possibility of a direct heterophilic interaction of a TIR domain with a protein kinase and the phosphorylation of both is novel, as is the idea that a phosphorylated or inactive kinase may serve as a scaffold for the assembly of downstream signalling components. These concepts may be particularly relevant to Pto signalling.

5.3 Do the NBS-LRR plant resistance proteins lacking an N-terminal TIR domain form a distinct group of related proteins?

There are two subgroups of NBS-LRR resistance proteins that do not contain TIR domains. One group, which includes proteins encoded by the *Arabidopsis RPS2* (Bent *et al.*, 1994; Mindrinos *et al.*, 1994), *RPS5* (Warren *et al.*, 1998),

RPM1 (Grant *et al.*, 1995) and *RPP8* (McDowell *et al.*, 1998) genes, the tomato *Prf* (Salmeron *et al.*, 1996) and *Mi* (Milligan *et al.*, 1998; Vos *et al.*, 1998) genes, and the potato *Rx* gene (Bendahmane *et al.*, 1999), is reported to contain an N-terminal LZ domain (usually a minimum of four consecutive leucines with heptad spacing). A second group is emerging, which has not been reported to have either a TIR or LZ domain. This group includes proteins encoded by the tomato *I2* (Ori *et al.*, 1997; Simons *et al.*, 1998), wheat *Cre3* (Lagudah *et al.*, 1997), lettuce *Dm3* (Meyers *et al.*, 1998a,b), rice *Xa1* (Yoshimura *et al.*, 1998) and *Pi-b* genes (Wang *et al.*, 1999), and maize *Rp1-D* gene (Collins *et al.*, 1999).

Two questions arise: firstly, whether the LZ-NBS-LRR proteins really form a group related by their LZ domains; and secondly, whether other non-TIR NBS-LRR proteins lacking an obvious LZ domain fall into the same or a different group. To answer these questions PSI BLAST searches (Altschul *et al.*, 1997; http://www.ncbi.nlm.nih.gov; an iterative search method using a single sequence for the first search but a cumulative matrix of homologous sequences for each subsequent search) were conducted against the Genbank database, using the N-terminal domain of each of the above non-TIR NBS-LRR resistance proteins together with those of a number of resistance gene homologues (RGHs). Many of these proteins were found to contain stretches of detectable homology immediately adjacent to the N-terminus which included the previously described LZ motifs. In the case of Mi and Prf, this homology was in a similar position relative to the NBS domain, but some distance downstream of the extended N-terminus. In the case of Xa1 and Pi-b, this was in a similar position relative to the N-terminus but some distance upstream of the NBS domain. In the case of Cre3, this homology was absent. Cre3 has only a short N-terminus of 93 amino acids preceding the kinase 1a motif, and this was found to be homologous to the equivalent region adjacent to the kinase 1a motif of Rp1-D and, therefore, well downstream of the homology between Rp1-D and the other resistance proteins. This suggests that the Cre3 sequence may be incomplete at the N-terminus.

Two mutually exclusive homology groups were found with no significant homology detectable between them. Group 1 comprised the LZ proteins, RPM1, RPP8, Prf, Mi and Rp1-D, as well as the non-LZ proteins, I2, Xa1 and Pi-b, and a number of RGHs. Group 2 comprised the LZ proteins, RPS2 and RPS5, as well as the non-LZ protein, Dm3, and a number of RGHs. No homology could be detected using Rx as a seed sequence and, at the time the analysis was carried out, Rx was not present in the database to determine whether it could be detected using other resistance sequences as seeds. Nevertheless, it is clear that Rx belongs to Group 1, albeit as a possible outlier. These groupings prompted an attempt to align the sequences within each group based on the PSI BLAST analysis and using the multiple sequence alignment program, Macaw (Schuler *et al.*, 1991; ftp://ncbi.nlm.nih.gov/pub/macaw). The results of these alignments are presented in Figure 5.3.

Domain 1

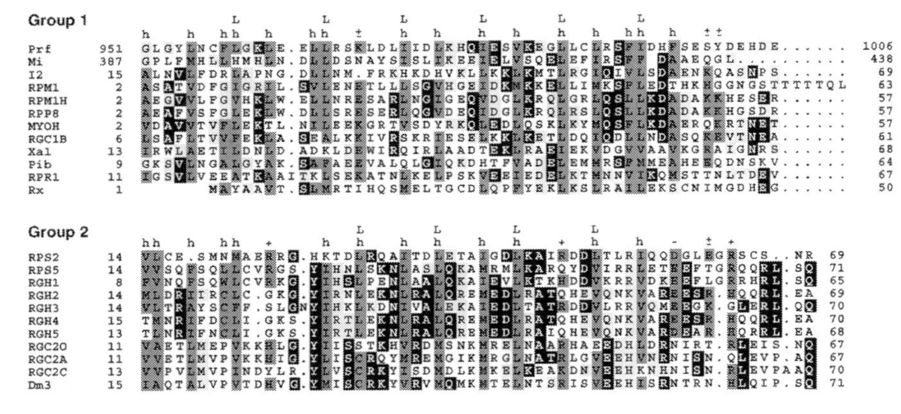

Domain 2

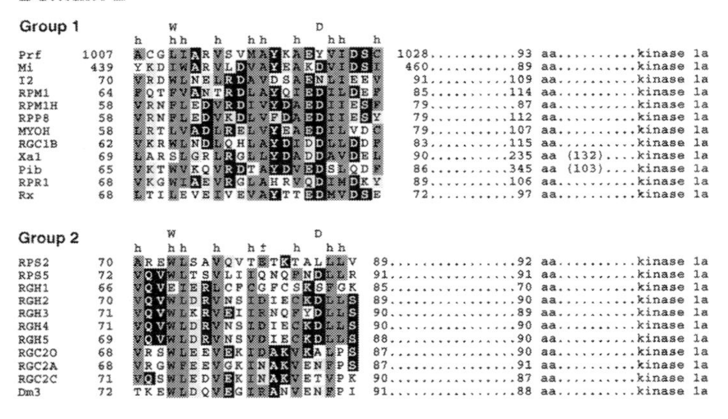

Figure 5.3 Multiple sequence alignment of the two consecutive, amino-terminal, $(hxxhxxx)_n$ domains of the non-TIR NBS-LRR proteins. The sequences are shown using the single letter amino acid code and the numbers to the left and right of each sequence indicate the position in the overall sequence. Resistance proteins included in this analysis are: Prf (Genbank Accession AAC49408); Mi (AAC67238); I2 (CAA03554); RPM1 (CAA61131); RPP8 (AAC83165); Xa1 (BAA25068); Pi-b (BAA76281); Rx (Bendahmane *et al.*, 1999); RPS2 (AAA21874); RPS5 (AAC26126); and Dm3 (published as RGC2B; AAD03156). NBS-LRR proteins without defined resistance functions included in this analysis are: RPM1H, an *Arabidopsis* RPM1 homologue (AAB65485); MYOH, an *Arabidopsis* myosin heavy chain homologue (a misnomer as it is clearly a resistance gene homologue; AAA63149; the methionine at position 55 of the published sequence is the presumed start methionine); lettuce Resistance Gene Candidates, RGC1B (AAC02202), RGC2A (AAC02203), RGC2C (AAD04191) and RGC2O (AAD03156); rice RPR1 (BAA75812); and *Arabidopsis* Resistance Gene Homologues, RGH1 (CAB10240), RGH2 (AAB71476), RGH3 (AAC35528), RGH4 (AAC13910) and RGH5 (AAB71477). Four or more identities are highlighted in black. Similarities are highlighted in light grey. The hydrophobic residues A, F, I, L, M or V (or occasionally C or W) forming a $(hxxhxxx)_n$ or a $(hxxhxxh)_n$ consensus are highlighted in dark grey, with the letter h above the alignments. Positions in which the $(hxxhxxx)_n$ or $(hxxhxxh)_n$ consensus is consistently disrupted are also indicated in dark grey, but with + for positively charged

From these comparisons, it is quite clear that the previously described LZ regions represent only a limited description of an overall pattern of hydrophobic residues common to all of the non-TIR proteins. The consensus for this pattern is $(hxxhxxx)_n$, where h = A, F, I, L, M or V (or occasionally W or C), x is any residue (with a tendency towards polar or charged residues), and n is the number of repetitions of the heptad motif. There is a further tendency towards a $(hxxhxxh)_n$ consensus, but the conservation of the hydrophobic residues at position 7 is not so great as that at positions 1 or 4. As suggested previously (Jones and Jones, 1997), the LZ motifs of RPM1 and Prf (Group 1), which have leucines or isoleucines at position 1, appear to be out of phase with those of RPS2 (Group 2), which have leucines or isoleucines at position 4, with respect to this consensus. This is a further indication that the LZ motif is an inadequate description of the similarity between these proteins. Resistance proteins in both groups appear to have two $(hxxhxxx)_n$ domains in tandem, with the boundaries between domains defined by charged or polar residues in hydrophobic consensus positions and by spacer regions, of varied length, which do not show the consensus pattern. Charged or polar residues also occur at hydrophobic consensus positions within both domains 1 and 2, but at different positions in the two groups. However, the hydrophobic consensus resumes in phase following these residues.

The $(hxxhxxx)_n$ pattern of hydrophobic residues (which includes the LZ pattern) is consistent with the formation of an amphipathic alpha-helical domain able to form a coiled-coil structure involved in protein homo- or hetero-oligomerisation, as suggested previously for RPS2 (Bent *et al.*, 1994). These sequences were analysed for the potential to form CC structures using the COILS program (Lupas *et al.*, 1991; http://www.ch.embnet.org/software/COILS_form.html). In Group 1, the RPP8, RPM1H, MYOH and RGC1B proteins gave predicted CC domains with a probability greater than 0.95 in domain 1. Prf, Mi, I2 and Xa1 gave probabilities greater than 0.5, but RPM1 and Rx gave low probabilities and Pi-b a very low probability. In Group 2, RPS2, RPS5, RGH1, RGH2, RGH3, RGH4, RGH5 and RGC2C gave predicted CC domains with probabilities greater than 0.95 in domain 1. RGC2O gave a probability greater than 0.5, but Dm3 and RGC2A gave very low probabilities. Despite some variation in probability for predicted CC domains,

residues, ± for positively or negatively charged residues, or − for negatively charged residues, above the alignments. The leucine or isoleucine residues suggested previously to form leucine zipper (LZ) domains in Prf, Mi, RPM1, RPP8, RPS2 and RPS5 are indicated with the letter L above the alignments. The position of a conserved tryptophan residue (or conservative phenylalanine substitution) is indicated with the letter W above the consensus. The position of a conserved aspartic acid residue (or conservative asparagine substitution) is indicated with the letter D above the consensus. The number of intervening amino acids (aa) between domain 2 and the first conserved glycine of the kinase 1a motif in the nucleotide binding site (NBS) domain is indicated for each sequence.

the majority trend towards predicted CC domains suggests that domain 1 probably does form a coiled-coil. Proteins in both Groups 1 and 2 all gave very low probabilities for predicted CC domains in domain 2, apart from I2 which had a probability greater than 0.95 and RGC2O which had a probability greater than 0.5. In the light of a majority trend against predicted CC domains, domain 2 probably does not form a coiled-coil, despite showing the $(hxxhxxx)_n$ pattern of hydrophobic residues. Domain 2 is shorter than domain 1 and shows a greater proportion of the $(hxxhxxh)_n$ pattern, which probably militates against a CC prediction. However, domain 2 shows greater conservation within Groups 1 and 2 than domain 1, as reflected by higher homology scores. Interestingly, there is a conserved tryptophan residue and a conserved aspartic acid residue in both Groups 1 and 2, suggesting that the two groups may be distantly related. This notion is further supported by the similar length of the two domains and their position relative to the N-terminus and to the NBS domain.

The conservation of distance from the NBS domain may be functionally important. The Group 2 proteins all have approximately 90 amino acids between the end of domain 2 and the kinase 1a motif of the NBS region, except for RGH1, which shows a 70 amino acid spacing (Figure 5.3). Apart from the first three residues, RGH1 seems to fit domain 2 quite poorly, raising the possibility that part of domain 2 has been deleted. If domain 2 is important for function, then RGH1 might be predicted to function poorly or differently from the other resistance genes and RGHs. The Group 1 proteins show a bimodal distribution in spacing between domain 2 and the kinase 1a domain. Prf, Mi, RPM1H and Rx all have about 90 amino acid spacers, whereas I2, Rpm1, RPP8, MYOH, RGC1B and RPR1 all have about 110 amino acid spacers. Xa1, Pi-b and Rp1- D are exceptions, with larger spacers. Pi-b has a degenerate proximal duplication of the kinase 1a, 2 and 3a motifs of the NBS region, which reduces the spacer distance from 345 amino acids to 103. There are two introns in the coding sequence of Xa1, lying between domain 2 and the kinase 1a domain, flanking a 309 base-pair (bp) exon, which may reflect an exon insertion event. This exon shows transposase homology in BLAST searches, which suggests that an alternative hypothesis of transposon insertion is also feasible. Subtracting the 103 amino acids encoded by this exon gives a 132 amino acid spacing between domain 2 and the kinase 1a domain, which is still larger than the other resistance proteins. However, Rp1-D has a 134 amino acid spacer, raising the possibility of a trimodal distribution with 20 amino acid increments.

From this overall analysis, it would seem that the non-TIR NBS-LRR resistance proteins are probably related by their N-terminal domains and evolved from a single ancestral sequence. It also seems that they might be more appropriately described as CC-NBS-LRRs rather than LZ-NBS-LRRs, but clearly they form two subgroups which could be described as CC1- and CC2-NBS-LRRs.

5.4 The NBS domain is a regulatory motif shared by resistance proteins in plants, apoptotic proteins in animals and pleiotropic regulatory proteins in Gram-positive bacteria

Recently, Chinnaiyan *et al.* (1997a) and van der Biezen and Jones (1998) drew attention to the homology between the NBS domains in plant disease resistances proteins and those in the apoptosis proteins, Apaf-1 (apoptotic protease-activating factor-1) in humans and CED-4 (*Caenorhabditis elegans* death regulating protein) in the nematode, *Caenorhabditis elegans*. This homology extends beyond the kinase 1a (P-loop), kinase 2 and kinase 3a motifs, which are common to many nucleotide binding proteins, to include several other unique domains which are conserved among the NBS-LRR resistance proteins.

Apaf-1 contains a central NBS domain flanked by an N-terminal CARD (caspase activation and recruitment domain) and a C-terminal domain carrying 12 WD-40 repeats (containing a conserved GH, glycine-histidine, motif separated from a conserved WD, tryptophan-aspartic acid, motif by approximately 40 amino acids) (Zou *et al.*, 1997). CED-4 is a homologue of Apaf-1 that also carries CARD and NBS domains but lacks a WD-40 domain (Yuan and Horvitz, 1992). Apaf-1 interacts with and activates procaspase 9 (Figure 5.4), and similarly CED-4 interacts with and activates CED-3 (a homologue of procaspase 9) (Chinnaiyan *et al.*, 1997b; Wu *et al.*, 1997b). These activated caspases (cysteine proteases that cleave after specific aspartic acid residues) trigger a caspase cascade that results in apoptosis. Apaf-1 also interacts with and is inhibited by Bcl-2 and other anti-apoptotic members of the Bcl-2 gene family, such as Bcl-X_L (Hu *et al.*, 1998a). Similarly, CED-9 (a homologue of Bcl-2) interacts with and inhibits CED-4 (Chinnaiyan *et al.*, 1997b; Wu *et al.*, 1997a). Bcl-X_L, Apaf-1 and procaspase-9 are thought to form either an inactive ternary complex anchored to the mitochondrial outer membrane by Bcl-X_L (Pan *et al.*, 1998) or an inactive binary complex with procaspase 9 recruited following activation (Li *et al.*, 1997; Zou *et al.*, 1999). CED-3, CED-4 and CED-9 are thought to interact in a similar way (Chinnaiyan *et al.*, 1997b; Wu *et al.*, 1997b).

Activation of either complex requires the binding of ATP or dATP and hydrolysis to ADP or dADP (Li *et al.*, 1997; Kanuka *et al.*, 1999; Zou *et al.*, 1999), together with a second activating signal. The production of pro-apoptotic members of the Bcl-2 family, which prevent Bcl-X_L and CED-9 interaction with Apaf-1 and CED-4 (del Peso *et al.*, 1998; Hegde *et al.*, 1998), or directly activate Apaf-1 or CED-4 (Inohara *et al.*, 1998b) is one possible signal. In the case of Apaf-1, another signal is thought to be the release of cytochrome *c* from the mitochondrion (Li *et al.*, 1997), which can be triggered by pro-apoptotic members of the Bcl-2 family, such as Bax or Bak (Desagher *et al.*, 1999; Finucane *et al.*, 1999; Shimizu *et al.*, 1999). Cytochrome *c* binds to Apaf-1 (Li *et al.*, 1997; Zou *et al.*, 1997, 1999) and causes its dissociation from

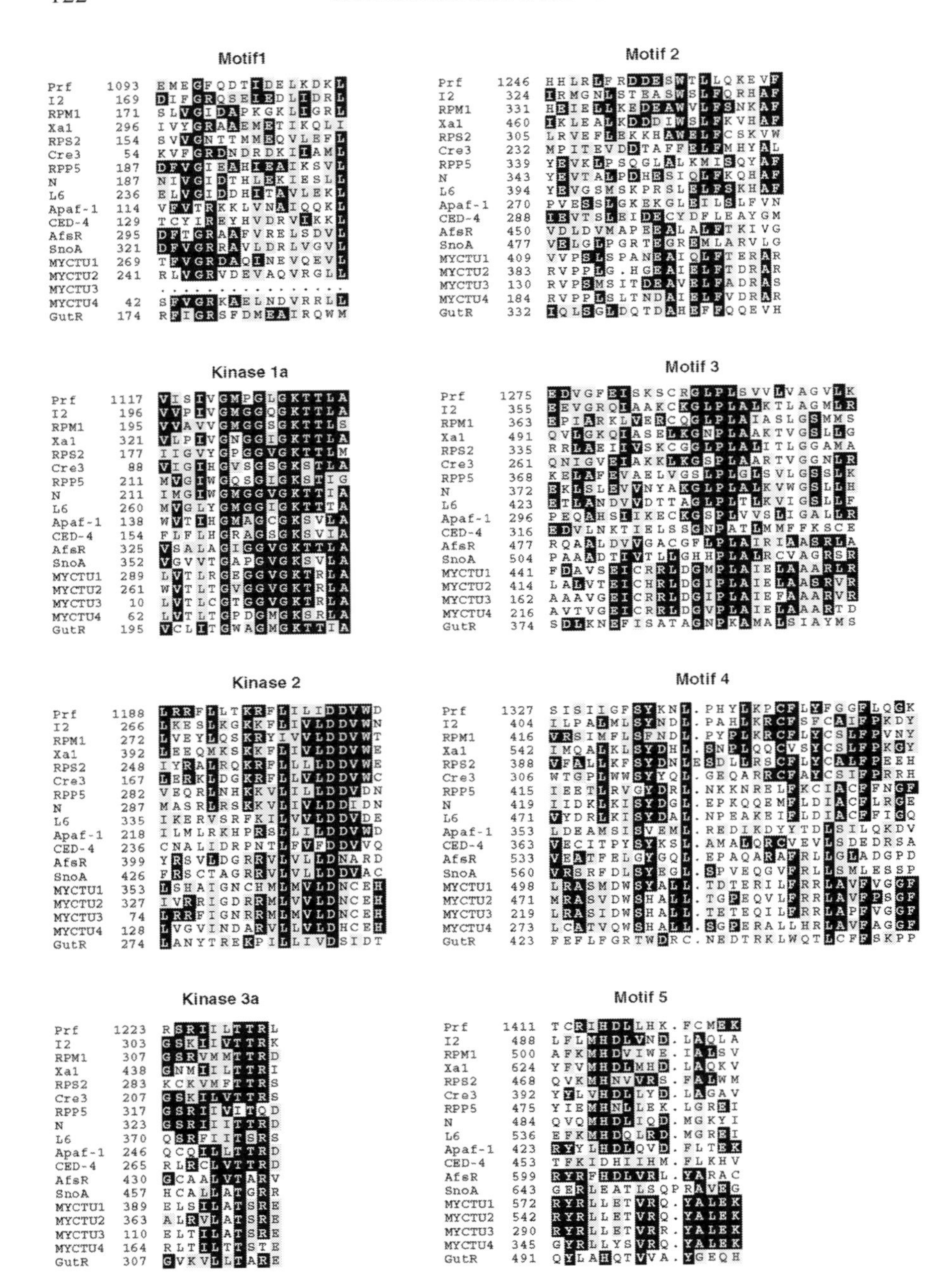
Motif1
Motif 2
Kinase 1a
Motif 3
Kinase 2
Motif 4
Kinase 3a
Motif 5
Prf
I2
RPM1
Xa1
RPS2
Cre3
RPP5
N
L6
Apaf-1
CED-4
AfsR
SnoA
MYCTU1
MYCTU2
MYCTU3
MYCTU4
GutR

Bcl-2 (or Bcl-X_L). Dissociation of Apaf-1 or CED-4 from Bcl-2 (or Bcl-X_L) or CED-9 allows oligomerisation of the complex via the NBS domain, which in turn triggers the auto-proteolytic self-activation and release of caspase 9 or CED-3 (Hu *et al.*, 1998b; Srinivasula *et al.*, 1998; Yang *et al.*, 1998b; Zou *et al.*, 1999). Apaf-1 and procaspase 9 interact via their N-terminal CARD domains (Li *et al.*, 1997; Pan *et al.*, 1998). Similarly, CED-4 and CED-3 also interact via their N-terminal CARD domains, but the CARD domain of CED-3 has also been shown to interact with the kinase 1a motif of the NBS domain of CED-4, and the protease domain of CED-3 has been shown to interact with both the NBS and CARD domains of CED-4 (Chaudhary *et al.*, 1998a). Bcl-2 (or Bcl-X_L) or CED-9 interact with the NBS domain of Apaf-1 or CED-4 (Chaudhary *et al.*, 1998a; Pan *et al.*, 1998). The WD-40 domain of Apaf-1 interacts with both Bcl-X_L and the NBS domain and is inhibitory to procaspase-9 activation (Hu *et al.*, 1998a,b). Cytochrome *c* binding to the WD-40 domain and hydrolysis of ATP or dATP are thought to trigger a conformational change that relieves inhibition by the WD-40 domain as well as the dissociation of Bcl-2 (or Bcl-X_L).

It is tempting to speculate that the LRR domain of NBS-LRR resistance genes performs a similar function in response to avirulence ligand binding. It is also tempting to speculate that the NBS domain may regulate resistance in plants in a similar fashion to apoptosis in animals, so that the NBS domain is not simply a site for nucleotide binding and hydrolysis, but may also bind negative regulators analogous or even homologous to Bcl-2, and may oligomerise following activation to effect the secondary oligomerisation and activation of downstream effectors via N-terminal adaptor domains. No Bcl-2 homologue has yet been reported in plants, although there is one report which presents immunological data to suggest they may exist (Dion *et al.*, 1997). Similarly, no caspase homologues have yet been reported in plants, but there is one report

←

Figure 5.4 Eight conserved motifs in the NBS region. Eight conserved motifs in the nucleotide binding site (NBS) region of: tomato Prf (GenBank accession U65391); tomato I2 (CAA03554); *Arabidopsis* RPM1 (X87851); rice Xa1 (AB002266); *Arabidopsis* RPS2 (U14158); wheat Cre3 (AF052641); *Arabidopsis* RPP5 (U97106); tobacco N (U15605); flax L6 (U27081); human Apaf-1 (AF013263); *Caenorhabditis elegans* CED-4 (X69016); *Streptomyces coelicolor* AfsR (D90155); *Streptomyces nogalater* SnoA (AJ224512); predicted *Mycobacterium tuberculosis* proteins, designated MYCTU1-4 (Z75555 [G1419061], AL021246 [G2791528], Z73101 [G1314029] and Z73101 [G1314033]); and *Bacillus subtilis* GutR (L19113). These include all the motifs described by van der Biezen and Jones (1998) and a new additional motif on the N-terminal side of the kinase 1a motif. The motifs have been numbered sequentially rather than historically, so that Motif 2 corresponds to Motif 4 of van der Biezen and Jones (1998), Motif 3 to Motif 3 plus the HD (sic) motif and Motif 4 to Motif 2. The motifs are shown sequentially from top to bottom and then left to right. Motif alignments were generated using Macaw (Schuler, *et al.*, 1991; ftp://ncbi.nlm.nih.gov/pub/macaw). Black boxes indicate four or more identical amino acids and grey boxes indicate similar amino acids. Numbers to the left of the sequences indicate the position in the overall sequence of the first amino acid shown. The sequences are shown using the single letter amino acid code.

of biochemical evidence obtained using specific caspase inhibitors to suggest there may be caspase activity in plants (del Pozo and Lam, 1998).

Homology in the NBS domain is not limited to plant resistance and animal apoptosis proteins, but also appears in a number of pleiotropic regulatory proteins in several species of Gram-positive bacteria (Figure 5.5) and at least

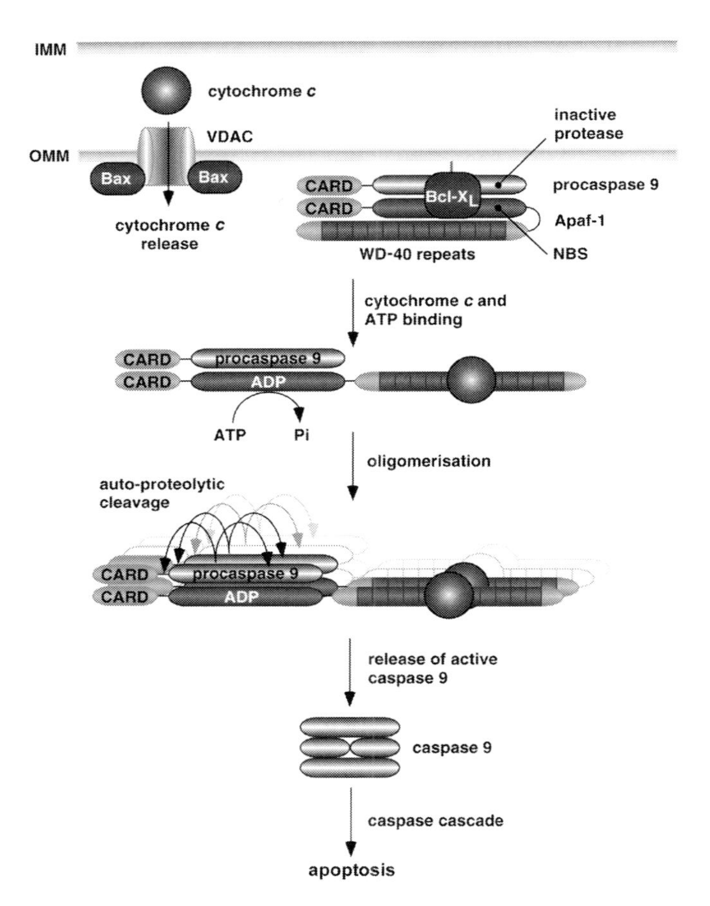

Figure 5.5 A model for the Apaf-1-regulated activation of apoptosis in human cells (adapted from Reed, 1997). The model depicts events that occur after signal-induced activation of Bax, which binds to the voltage-dependent anion channel on the mitochondrial outer membrane, causing the release of cytochrome. Abbreviations: IMM, inner mitochondrial membrane; OMM, outer mitochondrial membrane; VDAC, voltage-dependent anion channel; CARD, caspase activation and recruitment domain; Apaf-1, apoptotic protease-activating factor-1; NBS, nucleotide binding site; WD-40, a repeated amino acid sequence with a conserved GH (glycine-histidine) motif separated from a conserved WD (tryptophan-aspartic acid) motif by ~40 amino acids; Pi, inorganic phosphate; caspase, cysteine protease that cleaves after specific aspartic acid residues.

one archaean (*Pyrococcus horikoshii*; Genbank accession 3131196). This homology was detected using the NBS regions of Apaf-1, CED-4 or the NBS-LRR plant disease resistance proteins as seed sequences in PSI BLAST searches (Altschul *et al.*, 1997; http://www.ncbi.nlm.nih.gov) and has also been detected independently by Aravind *et al.* (1999). These proteins include the AfsR, SnoA and CdaR proteins in *Streptomyces* (Horinouchi *et al.*, 1990; Ylihonko *et al.*, 1996; Chong, 1999), GutR in *Bacillus subtilis* (Ye *et al.*, 1994), and several putative regulatory proteins identified by the recently completed *Mycobacterium tuberculosis* sequencing project (Cole *et al.*, 1998). It is also worth examining the way these proteins function for further clues to the way resistance proteins might function.

AfsR activates the expression of biosynthetic genes in a number of pathways involved in the production of various antibiotics, including undecylprodigiosin (designated Red because of its colour), actinorhodin (Act) and calcium-dependent antibiotic in *Streptomyces coelicolor* and *S. lividans* (Horinouchi *et al.*, 1990), and probably also daunorubicin (Dnr) in *S. peucetius* (Floriano and Bibb, 1996). CdaR, a homologue of AfsR, also regulates production of calcium-dependent antibiotic in *S. coelicolor* and *S. lividans* (Chong, 1999). SnoA, another homologue of AfsR, activates the expression of biosynthetic genes involved in the production of the antibiotic, nogalamycin, in *S. nogalater* (Ylihonko *et al.*, 1996) and GutR activates the metabolism of glucitol in *B. subtilis* (Ye *et al.*, 1994). Little more is known about the roles of CdaR, SnoA, GutR or the *M. tuberculosis* proteins.

AfsR is a conditional regulator that is required for the activation of Red and Act biosynthesis under low phosphate conditions (Floriano and Bibb, 1996). The production of Red, Act and Dnr is also controlled by the pathway-specific regulators, RedD, ActII-ORF4 and DnrI (Floriano and Bibb, 1996), of which DnrI has been shown to bind to the promoter of the Dnr biosynthetic genes and is probably a transcription factor (Tang *et al.*, 1996a). AfsR cannot substitute for these regulators, nor is it required for their transcription (Floriano and Bibb, 1996). The N-terminal domains of AfsR, RedD, ActII-ORF4 and DnrI share homology with one another (Floriano and Bibb, 1996), and with the C-terminal, RNA-polymerase- and DNA-binding domains of the OmpR family of transcription factors (Wietzorrek and Bibb, 1997). This suggests a possible role for AfsR in the transcriptional activation of the Red, Act and Dnr biosynthetic genes by cooperative binding to repeated sequence motifs in the promoters of these genes (Wietzorrek and Bibb, 1997), which is likely to involve protein interaction with RedD, ActII-ORF4 and DnrI as well as DNA binding, since AfsR alone cannot initiate transcription.

Beside AfsR, the production of the antibiotics Red and Act in *S. coelicolor* also depends on the activity of AfsK, which encodes a serine/threonine protein kinase (PK) that phosphorylates AfsR (Matsumoto *et al.*, 1994). The involvement of AfsR, an NBS protein, and AfsK, a PK, in the activation of antibiotic biosynthesis is reminiscent of the involvement of Prf and Pto in plant defence activation. The tomato *Pto* resistance gene encodes a PK that is the receptor for the *Pseudomonas syringae* pv. *tomato* AvrPto avirulence protein (Scofield *et al.*, 1996; Tang *et al.*, 1996b). The NBS protein Prf is required for Pto/AvrPto mediated resistance, but the role of Prf in Pto signalling is not understood (Salmeron *et al.*, 1996). One model for the interaction between Prf and Pto could be that the C-terminal leucine-rich repeat domain of Prf recognises the Pto/AvrPto complex, thereby activating Prf (Jones and Jones, 1997). However, the interaction between AfsR and AfsK provides an alternative model for the interaction between Prf and Pto, in which Prf is activated following phosphorylation by Pto kinase, whose activation is in turn dependent upon interaction between AvrPto and Pto.

Despite their apparent similarities, there are a number of key differences between these NBS proteins and those of the plant resistance proteins or the animal apoptosis proteins. Rather than N-terminal TIR or CC domains in the case of the NBS resistance proteins or N-terminal CARDs in the case of CED-4 or Apaf-1, GutR and AfsR have N-terminal DNA-binding domains with a single helix-turn-helix (HTH) or HTH-like motif (Ye *et al.*, 1994; Wietzorrek and Bibb, 1997). Rather than a C-terminal protein-interaction domain thought to be involved in ligand recognition, like the LRR domain of the plant resistance proteins or the WD-40 repeat domain of Apaf-1, AfsR and GutR seem to have tetratricopeptide repeat domains (Ye *et al.*, 1994; Aravind *et al.*, 1999), suggesting an alternative protein interaction domain. CdaR and SnoA, like CED-4, lack an extended C-terminal domain (Ylihonko *et al.*, 1996; Chong, 1999). Similarly, there are differences between the serine/threonine kinase proteins, AfsK and Pto. AfsK has a C-terminal extension (Matsumoto *et al.*, 1994), with at least ten WD-40-like repeats likely to be involved in protein interaction, whereas Pto does not.

Despite these differences, it is curious that the Gram-positive bacteria (including both high and low G + C subgroups) and at least one archaean possess a number of NBS proteins that are otherwise characteristic of eukaryotes rather than prokaryotes. One possible explanation is that during evolution there may have been lateral gene transfer between the Gram-positive bacteria, archaeans and eukaryotes. Another possible explanation is that the NBS domain is a very ancient regulatory motif that has acquired divergent functions in some lineages and has been lost from others. The real explanation is likely to be a combination of both. Nevertheless, the function of the NBS domain in the Gram-positive bacteria clearly indicates that this domain can be involved in the regulation of pathways other than those that result in cell death, and it may provide a tool to help understand the way the NBS domain functions in animals and plants.

5.5 NBS-LRR proteins are involved in human innate immunity

A human NBS-LRR protein, the major histocompatibility complex class II transcriptional activator (CIITA) (Steimle *et al.*, 1993), was reported previously (Jones and Jones, 1997), but its role in human immunity did not immediately suggest a functional parallel with plant NBS-LRR resistance genes. However, the recent discovery of the CARD4/Nod1 protein (Bertin *et al.*, 1999; Inohara *et al.*, 1999), homologous to CIITA, alters this perception dramatically. CARD4/Nod1 resembles Apaf-1 and CED-4 because it has an N-terminal CARD domain and a central NBS domain, and resembles plant resistance proteins because it has a central NBS domain and a C-terminal LRR domain. CARD4/Nod1 also has more obvious functional parallels than CIITA because, like the TIR domain of IL-1R, it is capable of triggering innate immune responses by activation of NF-κB and, like Apaf-1, it is capable of enhancing apoptotic cell death by activation of caspase-9. The CARD domain of CARD4/Nod1 is reported to most closely resemble that of RICK/RIP2/CARDIAK (CARD-containing interleukin-1-converting-enzyme associated kinase), a serine/threonine protein kinase that regulates both apoptosis and NF-κB activation (Inohara *et al.*, 1998a), and binds to the CARD domains of both procaspase-9 and RICK, suggesting their activation by CARD-mediated oligomerisation (Bertin *et al.*, 1999; Inohara *et al.*, 1999). CARD4/Nod1 appears to regulate caspase-9 mediated apoptosis and NF-κB activation independently (Inohara *et al.*, 1999) and, although CARD4/Nod1 does not alter the pro-apoptotic properties of RICK (Inohara *et al.*, 1999), activation of NF-κB is anti-apoptotic (Beg and Baltimore, 1996; Van Antwerp *et al.*, 1996; Kothny-Wilkes *et al.*, 1998; Wang *et al.*, 1998a). This suggests a dynamic balance between innate immunity and apoptosis, which provides a remarkable parallel to the emerging idea (discussed in section 5.9) of a similar balance between induction of PR proteins and activation of the hypersensitive response in resistance gene-mediated defence responses in plants.

Despite these remarkable parallels, it is clear that the NBS domains of CARD4/Nod1 and CIITA are not the same as those of Apaf-1, CED4 and the NBS-LRR plant disease resistance proteins. Despite the presence of kinase 1a, 2 and 3a motifs, CARD4/Nod1 and CIITA appear to lack the other five conserved motifs characteristic of the NBS domains of Apaf-1, CED-4 and the NBS-LRR plant disease resistance proteins. Furthermore, CARD4/Nod1 and CIITA are not detected using the NBS regions of Apaf-1, CED4 or the NBS-LRR plant disease resistance proteins as seed sequences in PSI BLAST searches (Altschul *et al.*, 1997; http://www.ncbi.nlm.nih.gov) and vice versa. This suggests that the two kinds of NBS domains have diverged considerably and probably function differently to one another. This difference is highlighted by the identification of a third NBS protein, NAIP (neuronal apoptosis inhibitory protein) (Roy *et al.*, 1995; Chen *et al.*, 1998), which was detected by PSI BLAST searches using CARD4/NodI or CIITA as seed sequences. Like CARD4/NodI and CIITA, NAIP also has a C-terminal LRR domain, but unlike these proteins it has three N-terminal BIR (baculovirus inhibition of apoptosis protein repeat) domains and functions as a suppressor of apoptosis (Liston *et al.*, 1996). These NBS proteins, therefore, seem to play more diverse roles than the apoptosis and disease resistance proteins, which only seem to be involved in the activation of cell death in animals and plants.

5.6 Do the LRR domains perform more than one function in resistance signalling?

The LRR domain has been discussed extensively in chapter 4 of this volume and in many other reviews (e.g. Jones and Jones, 1997). Essentially, the LRR domain is seen as providing a scaffold upon which residues involved in ligand recognition are presented and which allows recognition specificity to vary. There is considerable evidence that non-consensus residues in the xxLxLxx β turn – β strand – β turn motif within the LRRs of resistance genes are subject to diversifying selection consistent with a role in recognition of avirulence ligands (see chapter 4). Recent evidence for the resistance genes at the L locus in flax suggest that the LRRs are not the only determinants of resistance specificity, but that the TIR domain is also involved (Ellis *et al.*, 1999; see also chapter 4). Resistance specificity may be viewed in two ways, either as the ability to bind a ligand or as the ability to trigger a signal in response to ligand binding. The evidence obtained from domain swaps between L alleles indicates that there needs to be a match between part of the LRR domain and part of the TIR or NBS domain for a resistance signal to be initiated (Ellis *et al.*, 1999). This suggests that there may be an interaction between the LRR domain and the TIR or NBS domain and, therefore, that the LRR domain may not just play a role in recognition of the avirulence ligand. The L6 and L7 proteins differ only in the TIR region (Ellis *et al.*, 1999), suggesting that they either detect the corresponding avirulence ligands directly, which seems unlikely, or possibly via a conformational change in the LRR region that is unique to each of the corresponding avirulence ligands. If part of the LRR domain interacts with and represses the TIR domain (analogous to the repression of the NBS domain by the WD-40 domain in Apaf-1), then a match between the TIR domain and the conformationally altered LRR domain may be required for release from repression.

A mutation in the N-terminal portion of the LRR domain of RPS5 not only abolishes RPS5 resistance but also affects resistance conferred by *RPS2* and several uncharacterised genes for resistance to the oomycete, *Peronospora parasitica* (Warren *et al.*, 1998). This mutation, which occurs in a region of conservation between the LRRs of RPS5 and RPS2, may have generated a dominant negative inhibitor of shared signal transduction components, suggesting that the LRR domain of these CC-NBS-LRR proteins may play a dual role in interaction with both upstream and downstream signalling components. A similar idea has already been suggested for the LRR domain of the Cf proteins. The absence of an obvious signalling domain, combined with the observation that the N-terminal half of the LRR domain is variable and contains the recognition specificity for Cf-4 and Cf-9 as opposed to the C-terminal half which is conserved, suggests a dual function for the LRR domain, with the C-terminal half likely to be involved in interaction with a signalling partner (Jones and Jones, 1997; Thomas *et al.*, 1997). A similar prediction could be made for Xa21, given that receptor kinases frequently undergo ligand-mediated oligomerisation of their receptor domains in order to oligomerise their kinase domains and trigger their reciprocal phosphorylation. In this case however, there is also experimental evidence to support such a prediction. The Xa21D gene produces a protein truncated in the last LRR and so lacks the TM and PK domains of Xa21, yet it is still able to function, albeit weakly (Wang *et al.*, 1998b), suggesting possible oligomerisation between Xa21D and other members of the Xa21 family or with other components of a receptor complex. In TLR4 signalling, the CD14 LRR protein appears to have multiple capabilities of LBP, LPS and TLR4 binding, and the TLR4 LRR protein seems able to bind the CD14/LPS/LBP complex and oligomerise with itself (Figure 5.2). Furthermore, in the NBS protein, Apaf-1, there seems to be interaction between the C-terminal WD-40 repeat domain and the NBS domain to control apoptosis (Figure 5.4). All of this circumstantial evidence points to more than one role for the LRR domains of resistance proteins. The data presented by Ellis *et al.* (1999) and Warren *et al.* (1998) suggest that this could be a more positive role than simply repression of signalling in the absence of an avirulence ligand and activation of signalling in its presence.

5.7 Relationship with membranes

In animals, the TIR proteins of the Toll and IL-1R family and the NBS proteins, Apaf-1 and CED-4, all operate at membranes (Figures 5.1, 5.2 and 5.5). The operation of Cf and Xa21 resistance proteins at membranes is also obvious because they are integral membrane proteins. Pto contains a myristoylation site for membrane attachment suggesting a potential for membrane association, but this motif has been shown to be unnecessary for Pto to function (Loh *et al.*, 1998). However, this result is contradicted by the observation that Fen kinase (a member of the Pto family which gives a Prf-dependent hypersensitive response to the insecticide fenthion) (Salmeron *et al.*, 1994; Loh and Martin, 1995) does require the myristoylation motif to function (Rommens *et al.*, 1995). Perhaps myristoylation is a redundant function for Pto, which could be associated with a membrane in another way. RPM1 is the only NBS-LRR protein whose subcellular location has been examined so far, and it has been shown to be associated with the cytosolic face of the plasma membrane as a peripheral membrane protein (Boyes *et al.*, 1998). EDS1, shown to be necessary for TIR-NBS-LRR proteins to function in *Arabidopsis*, has lipase homology suggesting a possible functional relationship involving membrane alterations in downstream signalling (Falk *et al.*, 1999). NDR1, shown to be necessary for CC-NBS-LRR proteins, including RPM1, to function in *Arabidopsis* is a membrane-associated protein (Century *et al.*, 1997). Based on both demonstrated and circumstantial associations, the relationship between resistance proteins and membranes is clearly an area that should be investigated further. It is possible that in plant cells, as in apoptotic animal cells, decompartmentalisation triggered by alterations in membrane integrity may play a key role in initiating cell death.

5.8 Direct or indirect interaction with Avr gene products

Pto is the only resistance protein for which a direct interaction with the corresponding avirulence protein has been published (Scofield *et al.*, 1996; Tang *et al.*, 1996b). However, there have also been unpublished reports of a direct interaction between the protein encoded by the rice Pi^{ta} gene for resistance to the rice blast pathogen, *Magnaporthe grisea*, and the corresponding Avr2YAMO avirulence protein (Valent, unpublished). For many resistance proteins, the opportunity to show a direct interaction between resistance and avirulence gene products does not exist because the corresponding avirulence products have not been identified. The few host pathogen combinations for which both components have been identified are presented in Table 5.3. The question of direct Cf-9/Avr9 interaction has been examined in Avr9 binding studies (Kooman-Gersmann *et al.*, 1996); from which it is clear that tomato plasma membranes carry Avr9 binding sites, whether or not they carry Cf-9. This suggests that Avr9 may bind another protein that is a target for Avr9 in a role as a pathogenicity factor. This further suggests that Cf-9 could recognise a target protein/Avr9 complex rather than Avr9 alone. It is possible that at least some resistance proteins may recognise their avirulence counterparts indirectly rather than directly. Even the seemingly straightforward example of Pto/AvrPto interaction might support this idea.

Pto is the receptor for AvrPto (Scofield *et al.*, 1996; Tang *et al.*, 1996b) and it interacts with three transcription factors, Pti4, 5 and 6, in yeast two-hybrid analysis (Zhou *et al.*, 1997), as well as another serine/threonine protein kinase, Pti1 (Zhou *et al.*, 1995), but it does so in the absence of AvrPto. It has been suggested that Pti4, 5 and 6 activate the transcription of PR proteins (Zhou *et al.*, 1997). The role of Pti1 is unclear, but it can be phosphorylated by Pto *in vitro*

Table 5.3 Resistance/avirulence gene combinations for which both genes have been cloned

Host	*R* gene	Pathogen*	*Avr* gene	Reference
Tobacco	*N*	Tobacco mosaic virus	Helicase domain of replicase	Padgett *et al.*, 1997
Tomato	*Pto* *Prf*	*Pseudomonas syringae* pv. *tomato*	*AvrPto*	Salmeron and Staskawicz, 1993
	Cf-9	*Cladosporium fulvum*	*Avr9*	van Kan *et al.*, 1991
	Cf-4		*Avr4*	Joosten *et al.*, 1994
Potato	*Rx*	Potato virus X	Coat protein	Bendahmane *et al.*, 1995
Arabidopsis	*RPS2*	*P. s.* pv. *tomato*	*AvrRpt2*	Innes *et al.*, 1993
	RPS4	*P. s.* pv. *pisi*	*AvrRps4*	Hinsch and Staskawicz, 1996
	RPS5	*P. s.* pv. *phaseolicola*	*AvrPphB*	Jenner *et al.*, 1991
	RPM1	*P. s.* pv. *glycinea*	*AvrB*	Tamaki *et al.*, 1988
		P. s. pv. *maculicola*	*AvrRpm1*	Dangl *et al.*, 1992
Rice	Pi^{ta}	*Magnaporthe grisea*	*Avr2YAMO*	Valent, unpublished

*The *Pseudomonas syringae* avirulence genes were islolated from the sources shown but apart from *P. s.* pv. *maculicola* have usually been transferred to *P. s.* pv. *tomato* for host/pathogen studies.

in the absence of AvrPto (Zhou *et al.*, 1995). As suggested previously (Jones and Jones, 1997), Pto is a legitimate target for manipulation by the bacterium and the original purpose of AvrPto may have been to inhibit rather than activate defence mechanisms triggered by Pto. It may have played a role in bacterial pathogenicity by preventing activation of the defence mechanisms controlled by Pti4, 5 and 6. It has not yet been reported whether the flux of signalling through Pti4, 5 and 6 increases or decreases in response to AvrPto binding. However, there is evidence to suggest that Pti1 is unlikely to be phosphorylated by Pto after it binds AvrPto (Rathjen *et al.*, 1999). Even if the flux of signalling through Pti 4, 5 and 6 increases, it is clearly not sufficient, in the absence of functional Prf, to confer resistance. It therefore seems most likely that Prf interacts with Pto following AvrPto binding to trigger a hypersensitive response (HR). AvrPto has been shown to bind to the kinase activation loop of Pto (Frederick *et al.*, 1998; Rathjen *et al.*, 1999), which in the absence of AvrPto seems to be negatively regulated by an inhibitory tyrosine residue (Rathjen *et al.*, 1999). This suggests kinase activation following AvrPto binding. Consistent with this interpretation, Rathjen *et al.* (1999) have shown that kinase activity is not required for Pto to bind AvrPto, but is required for the activation of a Prf-dependent HR. They argue that an interaction between Prf and a Pto/AvrPto complex alone is not sufficient to trigger an HR, because co-expression of AvrPto and a kinase defective form of Pto able to bind AvrPto does not trigger an HR in Prf-competent tomato plants. However, this observation does not preclude such an interaction, because AvrPto-activated Pto may interact with and phosphorylate Prf to trigger a signal. This idea is reminiscent of the relationship between the AfsR NBS regulatory protein and the AfsK protein kinase discussed in section 5.4. Alternatively, AvrPto-activated Pto may phosphorylate another protein (an indirect target for the pathogenicty function of AvrPto perhaps), which is then recognised by Prf.

Like most resistance genes, *Pto* exists as part of a divergent multigene family at a complex locus, consistent with its role in specificity of recognition and the associated genetic structures needed to generate variation in specificity. Unlike most resistance genes, *Prf* exists as a single gene at a simple locus embedded within the complex *Pto* locus (Salmeron *et al.*, 1996). *RPM1* and *RPS2* also exist as single genes at simple loci and lack the complexity consistent with a role in specificity of recognition. Significantly, there have been no reports of the corresponding avirulence proteins, AvrRpm1 or AvrB, binding directly to RPM1 (despite attempts to demonstrate a direct interaction; Boyes *et al.*, 1998), nor of AvrRpt2 to RPS2; and the ability of RPM1 to respond to two dissimilar avirulence ligands is analogous to that of Prf with respect to Pto and Fen, which respond to AvrPto and fenthion. It is therefore possible that *RPM1* and *RPS2* and other simple resistance genes play a similar role to that of *Prf*. Arguing against this idea is the lack of evidence for genes analogous to the Pto gene family encoding the proposed target proteins. Perhaps these proteins are essential, and therefore less free to vary than the kinases of the Pto family, or they may be numerically redundant, and therefore not detectable in a mutation screen.

5.9 The relationship between the hypersensitive response and induction of PR proteins in plant disease resistance may parallel that of cell death and innate immunity in animals

Plant resistance genes trigger both cell death and resistance responses and these can be separable, suggesting at least two major signalling pathways. The potato *Rx* gene for resistance to potato virus X (PVX) encodes a NBS-LRR protein that confers resistance without an HR (Bendahmane *et al.*, 1999). Resistance without an HR can also occur in other systems. The tomato *Cf-9* gene for resistance to *Cladosporium fulvum* is capable of conferring resistance without an HR (Hammond-Kosack and Jones, 1994), and the *dnd1* mutation in *Arabidopsis* suppresses the HR without affecting *RPM1, RPS2* or *RPS4* mediated resistance to *Pseudomonas syringae* (Yu *et al.*, 1998).

Overexpression of the avirulence determinants corresponding to *Rx* and *Cf-9* leads to cell death, suggesting that an increased flux through the resistance signalling pathway can saturate the mechanism that suppresses the HR. Overexpression of the PVX coat protein, which is the avirulence determinant recognised by Rx (Bendahmane *et al.*, 1995), leads to an HR (Bendahmane *et al.*, 1999). Injection of a *Cf-9* plant with Avr9 peptide, or the transgenic expression of *Avr9* under the control of the CaMV 35S promoter in a Cf-9 plant, or infection of a *Cf-9* plant with PVX overexpressing *Avr9*, exposes the plant to a massive dose of Avr9 and causes an HR (Hammond-Kosack *et al.*, 1994, 1995). This massive dose is likely to produce a very high flux through the Cf-9 signalling pathway. Natural pathogen infection probably exposes plant cells to much lower levels of Avr9, likely to produce a much smaller flux through the signalling pathway, so that the plant can resist the pathogen without cell death necessarily being involved.

It is generally accepted that the outcome of pathogen challenge is a race against time in terms of success or failure in establishing an infection. Resistance itself may also be a race against time, with the plant attempting to respond without resorting to cell sacrifice if the level of pathogen inoculum, as measured by flux through resistance protein signalling pathways, is sufficiently low to allow the plant time to induce alternative defences. Overexpression of Prf leads to activation of plant defences without an HR (Oldroyd and Staskawicz, 1998), suggesting that there is a sub-threshold level of signalling through the Prf pathway, which triggers expression of PR proteins but not an HR. This kind of dynamic switching, if it really exists in plants, resembles that described earlier for signals that stimulate both apoptosis and NF-κB signalling pathways in mammals, with the latter providing negative feedback on the former.

5.10 Conclusion

This review attempts to examine closely the various resistance protein motifs, particularly of the NBS-LRR genes, and what can be learnt about them from other systems that use the same motifs. Some of these motifs function in non-obvious ways in other systems, and these functions suggest areas for future experimental investigation of resistance protein function that might not otherwise have been considered. It is unlikely that resistance proteins will function in exactly the same way, but the comparisons are useful, even if they only highlight mechanisms that need to be excluded. However, some themes, such as homophilic oligomerisation and adaptor proteins, seem to be pervasive and are perhaps instructive. Some biochemical data are beginning to emerge for resistance protein function, but it seems that the time-lag between gene isolation and an understanding of biochemical function may be long and the transition difficult. Fortunately, it seems likely that the time-lag will be nowhere near as long as that between the discovery of resistance genes and their eventual isolation.

Acknowledgements

The author thanks Jeff Ellis, Peter Dodds and Adrienne Hardham for their critical reading of the manuscript.

References

Altschul, S.F., Madden, T.L., Schäffer, A.A., Zhang, J., Zhang, Z., Miller, W. and Lipman, D.J. (1997) Gapped BLAST and PSI-BLAST: a new generation of protein database search programs. *Nucleic Acids Res.*, **25** 3389-402.

Anderson, P.A., Lawrence, G.J., Morrish, B.C., Ayliffe, M.A., Finnegan, E.J. and Ellis, J.G. (1997) Inactivation of the flax rust resistance gene, *M*, associated with loss of a repeated unit within the leucine-rich repeat coding region. *Plant Cell*, **9** 641-51.

Aravind, L., Dixit, V.M. and Koonin, E. (1999) The domains of death: evolution of the apoptosis machinery. *Trends Biochem. Sci.*, **24** 47-53.

Beg, A.A. and Baltimore, D. (1996) An essential role for NF-κB in preventing TNF-α-induced cell death. *Science*, **274** 782-84.

Belvin, M.P., Jin, Y. and Anderson, K.V. (1995) Cactus protein degradation mediates *Drosophila* dorsal-ventral signaling. *Genes Dev.*, **9** 783-93.

Bendahmane, A., Köhm, B.A., Dedi, C. and Baulcombe, D.C. (1995) The coat protein of potato virus X is a strain-specific elicitor of *Rx1*-mediated virus resistance in potato. *Plant J.*, **8** 933-41.

Bendahmane, A., Kanyuka, K. and Baulcombe, D.C. (1999) The *Rx* gene from potato controls separate virus resistance and cell death responses. *Plant Cell*, **11** 781-91.

Bent, A.F., Kunkel, B.N., Dahlbeck. D., Brown, K.L., Schmidt, R., Giraudat, J., Leung, J. and Staskawicz, B.J. (1994) *RPS2* of *Arabidopsis thaliana*: a leucine-rich repeat class of disease resistance genes. *Science*, **265** 1856-60.

Bertin, J., Nir, W.-J., Fischer, C.M., Tayber, O.V., Errada, P.R., Grant, J.R., Keilty, J.J., Gosselin, M.L., Robison, K.E., Wong, G.H.W., Glucksmann, M.A. and DiStefana, P.S. (1999) Human CARD4 protein is a novel CED-4/Apaf-1 cell death family member that activates NF-κB. *J. Biol. Chem.*, **274** 12955-58.

Botella, M.A., Parker, J.E., Frost, L.N., Bittner-Eddy, P.D., Beynon, J.L., Daniels, M.J., Holub, E.B. and Jones, J.D.G. (1998) Three gene of the *Arabidopsis RPP1* complex resistance locus recognize distinct *Peronospora parasitica* avirulence determinants. *Plant Cell*, **10** 1847-60.

Bowling, S.A., Clarke, J.D., Liu, Y.D., Klessig, D.F. and Dong, X.N. (1997) The *cpr5* mutant of *Arabidopsis* expresses both *NPR1*-dependent and *NPR1*-independent resistance. *Plant Cell*, **9** 1573-84.

Boyes, D.C., Nam, J. and Dangl, J.L. (1998) The *Arabidopsis thaliana RPM1* disease resistance gene product is a peripheral plasma membrane protein that is degraded coincident with the hypersensitive response. *Proc. Natl. Acad. Sci. USA*, **95** 15849-54.

Brown, K., Gerstberger, S., Carlson, L., Franzoso, G. and Siebenlist, U. (1995) Control of IκBα proteolysis by site-specific, signal-induced phosphorylation. *Science*, **267** 1485-88.

Burns, K., Martinon, F., Esslinger, C., Pahl, H., Schneider, P., Bodmer, J.-L., Di Marco, F., French, L. and Tschopp, J. (1998) MyD88, an adapter protein involved in interleukin-1 signaling. *J. Biol. Chem.*, **273** 12203-209.

Cai, D., Kleine, M., Kifle, S., Harloff, H.-J., Sandal, N.N., Marcker, K.A., Klein-Lankhorst, R.M., Salentijn, E.M.J., Lange, W., Stiekema, W.J., Wyss, U., Grundler, F.M.W. and Jung, C. (1997) Positional cloning of a gene for nematode resistance in sugar beet. *Science*, **275** 832-34.

Cao, Z., Henzel, W.J. and Gao, X. (1996) IRAK: a kinase associated with the interleukin-1 receptor. *Science*, **271** 1128-31.

Cao, H., Glazebrook, J., Clarke, J.D., Volko, S. and Dong, X. (1997) The *Arabidopsis NPR1* gene that controls systemic-acquired resistance encodes a novel protein containing ankyrin repeats. *Cell*, **88** 57-63.

Century, K.S., Shapiro, A.D., Repetti, P.P., Dahlbeck, D., Holub, E. and Staskawicz, B.J. (1997) *NDR1*, a pathogen-induced component required for *Arabidopsis* disease resistance. *Science*, **278** 1963-65.

Chaudhary, D., O'Rourke, K., Chinnaiyan, A.M. and Dixit, V.M. (1998a) The death inhibitory molecules, CED-9 and CED-4L, use a common mechanism to inhibit the CED-3 death protease. *J. Biol. Chem.*, **273** 17708-12.

Chaudhary, P.M., Ferguson, C., Nguyen, V., Nguyen, O., Massa, H.F., Eby, M., Jasmin, A., Trask, B.J., Hood, L. and Nelson, P.S. (1998b) Cloning and characterization of two Toll/interleukin-1 receptor-like genes, TIL3 and TIL4: evidence for a multigene receptor family in humans. *Blood*, **91** 4020-27.

Chen, Q.F., Baird, S.D., Mahadevan, M., Besner-Johnston, A., Farahani, R., Xuan, J.-Y., Kang, X.L., Lefebvre, C., Ikeda, J.-E., Korneluk, R.G. and MacKenzie, A.E. (1998) Sequence of a 131-kb region of 5q13.1 containing the spinal muscular atrophy candidate genes, SMN and NAIP. *Genomics*, **48** 121-27.

Chinnaiyan, A.M., Chaudhary, D., O'Rourke, K., Koonin, E.V. and Dixit, V.M. (1997a) Role of CED-4 in the activation of CED-3. *Nature*, **388** 728-29.

Chinnaiyan, A.M., O'Rourke, K., Lane, B.R. and Dixit, V.M. (1997b) Interaction of CED-4 with CED-3 and CED-9—a molecular framework for cell death. *Science*, **275** 1122-26.

Chong, P.P. (1999) Molecular genetic studies on the biosynthesis and regulation of calcium-dependent antibiotic (CDA) production by *Streptomyces coelicolor* A3(2). Genbank Accession AAD18045.

Chow, J.C., Young, D.W., Golenbock, D.T., Christ, W.J. and Gusovsky, F. (1999) Toll-like receptor-4 mediates lipopolysaccharide-induced signal transduction. *J. Biol. Chem.*, **274** 10689-92.

Cole, S.T., Brosch, R., Parkhill, J., Garnier, T., Churcher, C., Harris, D., Gordon, S.V., Eiglmeier, K., Gas, S., Barry, C.E., Tekaia, F., Badcock, K., Basham, D., Brown, D., Chillingworth, T., Connor, R., Davies, R., Devlin, K., Feltwell, T., Gentles, S., Hamlin, N., Holroyd, S., Hornsby, T., Jagels, K., Krogh, A., McLean, J., Moule, S., Murphy, L., Oliver, K., Osborne, J., Quail, M.A., Rajandream, M.-A., Rogers, J., Rutter, S., Seeger, K., Skelton, J., Squares, R., Squares, S., Sulston, J.E., Taylor, K., Whitehead, S. and Barrell, B.G. (1998) Deciphering the biology of *Mycobacterium tuberculosis* from the complete genome sequence. *Nature*, **393** 537-44.

Collins, N., Drake, J., Ayliffe, M., Sun, Q., Ellis, J., Hulbert, S. and Pryor, T. (1999) Molecular characterization of the maize *Rp1-D* rust resistance haplotype and its mutants. *Plant Cell*, **11** 1365-76.

Dangl, J.L., Ritter, C., Gibbon, M.J., Mur, L.A.J., Wood, J.R., Goss, S., Mansfield, J.W., Taylor, J.D. and Vivian, A. (1992) Functional homologs of the *Arabidopsis RPM1* disease resistance gene in bean and pea. *Plant Cell*, **4** 1359-69.

del Peso, L., González, V.M. and Núñez, G. (1998) *Caenorhabditis elegans* EGL-1 disrupts the interaction of CED-9 with CED-4 and promotes CED-3 activation. *J. Biol. Chem.*, **273** 33495-500.

del Pozo, O. and Lam, E. (1998) Caspase and programmed cell death in the hypersensitive response of plants to pathogens. *Curr. Biol.*, **8** 1129-32 (Errata ibid. R896).

Desagher, S., Osen-Sand, A., Nichols, A., Eskes, R., Montessuit, S., Lauper, S., Maundrell, K., Antonsson, B. and Martinou, J.C. (1999) Bid-induced conformational change of Bax is responsible for mitochondrial cytochrome *c* release during apoptosis. *J. Cell Biol.*, **144** 891-901.

Dion, M., Chamberland, H., St-Michel, C., Plante, M., Darveau, A., Lafontaine, J.G. and Brisson, L.F. (1997) Detection of a homologue of bcl-2 in plant cells. *Biochem. Cell Biol.*, **75** 457-61.

Dixon, M.S., Jones, D.A., Keddie, J.S., Thomas, C.M., Harrison, K. and Jones, J.D.G. (1996) The tomato *Cf-2* disease resistance locus comprises two functional genes encoding leucine-rich repeat proteins. *Cell*, **84** 451-59.

Dixon, M.S., Hatzixanthis, K., Jones, D.A., Harrison, K. and Jones, J.D.G. (1998) The tomato *Cf-5* disease resistance gene and six homologs show pronounced allelic variation in leucine-rich repeat copy number. *Plant Cell*, **10** 1915-25.

Dushay, M.S., Åsling, B. and Hultmark, D. (1996) Origins of immunity: *Relish*, a compound Rel-like gene in the antibacterial defense of Drosophila. *Proc. Natl. Acad. Sci. USA*, **93** 10343-47.

Edwards, D.N., Towb, P. and Wasserman, S.A. (1997) An activity-dependent network of interactions links the Rel protein, Dorsal, with its cytoplasmic regulators. *Development*, **124** 3855-64.

Eldon, E., Kooyer, S., D'Evelyn, D., Duman, M., Lawinger, P., Botas, J. and Bellen, H. (1994) The *Drosophila* 18-wheeler is required for morphogenesis and has striking similarities to Toll. *Development*, **120** 885-99.

Ellis, J. and Jones, D. (1998) Structure and function of proteins controlling strain-specific pathogen resistance in plants. *Curr. Opin. Plant Biol.*, **1** 288-93.

Ellis, J.G., Lawrence, G.J., Luck, J.E. and Dodds, P.N. (1999) Identification of regions in alleles of the flax rust resistance gene, *L*, that determine differences in gene-for-gene specificity. *Plant Cell*, **11** 495-506.

Falk, A., Feys, B.J., Frost, L.N., Jones, J.D.G., Daniels, M.J. and Parker, J.E. (1999) *EDS1*, an essential component of *R*-gene-mediated disease resistance in *Arabidopsis* has homology to eukaryotic lipases. *Proc. Natl. Acad. Sci. USA*, **96** 3292-97.

Feinstein, E., Kimchi, A., Wallach, D., Boldin, M. and Varfolomeev, E. (1995) The death domain—a module shared by proteins with diverse cellular functions. *Trends Biochem. Sci.*, **20** 342-44.

Ferrero, E. and Goyert, S.M. (1988) Nucleotide sequence of the gene encoding the monocyte differentiation antigen, CD14. *Nucleic Acids Res.*, **16** 4173.

Finucane, D.M., Bossy-Wetzel, E., Waterhouse, N.J., Cotter, T.G. and Green, D.R. (1999) Bax-induced caspase activation and apoptosis via cytochrome *c* release from mitochondria is inhibitable by Bcl-xL. *J. Biol. Chem.*, **274** 2225-33.

Floriano, B. and Bibb, M. (1996) *afsR* is a pleiotropic but conditionally required regulatory gene for antibiotic production in *Streptomyces coelicolor* A3(2). *Mol. Microbiol.*, **21** 385-96.

Frederick, R.D., Thilmony, R.L., Sessa, G. and Martin, G.B. (1998) Recognition specificity for the bacterial avirulence protein, AvrPto, is determined by Thr-204 in the activation loop of the tomato Pto kinase. *Mol. Cell*, **2** 241-45.

Gassmann, W., Hinsch, M.E. and Staskawicz, B.J. (1999) The Arabidopsis RPS4 bacterial-resistance gene is a member of the TIR-NBS-LRR family of disease-resistance genes. *Plant J.*, **20** 265-77.

Geisler, R., Bergmann, A., Hiromi, Y. and Nüsslein-Volhard, C. (1992) *Cactus*, a gene involved in dorsoventral pattern formation of *Drosophila*, is related to the IκB gene family of vertebrates. *Cell*, **71** 613-21.

Grant, M.R., Godiard, L., Straube, E., Ashfield, T., Lewald, J., Sattler, A., Innes, R.W. and Dangl, J.L. (1995) Structure of the *Arabidopsis RPM1* gene enabling dual specificity disease resistance. *Science*, **269** 843-46.

Greenfeder, S.A., Nunes, P., Kwee, L., Labow, M., Cizzonite, R.A. and Ju, G. (1995) Molecular cloning and characterization of a second subunit of the interleukin-1 receptor complex. *J. Biol. Chem.*, **270** 13757-65.

Großhans, J., Bergmann, A., Haffter, P. and Nüsslein-Volhard, C. (1994) Activation of the protein kinase, Pelle, by Tube in the dorsoventral signal transduction pathway of *Drosophila* embryo. *Nature*, **372** 563-66.

Großhans, J., Schnorrer, F. and Nüsslein-Volhard, C. (1999) Oligomerisation of Tube and Pelle leads to nuclear localisation of Dorsal. *Mech. Dev.*, **81** 127-38.

Hammond-Kosack, K.E. and Jones, J.D.G. (1994) Incomplete dominance of tomato *Cf* genes for resistance to *Cladosporium fulvum*. *Mol. Plant-Microbe Interact.*, **7** 58-70.

Hammond-Kosack, K.E., Harrison, K. and Jones J.D.G. (1994) Developmentally regulated cell death on expression of the fungal avirulence gene, *Avr9*, in tomato seedlings carrying the disease-resistance gene, *Cf-9*. *Proc. Natl. Acad. Sci. USA*, **91** 10445-49.

Hammond-Kosack, K.E., Staskawicz, B.J., Jones, J.D.G. and Baulcombe, D.C. (1995) Functional expression of a fungal avirulence gene from a modified potato virus X genome. *Mol. Plant-Microbe Interact.*, **8** 181-85.

Hegde, R., Srinivasula, S.M., Ahmad, M., Fernandes-Alnemri, T. and Alnemri, E.S. (1998) Blk, a BHS-containing mouse protein that interacts with Bcl-2 and Bcl-X_L, is a potent death agonist. *J. Biol. Chem.*, **273** 7783-86.

Hinsch, M. and Staskawicz, B.J. (1996) Identification of a new *Arabidopsis* disease resistance locus, *RPS4*, and cloning of the corresponding avirulence gene, *avrRps4*, from *Pseudomonas syringae* pv. *pisi*. *Mol. Plant-Microbe Interact.*, **9** 55-61.

Hoffmann, J.A., Kafatos, F.C., Janeway, C.A. Jr. and Ezekowitz, R.A.B. (1999) Phylogenetic perspectives in innate immunity. *Science*, **284** 1313-18.

Hopp, T.P. (1995) Evidence from sequence information that the interleukin-1 receptor is a transmembrane GTPase. *Protein Sci.*, **4** 1851-59.

Horinouchi, S., Kito, M., Nishiyama, M., Furuya, K., Hong, S.-K., Miyake, K. and Beppu, T. (1990) Primary structure of AfsR, a global regulatory protein for secondary metabolite formation in *Streptomyces coelicolor A3(2)*. *Gene*, **95** 49-56.

Hu, Y., Benedict, M.A., Wu, D., Inohara, N. and Núnez, G. (1998a) Bcl-X_L interacts with Apaf-1 and inhibits Apaf-1-dependent caspase-9 activation. *Proc. Natl. Acad. Sci. USA*, **95** 4386-91.

Hu, Y., Ding, L., Spencer, D.M. and Núnez, G. (1998b) WD-40 repeat region regulates Apaf-1 self-association and procaspase-9 activation. *J. Biol. Chem.*, **273** 33489-94.

Huang, J., Gao, X., Li, S. and Cao, Z. (1997) Recruitment of IRAK to the interleukin-1 receptor complex requires interleukin-1 receptor accessory protein. *Proc. Natl. Acad. Sci. USA*, **94** 12829-32.

Innes, R.W., Bent, A.F., Kunkel, B.N., Bisgrove, S.R. and Staskawicz, B.J. (1993) Molecular analysis of avirulence gene, *avrRpt2*, and identification of a putative regulatory sequence common to all known *Pseudomonas syringae* avirulence genes. *J. Bacteriol.*, **175** 4859-69.

Inohara, N., del Poso, L., Koseki, T., Chen, S. and Núñez, G. (1998a) RICK, a novel protein kinase containing a caspase recruitment domain, interacts with CLARP and regulates CD95-mediated apoptosis. *J. Biol. Chem.*, **273** 12296-300.

Inohara, N., Gourley, T.S., Carrio, R., Muñiz, M., Merino, M., Garcia, I., Koseki, T., Hu, Y., Chen, S. and Núñez, G. (1998b) Diva, a Bcl-2 homologue that binds directly to Apaf-1 and induces BH3-independent cell death. *J. Biol. Chem.*, **273** 32479-86.

Inohara, N., Koseki, T., del Peso, L., Hu, Y., Yee, C., Chen, S., Carrio, R., Merino, J., Liu, D., Ni, J. and Núñez, G. (1999) Nod1, an Apaf-1-like activator of caspase-9 and nuclear factor-κB. *J. Biol. Chem.*, **274** 14560-67 (Erratum ibid. 18675).

Jenner, C., Hitchin, E., Mansfield, J., Walters, K., Betteridge, P., Teverson, D. and Taylor, J. (1991) Gene-for-gene interactions between *Pseudomonas syringae* pv. *phaseolicola* and *Phaseolus*. *Mol. Plant-Microbe Interact.*, **4** 553-62.

Jones, D.A. and Jones, J.D.G. (1997) The role of leucine-rich repeat proteins in plant defences. *Adv. Bot. Res. Adv. Plant Pathol.*, **24** 89-167.

Jones, D.A., Thomas, C.M., Hammond-Kosack, K.E., Balint-Kurti, P.J. and Jones, J.D.G. (1994) Isolation of the tomato *Cf-9* gene for resistance to *Cladosporium fulvum* by transposon tagging. *Science*, **266** 789-93.

Joosten, M.H.A.J., Cozijnsen, T.J. and De Wit, P.J.G.M. (1994) Host resistance to a fungal tomato pathogen lost by a single base-pair change in an avirulence gene. *Nature*, **367** 384-86.

Kanuka, H., Hisahara, S., Sawamoto, K., Shoji, S., Okano, H. and Miura, M. (1999) Pro-apoptotic activity of *Caenorhabditis elegans* CED-4 protein in *Drosophila*: implicated mechanisms for caspase activation. *Proc. Natl. Acad. Sci. USA*, **96** 145-50.

Kidd, S. (1992) Characterization of the *Drosophila cactus* locus and analysis of interactions between cactus and dorsal proteins. *Cell*, **71** 623-35.

Kirschning, C.J., Wesche, H., Ayres, T.M. and Rothe, M. (1998) Human Toll-like receptor 2 confers responsiveness to bacterial lipopolysaccharide. *J. Exp. Med.*, **188** 2091-97.

Knop, J. and Martin, M.U. (1999) Effects of IL-1 receptor-associated kinase (IRAK) expression on IL-1 signaling are independent of its kinase activity. *FEBS Lett.*, **448** 81-85.

Kobe, B. and Deisenhofer, J. (1994) The leucine-rich repeat—a versatile binding motif. *Trends Biochem. Sci.*, **19** 415-21.

Kooman-Gersmann, M., Honee, G., Bonnema, G. and de Wit, P.J.G.M. (1996) A high-affinity binding site for the avr9 peptide elicitor of *Cladosporium fulvum* is present on plasma membranes of tomato and other solanaceous plants. *Plant Cell*, **8** 929-38.

Kothny-Wilkes, G., Kulms, D., Pöppelmann, B., Luger, T.A., Kubin, M. and Scharwz, T. (1998) Interleukin-1 protects transformed keratinocytes from tumor necrosis factor-related apoptosis-inducing ligand. *J. Biol. Chem.*, **273** 29247-53.

Lagudah, E.S., Moullet, O. and Appels, R. (1997) Map-based cloning of a gene sequence encoding a nucleotide binding domain and a leucine-rich region at the *Cre3* nematode resistance locus of wheat. *Genome*, **40** 650-65.

Lawrence, G.J., Finnegan, E.J., Ayliffe, M.A. and Ellis, J.G. (1995) The *L6* gene for flax rust resistance is related to the *Arabidopsis* bacterial resistance gene, *RPS2*, and the tobacco viral resistance gene, *N*. *Plant Cell*, **7** 1195-206.

Lemaitre, B., Nicolas, E., Michaut, L., Reichhart, J.M. and Hoffmann, J.A. (1996) The dorsoventral regulatory gene cassette, *Spatzle/Toll/Cactus*, controls the potent antifungal response in *Drosophila* adults. *Cell*, **86** 973-83.

Letsou, A., Alexander, S., Orth, K. and Wasserman, S.A. (1991) Genetic and molecular characterization of *tube*, a *Drosophila* gene maternally required for embryonic dorsoventral polarity. *Proc. Natl. Acad. Sci. USA*, **88** 810-14.

Levashina, E.A., Ohresser, S., Lemaitre, B. and Imler, J.-L. (1998) Two distinct pathways can control expression of the gene encoding the *Drosophila* antimicrobial peptide, metchnikowin. *J. Mol. Biol.*, **278** 515-27.

Li, P., Nijhawan, D., Budihardjo, J., Srinivasula, S.M., Ahmad, M., Alnemri, E.S. and Wang, X.D. (1997) Cytochrome *c* and dATP-dependent formation of Apaf-1/caspase-9 complex initiates an apoptotic protease cascade. *Cell*, **91** 479-89.

Liston, P., Roy, N., Tamai, K., Lefebvre, C., Baird, S., Cherton-Horvat, G., Farahani, R., McLean, M., Ikeda, J.-E., MacKenzie, A. and Korneluk, R.G. (1996) Suppression of apoptosis in mammalian cells by NAIP and a related family of IAP genes. *Nature*, **379** 349-53.

Loh, Y.T. and Martin, G.B. (1995) The disease-resistance gene, *Pto*, and the fenthion-sensitivity gene, *Fen*, encode closely-related functional protein kinases. *Proc. Natl. Acad. Sci. USA*, **92** 4181-84.

Loh, Y.T., Zhou, J.M. and Martin, G.B. (1998) The myristylation motif of Pto is not required for disease resistance. *Mol. Plant-Microbe Interact.*, **11** 572-76.

Lupas, A., Van Dyke, M. and Stock, J. (1991) Predicting coiled-coils from protein sequences. *Science*, **252** 1162-64.

Malek, S., Huxford, T. and Ghosh, G. (1998) IκBα functions through direct contacts with the nuclear localization signals and the DNA binding sequences of NF-κB. *J. Biol. Chem.*, **273** 25427-35.

Malinin, N.L., Boldin, M.P., Kovalenko, A.V. and Wallach, D. (1997) MAP3K-related kinase involved in NF-κB induction by TNF, CD95 and IL-1. *Nature*, **385** 540-44.

Martin, G.B., Brommonschenkel, S.H., Chunwongse, J., Frary, A., Ganal, M.W., Spivy, R., Wu, T., Earle, E.D. and Tanksley, S.D. (1993) Map-based cloning of a protein kinase gene conferring disease resistance in tomato. *Science*, **262** 1432-36.

Matsumoto, A., Hong, S.-K., Ishizuka, H., Horinouchi, S. and Beppu, T. (1994) Phosphorylation of the AfsR protein involved in secondary metabolism in *Streptomyces* species by a eukaryotic-type protein kinase. *Gene*, **146** 47-56.

McDowell, J.M., Dhandaydham, M., Long, T.A., Aarts, M.G.M., Goff, S., Holub, E.B. and Dangl, J.E. (1998) Intragenic recombination and diversifying selection contribute to the evolution of downy mildew resistance at the *RPP8* locus of *Arabidopsis*. *Plant Cell*, **10** 1861-74.

Medzhitov, R., Preston-Hurlburt, P. and Janeway, C.A. Jr. (1997) A human homologue of the *Drosophila* Toll protein signals activation of adaptive immunity. *Nature*, **388** 394-97.

Medzhitov, R., Preston-Hurlburt, P., Kopp, E., Stadlen, A., Chen, C.Q., Ghosh, S. and Janeway, C.A. Jr. (1998) MyD88 is an adaptor protein in the hToll/IL-1 receptor family signaling pathways. *Mol. Cell*, **2** 253-58.

Meng, X., Khanuja, B.S. and Ip, Y.T. (1999) Toll-receptor-mediated *Drosophila* immune response requires Dif, an NF-κB factor. *Genes Dev.*, **13** 792-97.

Meyers, B.C., Chin, D.B., Shen, K.A., Sivaramakrishnan, S., Lavelle, D.O., Zhang, Z. and Michelmore, R.W. (1998a) The major resistance gene cluster in lettuce is highly duplicated and spans several megabases. *Plant Cell*, **10** 1817-32.

Meyers, B.C., Shen, K.A., Rohani, P., Gaut, B.S. and Michelmore, R.W. (1998b) Receptor-like genes in the major resistance locus of lettuce are subject to divergent selection. *Plant Cell*, **10** 1833-46.

Milligan, S.B., Bodeau, J., Yaghoobi, J., Kaloshian, I., Zabel, P. and Williamson, V.M. (1998) The root knot nematode resistance gene, *Mi*, from tomato is a member of the leucine zipper, nucleotide binding, leucine-rich repeat family of plant genes. *Plant Cell*, **10** 1307-19.

Mindrinos, M., Katagiri, F., Yu, G.-L. and Ausubel, F.M. (1994) The *A. thaliana* disease resistance gene, *RPS2*, encodes a protein containing a nucleotide-binding site and leucine-rich repeats. *Cell*, **78** 1089-99.

Muzio, M., Ni, J., Feng, P. and Dixit, V.M. (1997) IRAK (Pelle) family member, IRAK-2, and MyD88 as proximal mediators of IL-1 signaling. *Science*, **278** 1612-15.

Muzio, M., Natoli, G., Saccani, S., Levrero, M. and Mantovani, A. (1998) The human Toll signaling pathway: divergence of nuclear factor-κB and JNK/SAPK activation upstream of tumor necrosis factor receptor-associated factor 6 (TRAF6). *J. Exp. Med.*, **187** 2097-101.

Norris, J.L. and Manley, J.L. (1995) Regulation of dorsal its cultured cells by Toll and tube–tube function involves a novel mechanism. *Genes Dev.*, **9** 358-69.

Oldroyd, G.E.D. and Staskawicz, B.J. (1998) Genetically engineered broad-spectrum disease resistance in tomato. *Proc. Natl. Acad. Sci. USA*, **95** 10300-305.

Ori, N., Eshed, Y., Paran, I., Presting, G., Aviv, D., Tanksley, S., Zamir, D. and Fluhr, R. (1997) The *I2C* family from the wilt disease resistance locus, *I2*, belongs to the nucleotide binding, leucine-rich repeat superfamily of plant resistance genes. *Plant Cell*, **9** 521-32.

Padgett, H.S., Watanabe, Y. and Beachy, R.N. (1997) Identification of the TMV replicase sequence that activates the *N* gene-mediated hypersensitive response. *Mol. Plant-Microbe Interact.*, **10** 709-15.

Pan, G., O'Rourke, K. and Dixit, V.M. (1998) Caspase-9, Bcl-X_L and Apaf-1 form a ternary complex. *J. Biol. Chem.*, **273** 5841-45.

Parker, J.E., Coleman, M.J., Szabo, V., Frost, L.N., Schmidt, R., van der Biezen, E., Moores, T., Dean, C., Daniels, M.J. and Jones, J.D.G. (1997) The *Arabidopsis* downy mildew resistance gene, *RPP5*, shares similarity to the Toll and interleukin-1 receptors with *N* and *L6*. *Plant Cell*, **9** 879-94.

Pennickx, I.A.M.A., Thomma, B.P.H.J., Buchala, A., Metraux, J.P. and Broekaert, W.F. (1998) Concomitant activation of jasmonate and ethylene response pathways is required for induction of a plant defensin gene in *Arabidopsis*. *Plant Cell*, **10** 2103-13.

Petersen, U.-M., Björklund, G., Ip, Y.T. and Engström, Y. (1995) The *dorsal*-related immunity factor, Dif, is a sequence-specific *trans*-activator of *Drosophila Cecropin* gene expression. *EMBO J.*, **14** 3146-58.

Rathjen, J.P., Chang, J.H., Staskawicz, B.J. and Michelmore, R.W. (1999) Constitutively active *Pto* induces a *Prf*-dependent hypersensitive response in the absence of *avrPto*. *EMBO J.*, **18** 3232-40.

Reed, J.C. (1997) Cytochrome *c*: can't live with it—can't live without it. *Cell*, **91** 559-62.

Rock, F.L., Hardiman, G., Timans, J.C., Kastelein, R.A. and Bazan, J.F. (1998) A family of human receptors structurally related to *Drosophila* Toll. *Proc. Natl. Acad. Sci. USA*, **95** 588-93.

Rommens, C.M.T., Salmeron, J.M., Baulcombe, D.C. and Staskawicz, B.J. (1995) Use of a gene expression system based on potato virus X to rapidly identify and characterize a tomato Pto homolog that controls fenthion sensitivity. *Plant Cell*, **7** 249-57.

Rossi, M., Goggin, F.L., Milligan, S.B., Kaloshian, I., Ullman, D.E. and Williamson, V.M. (1998) The nematode resistance gene, *Mi*, of tomato confers resistance against the potato aphid. *Proc. Natl. Acad. Sci. USA*, **95** 9750-54.

Roy, N., Mahadevan, M.S., McLean, M., Shutler, G., Yaraghi, Z., Farahini, R., Baird, S., Besner-Johnston, A., Lefebvre, C., Kang, X.L., Salih, M., Aubry, H., Tamai, K., Ioannou, P., Crawford, T.O., de Jong, P.J., Surh, L., Ikeda, J., Korneluk, R.G. and MacKenzie, A. (1995) The gene for neuronal apoptosis inhibitory protein is partially deleted in individuals with spinal muscular atrophy. *Cell*, **80** 167-78.

Ryals, J., Weymann, K., Lawton, K., Friedrich, L., Ellis, D., Steiner, H.-Y., Johnson, J., Delaney, T.P., Jesse, T., Vos, P. and Uknes, S. (1997) The *Arabidopsis NIM1* protein is homologous to the mammalian transcription factor inhibitor, IκB. *Plant Cell*, **9** 425-39.

Salmeron, J.M. and Staskawicz, B.J. (1993) Molecular characterization and *hrp* dependence of the avirulence gene, *avrPto*, from *Pseudomonas syringae* pv. *tomato*. *Mol. Gen. Genet.*, **239** 6-16.

Salmeron, J.M., Barker, S.J., Carland, F.M., Mehta, A.Y. and Staskawicz, B.J. (1994) Tomato mutants altered in bacterial disease resistance provide evidence for a new locus controlling pathogen recognition. *Plant Cell*, **6** 511-20.

Salmeron, J.M., Oldroyd, G.E., Rommens, C.M., Scofield, S.R., Kim, H.S., Lavelle, D.T., Dahlbeck, D. and Staskawicz, B.J. (1996) Tomato *Prf* is a member of the leucine-rich repeat class of plant disease resistance genes and lies embedded within the *Pto* kinase gene cluster. *Cell*, **86** 123-33.

Schuler, G.D., Altschul, S.F. and Lipman, D.J. (1991) A workbench for multiple alignment construction and analysis. *Proteins: Structure, Function and Genetics*, **9** 180-90.

Scofield, S.R., Tobias, C.M., Rathjen, J.P., Chang, J.H., Lavelle, D.T., Michelmore, R.W. and Staskawicz, B.J. (1996) Molecular basis of gene-for-gene specificity in bacterial speck disease of tomato. *Science*, **274** 2063-65.

Shah, J., Tsui, F. and Klessig, D.F. (1997) Characterization of a salicylic acid-insensitive mutant (*sai1*) of *Arabidopsis thaliana* identified in a selective screen utilizing the SA-inducible expression of the *tms2* gene. *Mol. Plant-Microbe Interact.*, **10** 69-78.

Shelton, C.A. and Wasserman, S.A. (1993) *Pelle* encodes a protein kinase required to establish dorsoventral polarity in the *Drosophila* embryo. *Cell*, **72** 515-25.

Shen, B.H. and Manley, J.L. (1998) Phosphorylation modulates direct interaction between the Toll receptor, Pelle kinase, and Tube. *Development*, **125** 4710-28.

Shimizu, S., Narita, M. and Tsujimoto, Y. (1999) Bcl-2 family proteins regulate the release of apoptogenic cytochrome *c* by the mitochondrial channel, VDAC. *Nature*, **399** 483-87.

Simons, G., Groenendijk, J., Wijbrandi, J., Reijans, M., Groenen, J., Diergaarde, P., Vanderlee, T., Bleeker, M., Onstenk, J., Deboth, M., Haring, M., Mes, J., Cornelissen, B., Zabeau, M. and Vos, P. (1998) Dissection of the Fusarium *I2* gene cluster in tomato reveals six homologs and one active gene copy. *Plant Cell*, **10** 1055-68.

Sims, J.E., March, C.J., Cosman, D., Widmer, M.B., MacDonald, H.R., McMahan, C.J., Grubin, C.E., Wignall, J.M., Jackson, J.L., Call, S.M., Friend, D., Alpert, A.R., Gillis, S., Urdal, D.L. and Dower, S.K. (1988) DNA expression cloning of the IL-1 receptor, member of the immunoglobulin superfamily. *Science*, **241** 585-89.

Singh, P.K., Jia, H.P., Wiles, K., Hesselberth, J., Liu, L.D., Conway, B.-A.D., Greenberg, E.P., Valore, E.V., Welsh, M.J., Ganz, T., Tack, B.F. and McCray, P.B. (1998) Production of β-defensins by human airway epithelia. *Proc. Natl. Acad. Sci. USA*, **95** 14961-66.

Song, W.-Y., Wang, G.-L., Chen. L.-L., Kim, H.-S., Pi, L.-Y., Holsten, T., Gardner, J., Wang, B., Zhai, W.-X., Zhu, L.-H., Fauquet, C. and Ronald, P. (1995) A receptor kinase-like protein encoded by the rice disease resistance gene, *Xa21*. *Science*, **270** 1804-806.

Song, H.Y., Régnier, C.H., Kirschning, C.J., Goeddel, D.V. and Rothe, M. (1997) Tumor necrosis factor (TNF)-mediated kinase cascades: birfurcation of nuclear factor-κB and c-jun N-terminal kinase (JNK/SAPK) pathways at TNF-receptor-associated factor 2. *Proc. Natl. Acad. Sci. USA*, **94** 9792-96.

Srinivasula, S.M., Ahmad, M., Fernandes-Alnemri, T. and Alnemri, E.S. (1998) Autoactivation of procaspase-9 by Apaf-1-mediated oligomerization. *Mol. Cell*, **1** 949-57.

Steimle, V., Otten, L.A., Zufferey, M. and Mach, B. (1993) Complementation cloning of an MHC class II transactivator mutated in hereditary MHC class II deficiency (or bare lymphocyte syndrome). *Cell*, **75** 135-46.

Steward, R. (1987) *Dorsal*, an embryonic polarity gene in *Drosophila*, is homologous to the vertebrate proto-oncogene, *c-rel*. *Science*, **238** 692-94.

Tamaki, S., Dahlbeck, D., Staskawicz, B. and Keen, N.T. (1988) Characterization and expression of two avirulence genes cloned from *Pseudomonas syringae* pv. *glycinea*. *J. Bacteriol.*, **170** 4846-54.

Tang, L., Grimm, A., Zhang, Y.X. and Hutchinson, C.R. (1996a) Purification and characterization of the DNA-binding protein, DnrI, a transcriptional factor of daunorubicin biosynthesis in *Streptomyces peucetius*. *Mol. Microbiol.*, **22** 801-13.

Tang, X.Y., Frederick, R.D., Zhou, J.M., Halterman, D.A., Jia, Y.L. and Martin, G.B. (1996b) Initiation of plant disease resistance by physical interaction of avrPto and Pto kinase. *Science*, **274** 2060-63.

Terras, F.R.G., Eggermont, K., Kovaleva, V., Raikhel, N.V., Osborn, R.W., Kester, A., Rees, S.B., Torrekens, S., Van Leuven, F., van der Leyden, J., Cammune, B.P.A. and Broekaert, W.F. (1995) Small cysteine-rich antifungal proteins from radish—their role in host defense. *Plant Cell*, **7** 573-88.

Thomas, C.M., Jones, D.A., Parniske, M., Harrison, K., Balint-Kurti, P., Hatzixanthis, K. and Jones, J.D.G. (1997) Characterization of the tomato *Cf-4* gene for resistance to *Cladosporium fulvum* identifies sequences that determine recognitional specificity in *Cf-4* and *Cf-9*. *Plant Cell*, **9** 2209-24.

van Antwerp, D.J., Martin, S.J., Kafri, T., Green, D.R. and Verma, I.M. (1996) Suppression of TNF-α-induced apoptosis by NF-κB. *Science*, **274** 787-89.

van der Biezen, E.A. and Jones, J.D.G. (1998) The NB-ARC domain: a novel signalling motif shared by plant resistance gene products and regulators of cell death in animals. *Curr. Biol.*, **8** R226-27.

van Kan, J.A.L., van den Ackerveken, G.F.J.M. and de Wit, P.J.G.M. (1991) Cloning and characterization of cDNA of avirulence gene, *avr9*, of the fungal pathogen *Cladosporium fulvum*, causal agent of tomato leaf mold. *Mol. Plant-Microbe Interact.*, **4** 52-59.

Vig, E., Green, M., Liu, Y., Donner, D.B., Mukaida, N., Goebl, M.G. and Harrington, M.A. (1999) Modulation of tumor necrosis factor and interleukin-1-dependent NF-κB activity by mPLK/IRAK. *J. Biol. Chem.*, **274** 13077-84.

Volpe, F., Clatworthy, J., Kaptein, A., Maschera, B., Griffin, A.-M. and Ray, K. (1997) The IL-1 receptor accessory protein is responsible for the recruitment of the interleukin-1 receptor associated kinase to the IL-1/IL-1 receptor I complex. *FEBS Lett.*, **419** 41-44.

Vos, P., Simons, G., Jesse, T., Wijbrandi, J., Heinen, L., Hogers, R., Frijters, A., Groenendijk, J., Diergaarde, P., Reijans, M., Fierens-Onstenk, J., De Both, M., Peleman, J., Liharska, T., Hontelez, J. and Zabeau, M. (1998) The tomato *Mi-1* gene confers resistance to both root-knot nematodes and potato aphids. *Nature Biotechnol.*, **16** 1365-69.

Wang, C.-Y., Mayo, M.W., Korneluk, R.G., Goeddel, D.V. and Baldwin, A.S. Jr. (1998a) NF-κB antiapoptosis—induction of TRAF1 and TRAF2 and C-IAP1 and C-IAP2 to suppress caspase-8 activation. *Science*, **281** 1680-83.

Wang, G.-L., Ruan, D.-L., Song, W.-Y., Sideris, S., Chen, L.-L., Pi, L.-Y., Zhang, S., Zhang, Z., Fauquet, C., Gaut, B.S., Whalen, M.C. and Ronald, P.C. (1998b) *Xa21D* encodes a receptor-like molecule with a leucine-rich repeat domain that determines race-specific recognition and is subject to adaptive evolution. *Plant Cell*, **10** 765-79.

Wang, Z.X., Yano, M., Yamanouchi, U., Iwamoto, M., Monna, L., Hayasaka, H., Kuboki, Y. and Sasaki, T. (1999) Map-based cloning of the candidate of *Pi-b*, a rice blast resistance gene. *Plant J.*, **19** 55-64.

Warren, R.F., Henk, A., Mowery, P., Holub, E. and Innes, R.W. (1998) A mutation within the leucine-rich repeat domain of the *Arabidopsis* disease resistance gene, *RPS5*, partially suppresses multiple bacterial and downy mildew resistance genes. *Plant Cell*, **10** 1439-52.

Wesche, H., Korherr, C., Kracht, M., Falk, W., Resch, K. and Martins, M.U. (1997a) The interleukin-1 accessory protein (IL-1RAcP) is essential for IL-1-induced activation of interleukin-1 receptor-associated kinase (IRAK) and stress-activated protein kinases (SAP kinases). *J. Biol. Chem.*, **272** 7727-31.

Wesche, H., Henzel, W.J., Shillinglaw, W., Li, S. and Cao, Z.D. (1997b) MyD88—an adapter that recruits IRAK to the IL-1 receptor complex. *Immunity*, **7** 837-847.

Whitham, S., Dinesh-Kumar, S.P., Choi, D., Hehl, R., Corr, C. and Baker, B. (1994) The product of the tobacco mosaic virus resistance gene, *N*: similarity to Toll and the interleukin-1 receptor. *Cell*, **78** 1101-15.

Wietzorrek, A. and Bibb, M. (1997) A novel family of proteins that regulates antibiotic production in streptomycetes appears to contain an OmpR-like DNA-binding fold. *Mol. Microbiol.*, **25** 1181-84.

Williams, M.J., Rodriguez, A., Kimbrell, D.A. and Eldon, E.D. (1997) The 18-wheeler mutation reveals complex antibacterial gene regulation in *Drosophila* host defense. *EMBO J.*, **16** 6120-30.

Wu, D., Wallen, H.D. and Nuñez, G. (1997a) Interaction and regulation of subcellular localization of CED-4 by CED-9. *Science*, **275** 1126-29.

Wu, D., Wallen, H.D., Inohara, N. and Nuñez (1997b) Interaction and regulation of the *Caenorhabditis elegans* death protease, CED-3, by CED-4 and CED-9. *J. Biol. Chem.*, **272** 21449-54.

Wu, L.P. and Anderson, K.V. (1998) Regulated nuclear import of Rel proteins in the *Drosophila* immune response. *Nature*, **392** 93-97.

Yamin, T.-T. and Miller, D.K. (1997) The interleukin-1 receptor-associated kinase is degraded by proteasomes following its phosphorylation. *J. Biol. Chem.*, **272** 21540-47.

Yang, J. and Steward, R.A. (1997) Multimeric complex and the nuclear targeting of the *Drosophila* rel protein, Dorsal. *Proc. Natl. Acad. Sci. USA*, **94** 14524-29.

Yang, R.-B., Mark, M.R., Gray, A., Huang, A., Xie, M.H., Zhang, M., Goddard, A., Wood, W.I., Gurney, A.L. and Godowski, P.J. (1998a) Toll-like receptor-2 mediates lipopolysaccharide-induced cellular signalling. *Nature*, **395** 284-88.

Yang, X., Chang, H.Y. and Baltimore, D. (1998b) Essential role of CED-4 oligomerization in CED-3 activation and apoptosis. *Science*, **281** 1355-57.

Ye, R., Rehemtulla, S.N. and Wong, S.-L. (1994) Glucitol induction in *Bacillus subtilis* is mediated by a regulatory factor, GutR. *J. Bacteriol.*, **176** 3321-27.

Ylihonko, K., Tuikkanen, J., Jussila, S., Cong, L. and Mäntsälä, P. (1996) A gene cluster involved in nogalamycin biosynthesis from *Streptomyces nogalater*: sequence analysis and complementation of early-block mutations in the anthracycline pathway. *Mol. Gen. Genet.*, **251** 113-20.

Yoshimura, S., Yamanouchi, U., Katayose, Y., Toki, S., Wang, Z.-X., Kono, I., Kurata, N., Yano, M., Iwata, N. and Sasaki, T. (1998) Expression of *Xa1*, a bacterial blight-resistance gene in rice, is induced by bacterial inoculation. *Proc. Natl. Acad. Sci. USA*, **95** 1663-68.

Yu, I.-C., Parker, J. and Bent, A.F. (1998) Gene-for-gene disease resistance without the hypersensitive response in *Arabidopsis dnd1* mutant. *Proc. Natl. Acad. Sci. USA*, **95** 7819-24.

Yuan, J. and Horvitz, H.R. (1992) The *Caenorhabditis elegans* cell death gene, *CED-4*, encodes a novel protein and is expressed during the period of extensive programmed cell death. *Development*, **116** 309-20.

Zandi, E., Rothwarf, D.M., Delhase, M., Hayakawa, M. and Karin, M. (1997) The IκB kinase complex (IKK) contains two kinase subunits, IKKα and IKKβ, necessary for IκB phosphorylation and NF-κB activation. *Cell*, **91** 243-52.

Zhang, F.X., Kirschning, C.J., Mancinelli, R., Xu, X.-P., Jin, Y., Faure, E., Mantovani, A., Rothe, M., Muzio, M. and Arditi, M. (1999a) Bacterial lipopolysaccharide activates nuclear factor-κB through interleukin-1 signaling mediators in cultured human dermal endothelial cells and mononuclear phagocytes. *J. Biol. Chem.*, **274** 7611-14.

Zhang, Y., Fan, W., Kinkema, M., Li, X. and Dong, X. (1999b) Interaction of NPR1 with basic leucine zipper protein transcription factors that bind sequences required for salicylic acid induction of the *PR-1* gene. *Proc. Natl. Acad. Sci. USA*, **96** 6523-28.

Zhou, J.M., Loh, Y.T., Bressan, R.A. and Martin, G.B. (1995) The tomato gene, *Pti1*, encodes a serine/threonine kinase that is phosphorylated by Pto and is involved in the hypersensitive response. *Cell*, **83** 925-35.

Zhou, J.M., Tang, T. and Martin, G.B. (1997) The Pto kinase conferring resistance to tomato bacterial speck disease interacts with proteins that bind a *cis*-element of pathogenesis-related genes. *EMBO J.*, **16** 3207-18.

Zou, H., Henzel, W.J., Liu, X., Lutschg, A. and Wang, X. (1997) Apaf-1, a human protein homologous to *C. elegans* CED-4, participates in cytochrome *c*-dependent activation of caspase-3. *Cell*, **90** 405-13.

Zou, H., Li, Y., Liu, X. and Wang, X. (1999) An APAF-1-cytochrome *c* multimeric complex is a functional apoptosome that activates procaspase-9. *J. Biol. Chem.*, **274** 11549-56.

6 Signalling in plant disease resistance

Jane E. Parker

6.1 Introduction

In the natural environment, plants have evolved mechanisms to perceive and respond effectively to a vast range of biotic stresses. A plant's survival depends on efficient recognition of microbial and invertebrate pathogens and the timely activation of the local and systemic defence machinery. Unlike multicellular animals, plants lack a circulatory system for the surveillance and destruction of foreign material. However, they possess a functionally equivalent recognition system in the form of a basal resistance machinery and a repertoire of pathogen-specific resistance (*R*) genes. The cloning of *R* genes from several dicotyledonous and monocotyledonous species sheds some light on genetic and molecular mechanisms by which a plant may respond to pathogen selection pressures through the evolution of novel recognition molecules. Also, activation of local plant defences leads to the generation of systemic-acquired resistance (SAR), establishing a heightened level of immunity within the whole plant.

R gene-mediated disease resistance to pathogens is commonly characterized by rapid programmed plant cell death at the site of pathogen attack, a process known as the hypersensitive response (HR). This is accompanied by a battery of defence-related processes, including ion flux changes, an oxidative burst, the accumulation of potent signalling molecules, such as salicylic acid (SA) or jasmonic acid (JA), and the local and systemic transcriptional activation of sets of genes encoding pathogenesis-related (PR) proteins. Although little is understood about the precise nature of the plant-pathogen recognition events and consequent signalling processes that result in arrest of pathogen infection, the complexities of plant disease resistance signalling are now beginning to be unravelled by a combination of scientific approaches.

This chapter highlights some of the most recent advances made in biochemical and pharmacological studies and in the molecular genetic dissection of various plant-pathogen interactions. What emerges is a potentially highly-branched informational network with multiple signal amplification loops. Superimposed on this signalling machinery are mechanisms that establish the hierarchy of pathway utilization, depending on the nature of signal input. Genetic analyses of plant resistance pathways are also providing insights into processes that limit growth of virulent pathogens in compatible interactions.

6.2 Resistance proteins as signalling molecules

Some clues to the mechanisms involved in disease resistance signalling have been gained from analysis of cloned *R* gene products in a number of plant species (Hammond-Kosack and Jones, 1997; Ellis and Jones, 1998). The R protein must in some way fulfil a dual role as a recognition molecule and a signal transducer, and how these two functions are accomplished and integrated with other disease resistance signalling components is currently under intense scrutiny. R proteins specifying resistance to different pathogen types: viruses, bacteria, fungi, oomycetes, aphids and nematodes, share a limited number of motifs, suggesting that the mechanisms controlling recognition and signal emission are broadly conserved in the detection of different pathogens. The tomato *Pto* (Martin *et al.*, 1993) and rice *Xa21* (Song *et al.*, 1995) genes possess a serine/threonine protein kinase domain, implicating protein phosphorylation as a central process in these disease resistance pathways. The *Xa21* gene product also shares a leucine-rich repeat (LRR) domain with other characterized R proteins. Significantly, Pto requires a LRR-containing protein, Prf, to elicit the resistance response (Salmeron *et al.*, 1994, 1996). LRR domains are present in yeast, animal and plant signalling proteins and in some cases are known to mediate protein-protein interactions (Jones and Jones, 1997; Kajava, 1998). Thus, it is anticipated that the LRRs may promote dimerization and/or association with other defence components as part of or in addition to their demonstrated contribution to pathogen recognition specificity (Parniske *et al.*, 1997; Botella *et al.*, 1998; McDowell *et al.*, 1998; Meyers *et al.*, 1998; Ellis *et al.*, 1999). A major subclass of LRR-containing R proteins, which are predicted to reside freely in the cytoplasm or as cytoplasmic proteins associated with membrane compartments (Bent, 1996; Parker *et al.*, 1997; Botella *et al.*, 1998; Boyes *et al.*, 1998; Milligan *et al.*, 1998; Warren *et al.*, 1998), contain a nucleotide binding (NB) domain, indicating that hydrolysis of guanosine triphosphate (GTP) or adenosine triphosphate (ATP) is likely to contribute to their function.

6.2.1 Pto*-mediated resistance*

One of the best characterized pathosystems is *Pto*-mediated resistance to bacterial speck disease in tomato caused by the bacterial pathogen, *Pseudomonas syringae* pv. *tomato*. *Pto* encodes a functional cytoplasmic serine/threonine protein kinase that interacts directly with the avirulence protein, AvrPto, in a yeast two-hybrid assay (Scofield *et al.*, 1996; Tang *et al.*, 1996); and recent *in planta* analyses suggest that Pto kinase activity is necessary for AvrPto-mediated induction of the HR (Rathjen *et al.*, 1999). These findings strongly support the role of Pto as receptor for the bacterial AvrPto ligand in an incompatible interaction. The association is highly discriminatory, since a related kinase,

Fen, encoded within the Pto gene cluster, does not interact with AvrPto. Fen confers sensitivity to an organophosphorous insecticide, Fenthion. Generation of Pto/Fen chimaeras and site-directed mutants established that conserved amino acids, Thr-204 and Thr-207, residing within the Pto kinase activation loop, determine recognition specificity (Frederick *et al.*, 1998; Rathjen *et al.*, 1999). A number of potential signalling components in Pto-mediated defence were identified, using Pto as bait in the yeast two-hybrid system. Pto interacts with a second serine/threonine protein kinase, Pti1 (Zhou *et al.*, 1995), and Pti1 was shown to be a substrate for phosphorylation by Pto, suggesting that it may lie downstream of Pto in a defence-related kinase cascade. Furthermore, overexpression of *Pti1* in tobacco enhanced the plant's local necrotic response to *avrPto*, revealing a possible physiological role of Pti1 in elaboration of the HR. The yeast assay also revealed a family of defence-related transcription factors, Pti4, Pti5 and Pti6 of the EREBP (ethylene responsive element binding protein class) (Zhou *et al.*, 1997). These proteins recognize a core hexanucleotide sequence present in the promoters of a significant number of genes encoding PR proteins, suggesting that they may contribute to PR gene induction during the defence response. It remains to be shown whether such inducible events are essential for restriction of pathogen within the plant. What does emerge, however, is a plausible separation of pathways leading to the HR and to those regulating PR protein expression.

Comprehending the role of the NB-LRR-containing protein, Prf, in relation to Pto kinase activity is probably crucial to our understanding of *R*-gene-mediated signalling. Prf is required both for Pto- and Fen-mediated plant responses. Although direct association between Prf and these recognition molecules has not been demonstrated, an attractive hypothesis would engage Prf in a recognition complex with either Pto-AvrPto or Fen-Fenthion, depending on the particular input signal (Van der Biezen and Jones, 1998a). In this model, Pto, possibly initially functioning as a basal resistance molecule, has been captured by 'VirPto' (destined to become 'AvrPto') as a virulence target in order to suppress resistance. The Pto-AvrPto complex has subsequently been recruited by Prf to subvert its virulence to an avirulence capacity. In this context, Prf might be considered a bona fide resistance protein and it follows that genetically defined NB-LRR type R proteins that are structurally similar to Prf might recruit a protein kinase component for pathogen recognition. This model is consistent with recent data that demonstrate enhanced plant cell death and pathogen resistance in transgenic tomato plants overexpressing Pto (Tang *et al.*, 1999). More intriguing, however, is the finding that elevated Prf expression leads to induction of SAR without concomitant plant cell death (Oldroyd and Staskawicz, 1998). Significantly, Rathjen *et al.* (1999) showed that a constitutive gain-of-function *pto* mutation requires *Prf* for elicitation of the HR, indicating that Prf must lie coincident or downstream of Pto in the resistance signalling pathway. Their data also suggest that the role of AvrPto may be limited to activation of Pto, which could then be independently engaged in Prf-dependent downstream signalling.

6.2.2 *Intriguing homologies in NB-LRR type R proteins*

The predominant structural class of plant *R* genes isolated so far encodes large putatively cytoplasmic proteins that possess a carboxy (C)-terminal LRR domain and a central NB site. These predicted 'NB-LRR' proteins can be further distinguished into two subclasses, based on possession of different amino (N)-terminals. Class I (so-called 'TIR-NB-LRR') proteins have an N-terminal domain, TIR, similar to the cytoplasmic domains of the *Drosophila* Toll and mammalian interleukin-1 (IL-1R) transmembrane receptors that transduce signals in evolutionarily conserved insect and vertebrate innate immunity pathways (Whitham *et al.*, 1994; Medzhitov *et al.*, 1997; Rock *et al.*, 1998; Hoffmann *et al.*, 1999). Class II (abbreviated to LZ-NB-LRR) proteins possess a predicted N-terminal leucine zipper (LZ) domain, again suggesting protein-protein interactions through promotion of coiled-coil (CC) structures (Bent, 1996; Hammond-Kosack and Jones, 1997).

Increasingly refined sequence and motif alignment searches are now revealing some intriguing homologies between NB-LRR type plant R proteins and regulators of apoptosis (programmed cell death) in animal cells (Van der Biezen and Jones, 1998b; Aravind *et al.*, 1999; see also chapter 7 of this volume). Again, apoptosis is controlled by an evolutionarily conserved programme that involves activation of a cysteine protease (caspase) cascade, leading to cell suicide. Critically, the apoptotic machinery is assembled through adaptor proteins, CED-4 in *Caenorhabditis elegans*, or its human counterpart Apaf-1, that are normally held in an inactive conformation by CED-9 or its human equivalent, Bcl2 family proteins, in a ternary complex called an 'apoptosome'. Extracellular death-inducing stimuli promote the release of CED-4/Apaf-1 and their consequent association and activation of caspases CED-3/pro-caspase-9 to initiate the cell death programme.

The related modular structures of predicted plant NB-LRR type R proteins and CED-4/Apaf-1 are discussed in detail by D. Jones in chapter 5 of this volume. Importantly, their occurrence suggests that particular protein domains may have been selected through evolution to perform related functions in plants and animals (Aravind *et al.*, 1999). By analogy with the WD-40 repeat domain of Apaf-1, stimulation of the plant R protein LRRs through protein-protein interaction may evoke a conformational activation of the N-terminal effector (TIR or LZ) portion of the protein, possibly by ATP/GTP hydrolysis within the NB domain (Van der Biezen and Jones, 1998b). Some support for this model comes from analyses of domain swaps between highly sequence-related but functionally distinct *L* alleles of the flax rust resistance locus (Ellis *et al.*, 1999). Here, as in previous *R* gene studies (Parniske *et al.*, 1997;

Botella *et al.*, 1998; McDowell *et al.*, 1998), the LRR domain was shown to be the probable determinant of Avr recognition specificity. However, certain *L* inter-allelic combinations did not give resistance, suggesting a requirement for cooperation between the LRR and TIR domains. It is now established that the intracellular 'TIR' domains of certain members of the IL-1R family of receptors in humans associate with a similar domain of corresponding adaptor proteins, e.g. human myeloid differentiation antigen 88 (MyD88), which are then able to interact with a kinase through their respective 'death domains', triggering, among other events, the translocation of Rel-related transcription factors, such as nuclear factor-κB (NF-κB), to the nucleus, where they induce transcription of immune response genes (Hoffmann *et al.*, 1999). Similarly, in Apaf-1 and CED-4, and the recently identified human caspase activation and recruitment domain (CARD)4 protein (Bertin *et al.*, 1999), the 'CARD' mediates homophyllic interactions with other CARD-containing signalling components, including caspases and, significantly, an NF-κB-regulating serine-threonine protein kinase. The presence of LRRs in the C-terminus of CARD4 reinforces the notion that mechanisitic similarities may exist in the regulatory core of programmes controlling animal apoptosis and plant disease resistance. It remains to be seen whether this is reflected in broader conservation of apoptotic machinery. The fact that no plant homologues of animal death-related genes have been isolated tends to suggest that plants may execute the HR and other forms of progammed cell death by a different means. However, observations of caspase-like activities (del Pozo and Lam, 1998; Solomon *et al.*, 1999) and dose-dependent HR-suppressing effects of CED-9 and a related mammalian apoptotic repressor, Bcl-X_L (Mitsuhara *et al.*, 1999), have stimulated a new wave of experiments to investigate the precise mechanisms underlying plant hypersensitive cell death.

6.3 Early cellular reprogramming

Much attention has been focused on examining the sequence of early, local cellular changes within the plant that are contingent on R-Avr protein recognition or the perception of pathogen-derived elicitors. The results of these studies have highlighted links between changes in plant cell ion concentrations and the activation of protein kinases and oxidative metabolism.

6.3.1 Ion fluxes and calcium homeostasis

Some of the most rapid, measurable changes in plant cell cultures upon stimulation with pathogen-derived elicitors, are large transient ion fluxes across the plasma membrane (Blumwald *et al.*, 1998). Effluxes of Cl^- and K^- ions accompanied by influxes of H^+ and Ca^{2+} were registered within minutes of elicitor application in several plant systems (Nürnberger *et al.*, 1994; Tavernier *et al.*, 1995; Atkinson *et al.*, 1996; Jabs *et al.*, 1997). The role of the various ion fluxes in plant-pathogen recognition is unclear. However, an increasing body of pharmacological and biochemical data drawn from analysis of different plant-pathogen interactions identify Ca^{2+} cellular homeostasis as a central modulating component of early plant defence responses. Analyses in plants are consistent with Ca^{2+} provision to the cytosol both from extracellular and internal stores (Sanders *et al.*, 1999). Omission of Ca^{2+} from the culture medium prevented elicitor inducible defence-related responses in soybean (Stäb and Ebel, 1987), parsley (Nürnberger *et al.*, 1994) and tobacco (Tavernier *et al.*, 1995; Piedras *et al.*, 1998) cell cultures, and in some cases this requirement was correlated with Ca^{2+} influxes during elicitation. Further experiments in parsley cell cultures treated with a defined fungal-derived oligopeptide elicitor, revealed a probable sequence of plasma-membrane-localized events in the transcriptional induction of defence-related genes and accumulation of phytoalexins (Jabs *et al.*, 1997). Receptor-mediated activation of ion channels, accompanied by increased Ca^{2+} uptake, were shown to precede an oxidative burst, which itself was necessary for downstream induction of defences. The role of Ca^{2+} in this plant defence response is reinforced by identification of an elicitor-responsive inward Ca^{2+} channel in parsley plasma membranes in patch clamp experiments (Zimmermann *et al.*, 1997). Elevation of cytosolic Ca^{2+} was also detected in transgenic tobacco cell cultures expressing the Ca^{2+}–sensitive bioluminescent protein, aequorin, after treament with various non-specific elicitors (Chandra and Low, 1997). Again, the kinetics of Ca^{2+} accumulation and applications of Ca^{2+} ion channel blockers placed Ca^{2+} upstream of an oxidative burst.

Studies in cultured cell systems have been instructive in defining a plausible sequence of early defence-related signalling events but provide few clues to the temporal and spatial relationships of ion movements within cells of the intact plant. Here, exciting advances with *in situ* analyses of intracellular ion accumulation in intact plant-pathogen interactions are harnessing aequorin-based luminescence and Ca^{2+} reporter dye monitoring techniques (Read *et al.*, 1993). Using a Ca^{2+}-specific fluorescent reporter dye in conjuction with confocal laser scanning microscopy, Xu and Heath (1998) showed an extended elevation of intracellular Ca^{2+} levels in epidermal cells of resistant but not susceptible cowpea plants after inoculation with basidiospores of cowpea rust fungus, *Uromyces vignae*. The spike of intracellular Ca^{2+} occurred as the fungus penetrated the cell wall and preceded elaboration of an HR. Consistent with a proposed role for Ca^{2+} in resistance to this obligate biotroph, pharmacological tests showed that blocking Ca^{2+} channel activity delayed the onset of HR. The aequorin-based detection system has been applied in intact *Arabidopsis* plants (M. Grant and co-workers, personal communication) to monitor changes in intracellular Ca^{2+} levels upon elicitation with *Pseudomonas syringae* expressing AvrRpm1. Here, a biphasic intracellular Ca^{2+} signature was registered, comprising an early initial non-specific increase (also observed in interactions with a compatible *P. syringae* strain or *Escherichia coli*) in cytosolic calcium within the first 10 min, and a subsequent RPM1-specific sustained increase beginning almost immediately after delivery of the AvrRpm1 protein. Thus, calcium may serve as an initial post-recognition switch controlling downstream events. The proposed proximity of Ca^{2+} channel fluxes to recognition events in this interaction may be pertinent to recent immunological observations that RPM1, the corresponding R protein that probably associates directly or indirectly with AvrRpm1, localizes in the plant cell as a peripheral membrane protein (Boyes *et al.*, 1998).

The mechanisms by which calcium exerts control of plant disease resistance signalling components remain to be resolved. Multiple roles of calcium in regulating different steps of animal signal transduction pathways suggest numerous possibilities for Ca^{2+} modulation of downstream targets in plants (Sanders *et al.*, 1999). In animal cells, calmodulin-dependent protein kinases and protein kinase C isoforms have be characterized as modulators of Ca^{2+}. In plants, it is possible that their role may be fulfilled by calcium-dependent protein kinases (CDPKs) (Sheen, 1996). Recently, the rapid activation of a 68-kDa CDPK isoform was detected after elicitation of *Cf9*-transgenic tobacco cell cultures with the corresponding Avr9 peptide from the fungal pathogen, *Cladosporium fulvum* (Romeis *et al.*, 2000). Activation involved a phosphorylation event that caused a shift in protein mobility from 68 kDa in unchallenged cells to 70 kDa in elicited cells. Significantly, pharmacological data suggested placement of the enzyme upstream of reactive oxygen intermediate (ROI) accumulation. In this respect, it is notable that an oxidative burst (see above and section 6.3.2) in several plant-pathogen interactions was shown to be Ca^{2+} and/or phosphorylation dependent.

Another potential direct or indirect target of a CDPK is the H^+-ATPase. Modulation of plasma membrane H^+-ATPase activity has been implicated as a possible molecular switch in several plant resistance responses (Blumwald *et al.*, 1998). Interestingly, activation of a tomato plasma membrane H^+-ATPase by *C. fulvum* race-specific elicitors appears to be mediated by a heterotrimeric G protein that may function through phosphorylation-dephosporylation reactions (Xing *et al.*, 1996, 1997). Schaller and Oecking (1999) demonstrated a potentially exquisite modulating capacity of a tomato plasma membrane H^+-ATPase in opposing wound and pathogen responses. In tomato cell cultures, application of systemin or inhibitors of H^+-ATPase activity (such as erythrosin B and vanadate) caused alkalization of the medium and induced a characteristic set of wound response genes. In constrast, activation of the H^+-ATPase by fusicoccin acidified the growth medium and suppressed induction of the wound response. Similarly, in whole tomato plants, fusicoccin treatment suppressed the wound response but resulted in accumulation of SA and expression of *PR* genes. The authors further showed that the wound-associated H^+-ATPase inhibition was dependent on Ca^{2+} influx and protein phosphorylation. Thus, even

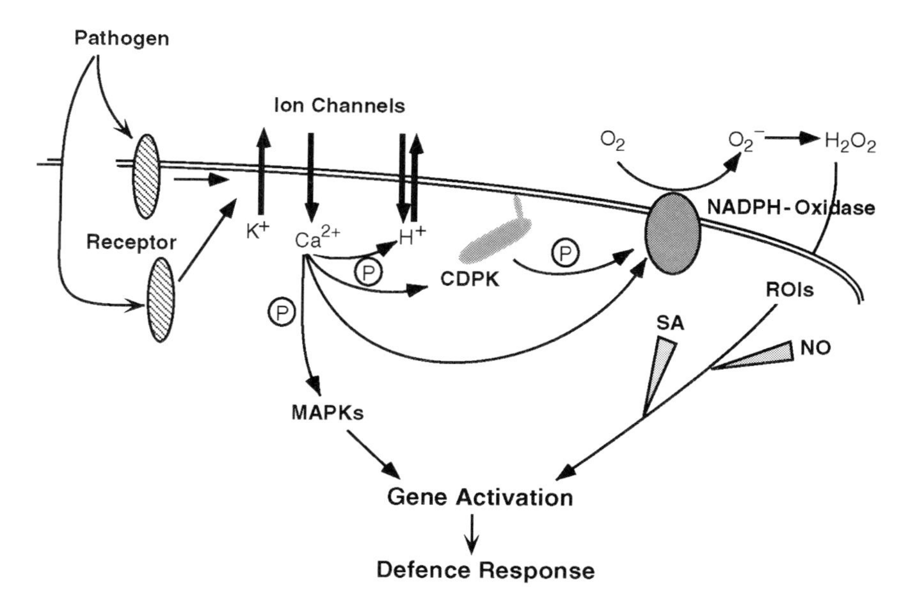

Figure 6.1 A scheme of early events in the plant cell in response to pathogen attack. In this model, recognition of the pathogen by a cytoplasmic or plasma-membrane-associated plant receptor leads initially to the activation of ion channels. Both Ca^{2+} influxes and phosphorylation events (P) appear to be necessary for generation of an oxidative burst. Ca^{2+} may exert an effect through stimulation of Ca^{2+}-dependent protein kinases (CDPK) or by interaction with other target proteins, such as the $gp91^{phox}$ subunit of a plasma membrane-localized NADPH-oxidase complex, which may contribute to generation of reactive oxygen intermediates (ROIs) at the cell surface. ROIs then cooperate with other signalling molecules, such as nitric oxide (NO) and salicylic acid (SA), to potentiate the plant defence response. Mitogen-activated protein kinase (MAPK) activation is placed downstream of ion fluxes and leads to defence gene induction by a process that is independent of the oxidative burst.

subtle alterations in the cellular partitioning of ions and pH gradients are likely to influence fundamental plant defence processes. A scheme depicting early plant cellular changes and the possible targets of calcium ions discussed in this section is presented in Figure 6.1.

6.3.2 *The oxidative burst*

Early generation of ROIs, most notably superoxide anions (O_2^-) and hydrogen peroxide (H_2O_2), is associated with many plant resistance responses (Lamb and Dixon, 1997). A massive oxidative burst at the plant cell surface is reminiscent of NADPH oxidase-mediated oxidative metabolism in primed mammalian neutrophils as part of the native immune response (Segal and Abo, 1993). Although the signalling roles of ROIs in plant defence or their relevance to elaboration of the HR are not entirely clear, an increasing number of studies point to a central position of ROIs in reinforcing the local containment zone at the site of pathogen ingress and as components of signal amplification in systemic resistance pathways (Lamb and Dixon, 1997). Moreover, the highly reactive oxidants, O_2^- and OH; may exert potent antimicrobial activity and contribute to lipid peroxidation of both plant and pathogen membranes. Analysis of ROI accumulation in cultured plant cells treated with avirulent or virulent bacteria revealed significant differences in the induction profiles of incompatible and compatible interactions (Levine *et al.*, 1994; Chandra *et al.*, 1996), suggesting physiological relevance of this oxidative reaction. Whilst a weak and transient (~1 h) accumulation of ROIs was common to both responses, a later more prolonged and extensive induction (~3–6 h) was specific to the incompatible interaction. Similar two-phase kinetics were observed in parsley cells treated with a fungal-derived peptide elicitor (Jabs *et al.*, 1997) and in *Cf9*-expressing tobacco cells responding specifically to *C. fulvum* Avr9 (Piedras *et al.*, 1998). A requirement for a second reinforcing signal to elicit a plant response may serve to prevent inadvertent cell damage from numerous benign stimuli that the plant undoubtedly encounters.

6.3.3 The origin and signalling role of ROIs

The origin of the pathogen-induced oxidative burst in plants is still controversial and ROIs may indeed be derived from multiple sources (Bolwell *et al.*, 1995; Allan and Fluhr, 1997; Martinez *et al.*, 1998). One anticipated mechanism of O_2^- production is by a plasma-membrane-localized NADPH oxidase complex. The mammalian neutrophil NADPH oxidase consists of two membrane proteins, $gp91^{phox}$ and $p22^{phox}$, that form an active complex with three cytosolic regulatory components, $p40^{phox}$, $p47^{phox}$ and $p67^{phox}$, upon stimulation. A small GTPase, Rac2, and other G proteins are also implicated in oxidase activation (Bokoch, 1994). Biochemical studies, including applications of the suicide substrate inhibitor of the neutrophil NADPH oxidase, diphenylene iodonium (DPI), implicate NADPH oxidase-based O_2^- generation as an essential component of resistance (Levine *et al.*, 1994; Jabs *et al.*, 1997; Piedras *et al.*, 1998). However, results of DPI-inhibition need to be interpreted cautiously, since its effects are not necessarily specific to NADPH oxidase but may influence other important enzyme systems, such as nitric oxide synthase (Stuehr *et al.*, 1991; see also section 6.3.4). What is apparent is that $gp91^{phox}$ homologues exist in plants that have been identified by sequence homology with mammalian counterparts (Groom *et al.*, 1996; Keller *et al.*, 1998; Torres *et al.*, 1998), and these provide targets for further functional analysis. Interestingly, structural and biochemical characterization of a putative 108-kDa $gp91^{phox}$ homologue in *Arabidopsis* reveals it to be a probable intrinsic plasma membrane protein (Keller *et al.*, 1998). The presence of an additional N-terminal portion containing two Ca^{2+}-binding 'EF' hand motifs and sequence similarity to a human

GTPase-activating protein, RanGAP1, highlights a possible mechanism for integration of early Ca^{2+} ion fluxes into the cytoplasm that have been shown to precede an oxidative burst (Figure 6.1; see also section 6.3.1).

Elucidating the precise roles of ROIs in plant disease resistance signalling remains an area of intense research. Pharmacological and biochemical studies utilizing cultured plant cells, although instructive with respect to fundamental cellular mechanisms, are unlikely to reveal the subtleties of local and systemic effects of ROIs and their associated signalling networks. Alvarez *et al.* (1998) demonstrated a capacity for the primary oxidative burst at the site of pathogen inoculation in *Arabidopsis* plants to induce systemic 'microbursts' that appear to be an essential component of systemic immunity. These and other studies on intact plants (Mur *et al.*, 1997) reinforce earlier work implicating ROIs as signals in the transcriptional activation of a subset of cellular protectant and defence genes (Levine *et al.*, 1994; Jabs *et al.*, 1997).

6.3.4 ROI, nitric oxide and SA cooperation in signal potentiation

In defence induction experiments with plant cell cultures, it was noted that physiological levels of ROIs were not sufficient to cause plant cell death (Glazener *et al.*, 1996; Jabs *et al.*, 1997), suggesting that other stimuli may be required to maximally elicit an HR. In mammalian macrophages, ROIs are known to cooperate with nitric oxide (NO) in the execution of bacterial pathogens (Nathan, 1995), and recent analyses suggest some exciting parallels in the plant defence response. Delledonne *et al.* (1998) and Durner *et al.* (1998) demonstrated in soybean and tobacco, respectively, that synthesis of NO was induced in an incompatible interaction but not in the compatible plant response. In soybean cells, detection of increased NO synthase activity was correlated with a measurable release of NO. Furthermore, Durner *et al.* (1998) showed that application of NO donors to tobacco induced expression of genes encoding pathogenesis-related protein-1 (PR1) and phenylanaline ammonia lyase (PAL); the latter being a key enzyme of the phenylpropanoid pathway and an anticipated component of salicylic acid biosynthesis (Mauch-Mani and Slusarenko, 1996). The induction of at least *PR1* appeared to be dependent on SA whereas PAL expression was SA-independent (Durner *et al.*, 1998). In the tobacco experiments, a combination of pharmacological data and enzyme assays implicated cyclic guanosine monophosphate (cGMP) as a second messenger in NO-mediated signalling. This further strengthens similarities with mammalian immune systems, in which NO is known to activate guanylate cyclase. Significantly, Delledonne *et al.* (1998) demonstrated a synergy between NO and ROI resulting in increased levels of soybean cell death. Cooperation was also observed between ROI and an SA-dependent process in the potentiation of the plant resistance response, extending earlier evidence of an SA-controlled amplification loop in plant defence against pathogens (Draper, 1997; Shirasu *et al.*, 1997).

Thus, a number of signalling elements are proposed that may constitute a cooperative and highly flexible signalling system, serving in the first instance to potentiate *R-avr* gene-dependent signals through ROI-NO collaboration, and secondly, to amplify the responses through the combined actions of ROI and SA (Figure 6.1).

The experiments outlined above provide an idea of the nature of possible plant resistance signalling molecules and the likely complexities of the informational networks involved. One of the many questions that remain to be answered is how, for example, ROIs exert their signalling effects. It is unlikely that they activate downstream defence genes directly. A more reasonable hypothesis is that they interact with other signalling components (NO and SA are prime candidates), possibly lowering the activation threshold of a signalling cascade. It is also vital for the plant's survival that the zone of plant cell death associated with restriction of the invading pathogen is delimited. Characterization of an expanding catalogue of mutations affecting plant resistance responses indeed reveals that induction and potentiation of plant defence systems, as well as the necessary negative controls to prevent runaway responses, are under strict genetic control (Heath, 1998; Richberg *et al.*, 1998; see also section 6.4).

6.3.5 Protein phosphorylation

The central role of the serine/threonine protein kinase, Pto, in specific recognition of AvrPto and consequent defence signalling was outlined in section 6.2. Evidence from *in vivo* phosphorylation and pharmacological experiments also suggests that protein phosphorylation cascades operate at several different levels in plant-pathogen interactions (Dietrich *et al.*, 1990; Felix *et al.*, 1994; Viard *et al.*, 1994), although is not clear which processes are instrumental in eliciting plant resistance.

6.3.5.1 MAP kinases

Mitogen-activated protein kinase (MAPK) signalling pathways are common to eukaryotes from yeast to humans and represent a pivotal regulatory apparatus for transducing external signals perceived by cell surface receptors to downstream cellular responses (Herskowitz, 1995; Machida *et al.*, 1997). They involve the activation of a serine/threonine MAPK enzyme through dual phosphorylation of specific tyrosine and threonine residues by an upstream MAPK. A large class of protein kinases with homology to animal MAPKs has been identified in various plant species and evidence points to the importance of MAPK cascades in plant responses to environmental stresses, such as wounding and drought, and to pathogen invasion (Hirt, 1997; Machida *et al.*, 1997).

In parsley cells, an elicitor-responsive MAPK (ERM kinase) was identified that acts downstream of elicitor-induced ion fluxes but upstream or independently of the oxidative burst (Ligterink *et al.*, 1997). Protein kinase activity was induced within 5 min of treatment with a specific oligopeptide elicitor, measured by an In-gel kinase assay using myelin basic protein as a substrate. The activity cross-reacted with specific polyclonal antisera corresponding to an alfalfa MAPK, MMK4, allowing subsequent cloning of the parsley ERM kinase complementary deoxyribonucleic acid (cDNA). The same antisera was used in immunofluorescence microscopy to establish that ERM kinase translocates to the nucleus upon elicitor treatment of cells. Since no nuclear localisation signals are detected in the ERM kinase sequence, it is envisaged that the kinase may associate with other protein(s) to direct movement. A plausible consequence of ERM kinase translocation to the nucleus may be phosphorylation of certain transcription factors that drive defence gene expression, and a number of candidate transcription factors have been identified (Rushton and Somssich, 1998).

A rather unexpected picture is now emerging from recent studies of plant MAPK in response to abiotic and biotic stresses. The Alfalfa MAPK, MMK4, was also shown to be associated with cold, drought and mechanical stresses (Bögre *et al.*, 1997). A SA-inducible tobacco MAPK, SIPK, originally identified by Zhang and Klessig (1997), is similarly stimulated in tobacco cell cultures by a carbohydrate cell wall elicitor and proteinaceous elicitins derived from *Phytophthora* spp. (Zhang *et al.*, 1998). Further studies established that wounding induced SIPK in tobacco (Zhang and Klessig, 1998), an activity that may previously have been attributed to a wound-induced MAPK, WIPK (Seo *et al.*, 1995), although activation of WIPK in wounded tobacco plants was confirmed in recent studies (Seo *et al.*, 1999). Furthermore, these authors demonstrated, in transgenic tobacco, a causative relationship between WIPK activity and accumulation of jasmonic acid, a central hormonal mediator of plant wounding. A convergence of signalling mechanisms leading to MAPK activation is further supported by studies in tomato using race-specific elicitation of plant defences (Romeis *et al.*, 1999). Here, *Cf9-avr9*-dependent activation of both SIPK and WIPK was detected. Pharmacological tests showed that stimulation of MAPK activity was dependent on Ca^{2+} influx and phosphorylation events, but not the oxidative burst. Also, extending previous experiments in elicitor-stimulated parsley cells (Ligterink *et al.*, 1997), a block in SIPK and WIPK activation did not abolish *Cf9-avr9*-dependent ROI production, suggesting that the MAPK pathway is likely to operate independently of the oxidative burst (see Figure 6.1).

It is apparent from these analyses that there is convergence between *R-avr*-gene-mediated resistance pathways and elicitor, wound and mechanical stress responses, at least at the level of MAPK activation. Many questions remain. MAPK activation requires tyrosine phosphorylation; how is this achieved? How is signal fidelity maintained in the various plant reactions? Following leads from studies of animal MAPK cascades (Madhani and Fink, 1998), it may be that signal specificity is imposed by selective recruitment of particular kinases into signalling complexes. Elicitation by different stimuli might also be registered in the cell by the extent and duration of MAPK activation rather than a simple on/off setting. Modes of transcriptional and/or post-translational regulation clearly differ in different plant responses and MAPK types. Romeis *et al.* (1999) have proposed that the multiplicity of MAPK inducers may reflect a function in early 'resetting' of the system after a stress stimulus rather than in the primary elicitation. A genetic approach, combining analysis of mutations, antisensed genes and overexpressing constructs may well help to resolve these outstanding questions.

6.4 Genetic dissection of disease resistance pathways

Molecular genetic approaches to the identification of components of plant disease resistance pathways through forward mutational screens, gene mapping and cloning, and genomic strategies, provide a powerful complement to the biochemical studies discussed in the previous section. Genetic dissection of resistance mechanisms immediately targets genes that have physiological importance in a particular defence response. Phenotypic analysis of mutant lines or combinations of mutants can then reveal crucial information about signalling hierarchies and pathway utilization. Ultimately, however, the precise function of genetically-defined components requires a full exploitation of molecular biology and biochemistry.

6.4.1 Components of R*-gene-mediated resistance*

Forward screens for mutations compromising resistance to a particular pathogen or creating defects in a specific plant defence response have identified genes that are essential for the full expression of *R*-gene-mediated resistance in a number of plant-pathogen interactions, as shown in Table 6.1. The most extensive mutational analyses have been performed in the model crucifer, *Arabidopsis thaliana*, as a tractable plant for genetics and gene cloning. Moreover, the anticipated completion of its ~120 Mb genome sequence by the year 2000, with associated inventory of gene and protein sequences, make it potentially the most informative plant system for unravelling signalling networks. Importantly, it is host to all the major pathogen classes: viruses, bacteria, fungi, oomycetes and nematodes, and in many of these interactions distinct *R* loci have been identified. *Arabidopsis* also responds appropriately to other stress stimuli, such as insect wounding and drought, providing a genetic framework for effective integration of information on different biotic and abiotic stress responses.

Pathology-based screens for suppression of *R*-gene-mediated resistance resulted in the isolation of: the *rcr1, rcr2* and *rcr3* (required for *Cladosporium* resistance) mutations in tomato; *rar1* and *rar2* (required for *Mla12* resistance) in barley; and the *ndr1* (non-specific disease resistance), *eds1* (enhanced disease susceptibility) and *pbs1, pbs2* and *pbs3* (*avrPphB* susceptible) mutations in

Table 6.1 Mutations suppressing plant *R*-gene-specified disease resistance

Mutation* (plant)	Structure of corresponding wild-type gene product	Affected *R* genes (pathogen)	Reference
Prf (tomato)	Putative cytoplasmic LZ-NB-LRR protein	*Pto* (*P. syringae*)	Salmeron *et al.*, 1994, 1996
rcr1, rcr2 (tomato)		*Cf9* (*C. fulvum*)**	Hammond-Kosack *et al.*, 1994
rcr3 (tomato)		*Cf2* (*C. fulvum*)	(J.D.G Jones, personal communication)
rar1, rar2 (barley)	Rar1: Zn-binding 'CHORD' protein	*Mla-6, Mla9, Mla-12, Mla-13, Mla-14, Mla-22 Mlat, Mlh, Mlk, Mlra* (*E. graminis* f.sp. *hordei*)	Jørgensen, 1996 Peterhänsel *et al.*, 1997 Shirasu *et al.*, 1999
ndr1 (*Arabidopsis*)	Novel protein with two putative membrane spanning domains	*RPS2, RPM1, RPS5* (*P. syringae*) *RPP4*** (*P. parasitica*)	Century *et al.*, 1995, 1997 Aarts *et al.*, 1998
eds1 (*Arabidopsis*)	Putative cytoplasmic protein with lipase motifs	*RPS4* (*P. syringae*) *RPP5, RPP1, RPP2 RPP4, RPP12, RPP21* (*P. parasitica*)	Parker *et al.*, 1996 Aarts *et al.*, 1998 Falk *et al.*, 1999
pad4 (*Arabidopsis*)	Putative cytoplasmic protein with EDS1-type lipase motifs	*RPS2* (*P. syringae*) *RPP2, RPP4, RPP19 RPP5***, *RPP21* (*P. parasitica*)	Glazebrook *et al.*, 1997 Zhou *et al.*, 1999
pad1, pad2, pad3 (*Arabidopsis*)	PAD3: cytochrome P450 monooxygenase	*RPP2, RPP4, RPP19*** (*P. parasitica*)	Glazebrook *et al.*, 1997 Jirage *et al.*, 1999
pbs1 (*Arabidopsis*)		*RPS5* (*P. syringae*)	Warren *et al.*, 1999
pbs2 (*Arabidopsis*)		*RPS5, RPS2, RPM1* (*P. syringae*) *RPP4, RPP6, RPP19*** (*P. parasitica*)	Warren *et al.*, 1999
pbs3 (*Arabidopsis*)		*RPS5, RPS2, RPM1, RPS4*** (*P. syringae*) *RPP4, RPP19*** (*P. parasitica*)	Warren *et al.*, 1999
npr1 (*nim1*) (*Arabidopsis*)	Ankyrin repeat protein	*RPP1, RPP2, RPP4, RPP12, RPP19*** (*P. parasitica*)	Delaney *et al.*, 1995 Cao *et al.*, 1997 Warren *et al.*, 1999

*All mutations shown are recessive; **Indicates partial effects on resistance; in the cases of *pad1, pad2* and *pad3*, an effect on resistance to *P. parasitica* was observed only in double mutant combinations.

Arabidopsis. A broader assessment of mutant phenotypes with different pathogen races or species established that the majority of these signalling or effector genes are utilized by multiple *R* genes (Table 6.1), suggesting a function in common pathways downstream of R-Avr protein-mediated recognition events. A rather different strategy was employed to identify *Arabidopsis pad* (phytoalexin accumulation deficient) loci, *pad1, pad2* and *pad3*, by screening for defects in pathogen-induced accumulation of the indole phytoalexin, camalexin (Glazebrook and Ausubel, 1994). Of these mutations, *pad2* and a subsequently isolated mutant, *pad4*, were shown to significantly increase susceptibility to a compatible *P. syringae* strain in a direct pathology screen (Glazebrook *et al.*, 1996). The latter screen also revealed alleles of *npr1* (non-expressor of *PR* genes), a mutation that was originally isolated in a search for defects in responses to SA (Cao *et al.*, 1994) or an active SA analogue, 2,6-dichloroisonicotinic acid (INA) (Delaney *et al.*, 1995).

It is notable that several of the mutations shown in Table 6.1 cause only a partial suppression or 'relaxation' of particular *R*-gene-mediated resistance responses. In some cases, this may be due to partially functional mutant alleles and this can only be resolved once the corresponding wild-type gene is cloned and the mutations molecularly characterized. In other cases, an intermediate resistance phenotype may reflect partial dependence of the *R-avr* gene specified pathway on the corresponding signalling component. Analysis of an *ndr1* deletion mutation (Century *et al.*, 1995, 1997) is a case in point. The *ndr1* plants exhibited a slight but quantitative suppression of *RPP4*-specified resistance to the oomycete pathogen, *Peronospora parasitica*, that can only be due to intermediate dependence of *RPP4* on *NDR1* function (Century *et al.*, 1995, 1997; Aarts *et al.*, 1998). In different *Arabidopsis*-pathogen interactions, *NDR1* is fully employed (see discussion below).

Phenotypic analyses of other plant defence signalling mutants are beginning to reveal defects in SA/ROI-associated 'potentiation' mechanisms that have been implicated in biochemical and pharmacological studies on plant cell cultures and whole plants (see section 6.3.4). As an example, the *Arabidopsis pad4* mutation compromises accumulation of SA and camalexin after inoculation with a *P. syringae* pv. *maculicola* strain and allows enhanced pathogen growth (Zhou *et al.*, 1998). The results are consistent with PAD4 regulatory function upstream of SA and camalexin biosynthesis in certain *Arabidopsis*-pathogen interactions. In another *Arabidopsis* genetic background, comparison of a null *pad4* mutant and wild-type plant responses to *P. parasitica* inoculation showed an inability of the mutant to consolidate the initial plant resistance response, resulting in a characteristic trailing necrotic reaction as the pathogen repeatedly overcomes the host defence response (J. Parker, unpublished data). In a different study, phenotypic analyses of the cellular responses of *rar1* alleles and wild-type barley to powdery mildew infection revealed two temporally and spatially distinct phases to the resistance response (Shirasu *et al.*, 1999) that possibly mirror aspects of the biphasic oxidative reactions of elicited plant cell cultures (Lamb and Dixon, 1997). Resistant wild-type epidermal cells responded to fungal penetration with an early accumulation of H_2O_2 (monitored by diaminobenzidine deposition) (Thordal-Christensen *et al.*, 1997), which was confined to the area of the cell wall apposition. This was followed by more pronounced whole cell H_2O_2 production that accompanies epidermal cell death. Subsequently, cell death was observed (by lactophenol-trypan blue staining) in small clusters of mesophyll cells subtending individual penetrated epidermal cells, suggesting the emission of potentiating signals from the epidermis to the mesophyll to reinforce the primary resistance response. In the mutant alleles, *rar1-1* and *rar1-2*, the initial restricted H_2O_2 spike was maintained but the subsequent whole epidermal cell response severely diminished and mesophyll cell death almost abolished. Thus, both the *Arabidopsis* PAD4 and barley Rar1 proteins define functions that are necessary for reinforcement of local, and possibly systemic, resistance processes, leading to effective containment of the pathogen upon *R-avr* gene-mediated recognition.

It is also apparent from various studies that mutations such as *eds1, npr1* and *pad4*, suppressing or partially suppressing *R*-gene-conditioned responses, cause enhanced disease susceptibility in compatible *Arabidopsis*-pathogen interactions (Glazebrook *et al.*, 1997; Aarts *et al.*, 1998). Therefore, it is likely that some commonality exists between mechanisms evoked during *R-avr* gene-mediated resistance (normally associated with the HR) and the less well-defined restriction of compatible pathogens in disease. These mutations add to the expanding battery of '*eds*' mutant loci identified in *Arabidopsis*, demonstrating that growth of virulent pathogens is actively limited by host plant defences that are not necessarily associated with the HR (Rogers and Ausubel, 1997; Reuber *et al.*, 1998). Indeed, the importance of the HR in containment of invading pathogens has been scrutinized in a number of plant-pathogen interactions, and there are numerous examples of effective resistance in the absence of plant cell death (Morel and Dangl, 1997). In barley, *Mlg*-mediated restriction of powdery mildew fungal isolates is accomplished by formation of cell wall appositions well in advance of the plant HR (Görg *et al.*, 1993). In *Arabidopsis*, the *dnd1* mutation allows the plant to circumvent the HR whilst maintaining *R*-gene-specified resistance to *Pseudomonas syringae* (Yu *et al.*, 1998). In potato, *Rx*-specified resistance to potato virus X occurs in the absence of plant cell necrosis (Köhm *et al.*, 1993; Bendahmane *et al.*, 1999). Strikingly, this extreme resistance is epistatic to a plant HR that is normally elicited in *N*-gene-specified resistance to tobacco mosaic virus, suggesting a separation between processes leading to pathogen arrest and cell death. However, the plant can be driven into an *Rx*-dependent HR through overexpression of the potato virus X Avr determinant encoded within its coat protein (Bendahmane *et al.*, 1999). Altogether, these findings are consistent with the idea that the plant HR is a final manifestation of an incremental series of plant defences. It may well represent the 'coup de grâce' for any biotrophic pathogen that breeches a particular threshold of HR-induction, depriving it of nutrients and further means of colonization. In this model, a number of key host responses, such as protein phosphorylation or transcriptional activation of defence genes, could be induced without elicitation of the HR.

6.4.2 Definition of resistance signalling pathways

Comparison of barley and *Arabidopsis* wild-type and mutant defence responses to a range of pathogen stimuli have provided important clues to the utilization of particular signalling circuits, leading to the definition of distinct *R*-gene-specified signalling pathways.

In the barley-powdery mildew interaction, the *rar1* and *rar2* mutations define a pathway conditioned by *Mla12* and several other *R* loci that is genetically and phenotypically separate from resistance mediated by *Mlg* or *mlo* (Peterhänsel *et al.*, 1997). In *Arabidopsis*, mutations in *EDS1* fully suppress resistance conditioned by several *RPP* genes against the obligate biotroph, *P. parasitica*, as well as *RPS4*- but not *RPM1*-, *RPS2*- or *RPS5*-specified resistance to *P. syringae* (Parker *et al.*, 1996; Aarts *et al.*, 1998). Resistance could be rescued in *eds1* plants by application of SA, allowing placement of wild-type *EDS1* upstream of SA-dependent processes that are, at least in part, defined by the *NPR1* gene (Cao *et al.*, 1994, 1997). In contrast to *eds1*, the *ndr1* mutation strongly suppressed resistance conferred by *RPM1, RPS2* and *RPS5* but had a negligible effect on *EDS1*-dependent resistance responses. Moreover, utilisation of *EDS1* or *NDR1* was generally correlated with the predicted R protein structural type in the cases where these were cloned (Aarts *et al.*, 1998). Thus, TIR-NB-LRR type R proteins (see section 6.2.2) were strongly dependent on *EDS1*, whereas most LZ-NB-LRR type R proteins were dependent on *NDR1*. It was concluded from these studies that certain disease resistance pathways are likely to be governed by R protein structural class rather than pathogen type. One exception to this scheme was *RPP8* encoding a LZ-NB-LRR type protein (McDowell *et al.*, 1998) that exhibited no unique requirement for either *EDS1* or *NDR1* (Aarts *et al.*, 1998). This suggests the existence of a third resistance pathway. It also points to a divergence of signalling requirements between LZ-NB-LRR type R protein members, consistent with previous observations of differences in early defence gene induction profiles specified by two LZ-NB-LRR type R genes, *RPM1* and *RPS2*, in response to *P. syringae* infection (Reuber and Ausubel, 1996; Ritter and Dangl, 1996). Analyses of *RPP7*-specified resistance to *P. parasitica* also suggest a different resistance mechanism that appears to be independent of existing signalling components characterized so far (Warren *et al.*, 1999).

A model for *R* gene signalling preferences in *Arabidopsis* is illustrated in Figure 6.2. Incorporated here are the anticipated positions of three recently identified signalling components, *PBS1, PBS2* and *PBS3*, inferred from the responses of corresponding mutant lines to a range of *P. syringae* and

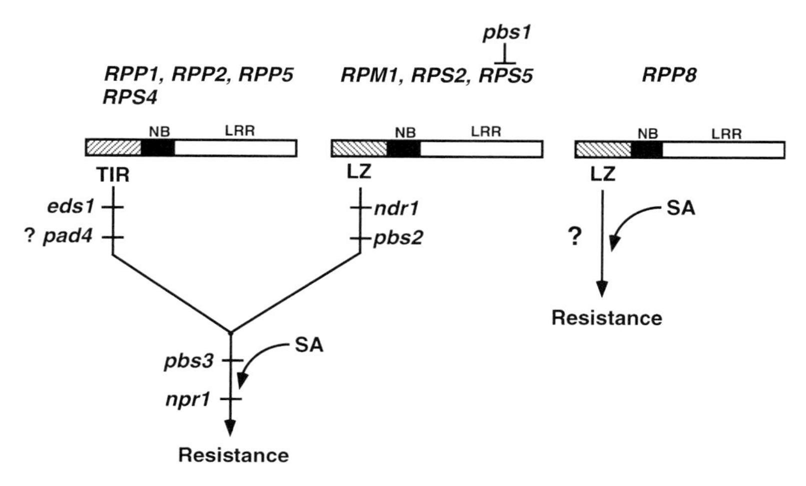

Figure 6.2 *R*-gene-mediated disease resistance pathways in *Arabidopsis*. Analysis of various mutations compromising *R*-gene-mediated responses reveals particular *R* gene signalling preferences. Whereas the *eds1* mutation defines a pathway conditioned by *R* genes encoding TIR-NB-LRR type proteins, *ndr1* and *pbs2* suppress resistance specified by most R proteins of the LZ-NB-LRR type. Analysis of *eds1* and *pad4* plants places their wild-type proteins upstream of the signalling molecule, SA. Perception of SA is, at least in part, mediated by *NPR1*. An exception to this scheme is *RPP8* encoding a LZ-NB-LRR protein that exhibits no unique requirement for *eds1* or *ndr1*, suggesting the existence of a third signalling pathway. *RPP8*-mediated resistance appears to be partially dependent on salicylic acid (SA) but not on *NPR1*. The relative positions of signalling components within each pathway are not known and the placement of *NDR1*, *PBS2* and *PBS3* upstream of SA accumulation is speculative (see text for more details). Abbreviations: *R*-gene, resistance gene; *eds*, enhanced disease susceptibility; TIR, Toll/interleukin-1 (IL-1) receptor; NB, nucleotide binding; LRR, leucine-rich repeat; *ndr*, non-specific disease resistance; *pbs*, avrPphB susceptible; LZ, leucine zipper.

P. parasitica isolates (Warren *et al.*, 1999). Interestingly, *pbs1* specifically affects *RPS5*-mediated resistance, highlighting the existence of highly discriminatory *R*-gene-associated molecules that may possibly interact with the R-Avr protein in a plant-pathogen recognition complex. The *pbs2* mutation affects the same spectrum of *R* genes as *ndr1*, whereas *pbs3* partially attenuates resistance conferred by both LZ- and TIR-type NB-LRR R genes, suggesting a function downstream of a point of convergence of distinct *R*-gene-specified pathways. The extent to which these pathways lead to separate downstream defences or converge to evoke common downstream events is unclear. The latter scenario is likely in a number of responses, as illustrated in Figure 6.2. For example, *RPS2*-mediated resistance to *P. syringae avrRpt2* does not require *PAD4* but still relies on accumulation of SA (Delaney *et al.*, 1994; Zhou *et al.*, 1998). Also, *RPP8*-mediated resistance to *P. parasitica* operates independently of the SA-response regulator, *NPR1* (G. Rairdan and T. Delaney, personal communication). In this context, NPR1 is emerging as a rather more versatile modulator of downstream plant defences than previously envisaged, as the complexities of interconnecting resistance pathways are revealed (see section 6.4.3).

Further insights into the mode of expression and function of resistance signalling components defined by mutations depend on cloning their respective wild-type genes and several recent advances have been made on this front. Barley *Rar1* was isolated using a positional cloning approach and found to encode a novel, small (25 kDa), putatively cytoplasmic protein containing two 60 amino acid cysteine and histidine rich domains, designated 'CHORD' (Shirasu *et al.*, 1999). A striking feature is a conserved tandem arrangement of two CHORD domains in *Rar1* homologues present in all eukaryotes examined except yeast. Shirasu *et al.* (1999) propose a signalling rather than effector function for Rar1, since it operates upstream of whole cell H_2O_2 accumulation in pathogen-attacked epidermal cells, possibly corresponding to the second potentiating phase of an oxidative burst that precedes mesophyll cell death (see sections 6.3.4 and 6.4.1). Interestingly, the CHORD I and II sequences that define a novel eukaryotic Zn^{2+}-binding motif are more highly related with corresponding domains I and II between phyla than between domains I and II in individual proteins, suggesting that each domain has evolved separately to perform a distinct function. RNA interference analyses of the single *C. elegans* CHORD homologue, *CHP*, caused embryo lethality and semi-sterility of the adult hermaphrodites, pointing to perturbation of vital processes in germ line development and embryogenesis. Whether there is a functional correlation between these developmental phenotypes and failures in an *R*-gene-mediated plant cell death programme of *rar1* plants is unknown. However, the fact that the two *rar1* alleles, *rar1-1* and *rar1-2*, obtained after exhaustive pathology screens, are incomplete gene knockouts, suggests that Rar1 wild-type protein may have a crucial cellular role besides being recruited to a disease resistance pathway.

A rather different signalling scenario is envisaged based on the predicted products of the *Arabidopsis EDS1* and *PAD4* genes. Both genes encode proteins that have no obvious homologies in the databases except over discrete blocks of amino acids that comprise the catalytic site of eukaryotic lipases (Falk *et al.*, 1999; Jirage *et al.*, 1999). The lipases belong to a class of serine hydrolases that utilize a triad of the highly-conserved amino acids, serine, aspartic acid and histidine, in the hydrolytic cleavage of ester bonds. During evolution, the serine hydrolase conformation has been recruited several times independently to derive distinct hydrolytic enzymes (Schrag and Cygler, 1997), and therefore it is not clear whether EDS1 and PAD4 cleave a lipid or a non-lipid substrate. Further experiments will be needed to elucidate their anticipated enzymatic activities *in vivo* and ascribe a more precise function to these signalling components. The recruitment of a lipase domain in EDS1 and PAD4 does, however, raise exciting possibilities in plant fatty acid signalling. One candidate lipid signalling molecule that is an essential component of plant wound responses and resistance in certain plant-pathogen combinations is jasmonic acid, a 12-carbon cyclic fatty acid synthesized from linoleic acid by the octadecanoid pathway (see section 6.4.3). However, applications of JA failed to rescue disease resistance in *eds1* plants, whereas similar treament with SA did, suggesting that if EDS1 is a lipase it must process a different lipid metabolite leading to SA-dependent *PR1* gene induction (Falk *et al.*, 1999). Increasingly refined methods for fatty acid profiling and detection of lipid peroxidation products are already revealing other potential lipid-based signals in plant cells (Farmer *et al.*, 1998).

6.4.3 Interplay of low molecular weight signalling molecules

The phenolic signalling molecule, SA, has a central role in efficient generation of the local HR in many *R*-gene-specified resistance responses and in establishment of SAR (Gaffney *et al.*, 1993; Delaney *et al.*, 1994; Mur *et al.*, 1997). More recently, however, the plant hormones, JA, and the gaseous hydrocarbon, ethylene, are emerging as important modulators of certain plant-pathogen interactions, in addition to their established activities in the plant wound response (Reymond and Farmer, 1998; Pieterse and van Loon, 1999). The existence of *Arabidopsis* mutants deficient in SA, JA and ethylene biosynthesis or perception, and lines genetically engineered to deplete SA has been valuable in assessing the relative contributions of these diverse signalling molecules in defence pathways. Analyses have also been greatly assisted by identification of diagnostic pathogenesis-related (*PR*) gene markers. SA accumulation and activity almost invariably precedes expression of *PR-1* transcripts (Cao *et al.*, 1994; Delaney *et al.*, 1994; Penninckx *et al.*, 1996). In contrast, JA- and ethylene-mediated defence responses are associated with induced expression of an antifungal plant defensin, PDF1.2 (Penninckx *et al.*, 1996), and a thionin, Thi2.1 (Epple *et al.*, 1995).

Increased susceptibility to the fungal necrotroph, *Pythium*, was found in the fatty acid desaturase triple mutant line (*fad3-fad7-fad8*) depleted in JA accumulation and in *coi1* mutant plants that fail to respond to JA (Vijayan *et al.*, 1998). Wild-type levels of resistance could be restored by exogenous applications of methyl jasmonate (MeJA) in the triple *fad* but not the *coi1* mutant lines, indicating that JA promoted resistance through a signalling chain rather than by direct antifungal action. Staswick *et al.* (1998) similarly encountered enhanced susceptibility to *Pythium* infection in *jar1* mutants that exhibit reduced sensitivity to JA. Resistance to another necrotrophic fungal pathogen, *Alternaria brassicicola*, was found to be dependent on JA and not SA or *NPR1* (Thomma *et al.*, 1998), and the JA responsiveness was associated with local and systemic induction of *PDF1.2* but not *PR1* (Penninckx *et al.*, 1996, 1998). Interestingly, the plant defence process requires concomitant action of JA and ethylene, mirroring certain aspects of JA-ethylene synergism observed previously in the wound response of tomatoes (O'Donnell *et al.*, 1996). Further studies (Thomma *et al.*, 1998, 1999) have established that the camalexin biosynthetic mutant, *pad3*, significantly attenuated resistance to *Alternaria brassicicola*, although it was previously shown not to compromise resistance to *P. syringae* or *P. parasitica* (Glazebrook and Ausubel, 1994; Glazebrook *et al.*, 1997). Camalexin exhibits antifungal activity against *A. brassicicola*, suggesting that it may be an effective defence molecule against this pathogen. However, whilst application of JA failed to induce camalexin accumulation in wild-type or *pad3* plants, resistance to *A. brassicicola* was enhanced, pointing to the existence of a JA-dependent, camalexin-independent pathway in this particular plant response (Thomma *et al.*, 1999). The *PAD3* gene was recently cloned (Zhou *et al.*, 1999) and, consistent with its anticipated biosynthetic role, shown to encode a protein with homology to maize cytochrome P450 monooxygenases that directs biosynthesis of indole-based compounds, DIMBOA (2,4-dihydroxy-7-methoxy-1,4-benzoxaxin-3-one) and DIBOA (2,4-dihydroxy-1,4-benzoxaxin-3-one), implicated in disease resistance (Frey *et al.*, 1997).

A different form of systemic disease resistance is elicited by root-colonizing, non-pathogenic rhizobacteria that is referred to as induced systemic resistance (ISR) (Pieterse *et al.*, 1998). As in the *Alternaria*-triggered response, this mechanism requires JA and ethylene and is independent of SA. However, in contrast to the interaction with *Alternaria*, effective resistance appears to engage JA and ethylene sequentially and requires functional NPR1. The NPR1-mediated process is not associated with transcriptional induction of JA and ethylene response genes, suggesting that resistance may be conferred by increased sensitivity to these signalling molecules, as was implicated in rice plants infected with the fungal pathogen, *Magnaporthe grisea* (Schweizer *et al.*, 1997). Alternatively, accumulation of JA and ethylene is below the threshold needed to induce expression of these marker genes. These data are significant, not least because they reveal a capacity for *NPR1* to differentially regulate downstream processes. They also extend earlier studies that have dissected *NPR1* function in relation to SA and JA utilization in SAR constitutively expressing mutants, *cpr5* (Bowling *et al.*, 1997), *cpr6* (Clarke *et al.*, 1998) and *ssi1* (Shah *et al.*, 1999), that are upregulated in both SA- and JA-related defence gene expression.

These and other '*cpr*' type mutations probably reflect the existence of multiple negative or resistance 'dampening' circuits that may be brought into play during particular plant resistance responses (Dangl *et al.*, 1996; Richberg *et al.*, 1998). The recessive mutation, *cpr5*, enhances resistance to *P. syringae* and *P. parasitica*. In *cpr5*, depletion of SA or mutations in *NPR1* suppressed *PR1* expression and bacterial resistance but did not diminish *PDF1.2* messenger ribonucleic acid (mRNA) levels or the heightened resistance to *P. parasitica*, revealing a divergence between *NPR1*-dependent and *NPR1*-independent processes. In contrast, the dominant mutations, *cpr6* and *ssi1*, appear to uncouple NPR1 activity from SA and invoke NPR1 function in regulating the JA-response pathway. Interestingly, *cpr6*-induced *PR1* gene expression requires SA but not NPR1. However, NPR1 is necessary for bacterial resistance in a mechanism that must logically be independent of SA/PR1. The *ssi1* mutation completely bypasses *npr1* but depends on SA to induce both *PR1* and *PDF1.2* expression, suggesting that the wild-type Ssi1 protein may participate in signal communication betweeen SA and JA-dependent pathways. For a more detailed description see chapter 8 in this volume.

The capacity of NPR1 to participate in both SA-dependent and SA-independent processes points to a central signal modulating role in different systemic resistance pathways that is clearly influenced by the nature of the input signal, as shown in Figure 6.3. How this role is achieved is unclear but it might rely on selective association with other signalling components after sensing a particular pathogen-derived stimulus. Some clues to NPR1 function derive from characterization of the predicted protein structure, its localization, and potential associations within the plant cell. *NPR1* encodes a protein with ankyrin repeats that have been shown to mediate protein-protein interactions in several prokaryotic and eukaryotic proteins (Cao *et al.*, 1997). Recent studies reveal that NPR1 interacts in a yeast two-hybrid assay with two transcription factors of the basic leucine zipper family (Zhang *et al.*, 1999). Specificity of

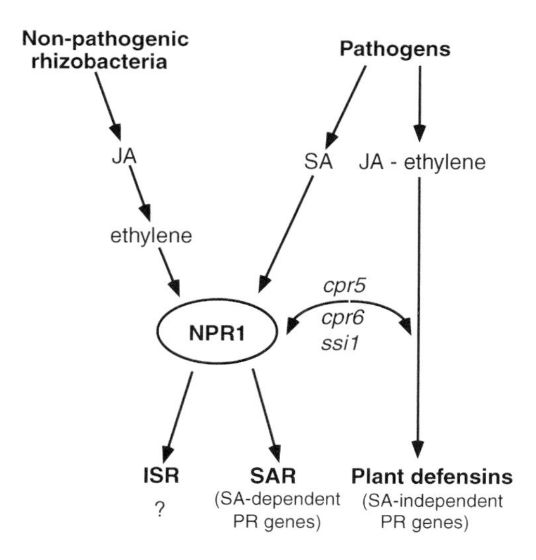

Figure 6.3 Involvement of NPR1 in systemic disease resistance pathways in *Arabidopsis*. A number of resistance pathways (SAR, ISR, and plant defensin induction) have been defined in *Arabidopsis* that have different requirements for the signalling molecules salicylic acid (SA), jasmonic acid (JA) and ethylene and lead to the induction of different sets of pathogenesis-related (*PR*) genes. Recruitment of NPR1 to these pathways appears to depend on the nature of the input signal and evidence points to NPR1 functioning as signal modulator, determining a particular signal output. Analysis of several resistance-upregulating mutations (*cpr5*, *cpr6* and *ssi1*) also points to cross-talk between SA- and JA/ethylene-responsive processes. Non-pathogenic rhizobacteria induce a JA/ethylene-dependent systemic response (ISR) that requires functional NPR1 but induces an as yet unknown set of downstream events (see text for more details). Abbreviations: SAR, systemic acquired resistance; ISR, induced systemic resistance.

this interaction was demonstrated using two *npr1* single amino acid exchange mutations that both destroy the protein interactions and NPR1 activity in the plant. Interestingly, gel mobility shift assays showed that one of the transcription factors, AHBP-1b, specifically bound an SA-responsive promoter element of the *Arabidopsis PR1* gene, suggesting a direct link between NPR1 activity and downstream *PR1* activation. This model is supported by cellular localization studies of NPR1-reporter fusions, in which NPR1 protein becomes mobilized from the cytoplasm to the nucleus after pathogen or SA treatment (Mark Kinkema and Xinnian Dong, personal communication).

What is the extent of positive and negative cross-talk between SA and JA signalling pathways? Several studies have demonstrated a clear antagonism between SA-associated plant-pathogen resistance and JA-mediated wound responses to insect feeding (Doherty *et al.*, 1988; Doares *et al.*, 1995; Felton *et al.*, 1999). Furthermore, mutual antagonism was observed between SA-dependent *PR* gene expression and defence responses induced by *Erwinia carotovora* derived elicitors in tobacco plants (Vidal *et al.*, 1997). It may be that opposing or synergistic interactions between SA and JA/ethylene signals reflect the plant's ability to prioritize responses very effectively through subtle differences in the kinetics of accumulation or distribution of particular signalling species. Reymond and Farmer (1998) illustrate this idea very appropriately in their 'turnable dial' model, in which different pathogen-derived input signals are responded to by fine tuning of the SA/JA/ethylene balance. Certainly, detection of JA and SA very early in tobacco plant cells undergoing an HR in response to *P. syringae* infiltration highlights opportunities for these molecules and their associated pathway components to interplay at an early stage of plant defence (Kenton *et al.*, 1999). This is an area primed for further investigation as more mutants are characterized in forward and reverse genetic screens and as the biochemical complexities of lipid-derived signals become resolved (Farmer *et al.*, 1998; Rancé *et al.*, 1998; Sanz *et al.*, 1998).

6.5 Concluding remarks

The last few years have witnessed some key advances in our understanding of plant defence signalling and an appreciation of the complexities of resistance pathway interactions. Based on the most recent biochemical and molecular genetic data described in this chapter, it seems most appropriate to view signalling in plant defence as a 'lattice' of interconnecting circuits in which pathway fluxes are constantly gauged and thresholds established to determine a suitable defence response. This may enable the plant to engage mechanisms that are effective against a particular pathogen type whilst suppressing inappropriate responses, and therefore conserving energy for the next encounter. It is clear that further progress in understanding these processes will rely on a full integration of complementary scientific approaches. Reverse genetics strategies should, for example, provide a means to address the physiological relevance of a number of pharmacologically resolved components or genes identified in genomics-based studies, such as the ion channel components, NADPH oxidase complexes or MAPK described here. Comprehension of the key regulatory mechanisms determining plant-pathogen recognition and defence pathway utilization will doubtless furnish plant breeders with increasingly refined methods of creating disease-resistant plants.

Acknowledgements

The author thanks all colleagues who have provided data prior to publication, and is also grateful to Erik Van der Biezen, Tina Romeis and Bart Feys at The Sainsbury Laboratory for helpful discussions. Work at The Sainsbury Laboratory is supported by The Gatsby Charitable Foundation.

References

Aarts, N., Metz, M., Holub, E., Staskawicz, B.J., Daniels, M.J. and Parker, J.E. (1998) Different requirements for *EDS1* and *NDR1* by disease resistance genes define at least two *R* gene-mediated signaling pathways in *Arabidopsis. Proc. Natl. Acad. Sci. USA*, **95** 10306-11.

Allan, A.C. and Fluhr, R. (1997) Two distinct sources of elicited reactive oxygen species in tobacco epidermal cells. *Plant Cell*, **9** 1559-72.

Alvarez, M.E., Pennell, R.I., Meijer, P.J., Ishikawa, A., Dixon, R.A. and Lamb, C. (1998) Reactive oxygen intermediates mediate a systemic signal network in the establishment of plant immunity. *Cell*, **92** 773-84.

Aravind, L., Dixit, V.M. and Koonin, E.V. (1999) The domains of death: evolution of the apoptosis machinery. *Trends Biochem. Sci.*, **24** 47-53.

Atkinson, M.M., Midland, S.L., Sims, J.J. and Keen, N.T. (1996) Syringolide 1 triggers Ca^{2+} influx, K^{+} efflux and extracellular alkalization in soybean cells carrying the disease-resistance gene, *Rpg4*. *Plant Physiol.*, **112** 297-302.

Bendahmane, A., Kanyuka, K. and Baulcombe, D.C. (1999) The *Rx* gene from potato controls separate virus resistance and cell death responses. *Plant Cell*, **11** 781-91.

Bent, A.F. (1996) Function meets structure in the study of plant disease resistance genes. *Plant Cell*, **8** 1757-71.

Bertin, J., Nir, W.J., Fischer, C.M., Tayber, O.V., Errada, P.R., Grant, J.R., Keilty, J.J., Gosselin, M.L., Robison, K.E., Wong, G.H.W., Glucksmann, M.A. and DiStefano, P.S. (1999) Human CARD4 protein is a novel CED-4/Apaf-1 cell death family member that activates NF-κB. *J. Biol. Chem.*, **274** 12955-58.

Blumwald, E., Aharon, G.S. and Lam, B.C.H. (1998) Early signal transduction pathways in plant-pathogen interactions. *Trends Plant Sci.*, **3** 342-46.

Bögre, L., Ligterink, W., Meskiene, I., Barker, P.J., HeberleBors, E., Huskisson, N.S. and Hirt, H. (1997) Wounding induces the rapid and transient activation of a specific MAP kinase pathway. *Plant Cell*, **9** 75-83.

Bokoch, G.M. (1994) Regulation of the human neutrophil NADPH oxidase by the Rac GTP-binding proteins. *Curr. Opin. Cell Biol.*, **6** 212-18.

Bolwell, G.P., Butt, V.S., Davies, D.R. and Zimmerlin, A. (1995) The origin of the oxidative burst in plants. *Free Rad. Res.*, **23** 517-32.

Botella, M.A., Parker, J.E., Frost, L.N., Bittner-Eddy, P.D., Beynon, J.L., Daniels, M.J., Holub, E.B. and Jones, J.D.G. (1998) Three genes of the *Arabidopsis RPP1* complex resistance locus recognize distinct *Peronospora parasitica* avirulence determinants. *Plant Cell*, **10** 1847-60.

Bowling, S.A., Clarke, J.D., Liu, Y.D., Klessig, D.F. and Dong, X.N. (1997) The *cpr5* mutant of *Arabidopsis* expresses both *NPR1*-dependent and *NPR1*-independent resistance. *Plant Cell*, **9** 1573-84.

Boyes, D.C., Nam, J. and Dangl, J.L. (1998) The *Arabidopsis thaliana RPM1* disease resistance gene product is a peripheral plasma membrane protein that is degraded coincident with the hypersensitive response. *Proc. Natl. Acad. Sci. USA*, **95** 15849-54.

Cao, H., Bowling, S.A., Gordon, S. and Dong, X. (1994) Characterization of an *Arabidopsis* mutant that is nonresponsive to inducers of systemic-acquired resistance. *Plant Cell*, **6** 1583-92.

Cao, H., Glazebrook, J., Clarke, J.D., Volko, S. and Dong, X. (1997) The *Arabidopsis NPR1* gene that controls systemic-acquired resistance encodes a novel protein containing ankyrin repeats. *Cell*, **88** 57-63.

Century, K.S., Holub, E.B. and Staskawicz, B.J. (1995) *NDR1*, a locus of *Arabidopsis thaliana* that is required for disease resistance to both a bacterial and a fungal pathogen. *Proc. Natl. Acad. Sci. USA*, **92** 6597-601.

Century, K.S., Shapiro, A.D., Repetti, P.P., Dahlbeck, D., Holub, E. and Staskawicz, B.J. (1997) *NDR1*, a pathogen-induced component required for *Arabidopsis* disease resistance. *Science*, **278** 1963-65.

Chandra, S. and Low, P.S. (1997) Measurement of Ca^{2+} fluxes during elicitation of the oxidative burst in aequorin-transformed tobacco cells. *J. Biol. Chem.*, **272** 28274-80.

Chandra, S., Martin, G.B. and Low, P.S. (1996) The Pto kinase mediates a signaling pathway leading to the oxidative burst in tomato. *Proc. Natl. Acad. Sci. USA*, **93** 13393-97.

Clarke, J.D., Liu, Y.D., Klessig, D.F. and Dong, X.N. (1998) Uncoupling *PR* gene expression from *NPR1* and bacterial resistance: characterization of the dominant *Arabidopsis cpr6-1* mutant. *Plant Cell*, **10** 557-69.

Dangl, J.L., Dietrich, R.A. and Richberg, M.H. (1996) Death don't have no mercy: cell death programs in plant-microbe interactions. *Plant Cell*, **8** 1793-807.

del Pozo, O. and Lam, E. (1998) Caspases and programmed cell death in the hypersensitive response of plants to pathogens. *Curr. Biol.*, **8** R896.

Delaney, T.P., Uknes, S., Vernooij, B., Friedrich, L., Weymann, K., Negrotto, D., Gaffney, T., Gut-Rella, M., Kessmann, H., Ward, E. and Ryals, J. (1994) A central role of salicylic acid in plant disease resistance. *Science*, **266** 1247-50.

Delaney, T.P., Friedrich, L. and Ryals, J.A. (1995) *Arabidopsis* signal transduction mutant defective in chemically and biologically induced disease resistance. *Proc. Natl. Acad. Sci. USA*, **92** 6602-606.

Delledonne, M., Xia, Y.J., Dixon, R.A. and Lamb, C. (1998) Nitric oxide functions as a signal in plant disease resistance. *Nature*, **394** 585-88.

Dietrich, A., Mayer, J.E. and Hahlbrock, K. (1990) Fungal elicitor triggers rapid, transient and specific protein phosphorylation in parsley cell suspension cultures. *J. Biol. Chem.*, **265** 6360-68.

Doares, S.H., Narvaezvasquez, J., Conconi, A. and Ryan, C.A. (1995) Salicylic acid inhibits synthesis of proteinase inhibitors in tomato leaves induced by systemin and jasmonic acid. *Plant Physiol.*, **108** 1741-46.

Doherty, H.M., Selvendran, R.R. and Bowles, D.J. (1988) The wound response of tomato plants can be inhibited by aspirin and related hydroxybenzoic acids. *Physiol. Mol. Plant Pathol.*, **33** 377-84.

Draper, J. (1997) Salicylate, superoxide synthesis and cell suicide in plant defence. *Trends Plant Sci.*, **2** 162-65.

Durner, J., Wendehenne, D. and Klessig, D.F. (1998) Defense gene induction in tobacco by nitric oxide, cyclic GMP and cyclic ADP-ribose. *Proc. Natl. Acad. Sci. USA*, **95** 10328-33.

Ellis, J. and Jones, D. (1998) Structure and function of proteins controlling strain-specific pathogen resistance in plants. *Curr. Opin. Plant Biol.*, **1** 288-93.

Ellis, J.G., Lawrence, G.J., Luck, J.E. and Dodds, P.N. (1999) Identification of regions in alleles of the flax rust resistance gene, *L*, that determine differences in gene-for-gene specificity. *Plant Cell*, **11** 495-506.

Epple, P., Apel, K. and Bohlmann, H. (1995) An *Arabidopsis thaliana* thionin gene is inducible via a signal transduction pathway different from that for pathogenesis-related proteins. *Plant Physiol.*, **109** 813-20.

Falk, A., Feys, B.J., Frost, L.N., Jones, J.D.G., Daniels, M.J. and Parker, J.E. (1999) *EDS1*, an essential component of *R*-gene-mediated disease resistance in *Arabidopsis* has homology to eukaryotic lipases. *Proc. Natl. Acad. Sci. USA*, **96** 3292-97.

Farmer, E., Weber, H. and Vollenweider, S. (1998) Fatty acid signaling in *Arabidopsis*. *Planta*, **206** 167-74.

Felix, G., Regenass, M., Spanu, P. and Boller, T. (1994) The protein phosphatase inhibitor calyculin-A mimics elicitor action in plant cells and induces rapid hyperphosphorylation of specific proteins as revealed by pulse labeling with [P-33] phosphate. *Proc. Natl. Acad. Sci. USA*, **91** 952-56.

Felton, G.W., Korth, K.L., Bi, J.L., Wesley, S.V., Huhman, D.V., Mathews, M.C., Murphy, J.B., Lamb, C. and Dixon, R.A. (1999) Inverse relationship between systemic resistance of plants to micro-organisms and to insect herbivory. *Curr. Biol.*, **9** 317-20.

Frederick, R.D., Thilmony, R.L., Sessa, G. and Martin, G.B. (1998) Recognition specificity for the bacterial avirulence protein, AvrPto, is determined by Thr-204 in the activation loop of the tomato Pto kinase. *Mol. Cell*, **2** 241-45.

Frey, M., Chomet, P., Glawischnig, E., Stettner, C., Grun, S., Winklmair, A., Eisenreich, W., Bacher, A., Meeley, R.B., Briggs, S.P., Simcox, K. and Gierl, A. (1997) Analysis of a chemical plant defense mechanism in grasses. *Science*, **277** 696-99.

Gaffney, T., Friedrich, L., Vernooij, B., Negrotto, D., Nye, G., Uknes, S., Ward, E., Kessmann, H. and Ryals, J. (1993) Requirement of salicylic acid for the induction of systemic-acquired resistance. *Science*, **261** 754-56.

Glazebrook, J. and Ausubel, F.M. (1994) Isolation of phytoalexin-deficient mutants of *Arabidopsis thaliana* and characterization of their interactions with bacterial pathogens. *Proc. Natl. Acad. Sci. USA*, **91** 8955-59.

Glazebrook, J., Rogers, E.E. and Ausubel, F.M. (1996) Isolation of *Arabidopsis* mutants with enhanced disease susceptibility by direct screening. *Genetics*, **143** 973-82.

Glazebrook, J., Zook, M., Mert, F., Kagan, I., Rogers, E.E., Crute, I.R., Holub, E.B., Hammerschmidt, R. and Ausubel, F.M. (1997) Phytoalexin-deficient mutants of *Arabidopsis* reveal that *PAD4* encodes a regulatory factor and that four *PAD* genes contribute to downy mildew resistance. *Genetics*, **146** 381-92.

Glazener, J.A., Orlandi, E.W. and Baker, C.J. (1996) The active oxygen response of cell suspensions to incompatible bacteria is not sufficient to cause hypersensitive cell death. *Plant Physiol.*, **110** 759-63.

Görg, R., Hollricher, K. and Schulze-Lefert, P. (1993) Functional analysis and RFLP-mediated mapping of the *Mlg* resistance locus in barley. *Plant J.*, **3** 857-66.

Groom, Q.J., Torres, M.A., Fordham-Skelton, A.P., Hammond-Kosack, K.E., Robinson, N.J. and Jones, J.D.G. (1996) *RbohA*, a rice homologue of the mammalian *gp91phox* respiratory burst oxidase gene. *Plant J.*, **10** 515-22.

Hammond-Kosack, K.E. and Jones, J.D.G. (1997) Plant disease resistance genes. *Annu. Rev. Plant Physiol. Plant Mol. Biol.*, **48** 575-607.

Hammond-Kosack, K.E., Jones, D.A. and Jones, J.D.G. (1994) Identification of two genes required in tomato for full *Cf9*-dependent resistance to *Cladosporium fulvum*. *Plant Cell*, **6** 2419-28.

Heath, M.C. (1998) Apoptosis, programmed cell death and the hypersensitive response. *Eur. J. Plant Pathol.*, **104** 117-24.

Herskowitz, I. (1995) MAP kinase pathways in yeast—for mating and more. *Cell*, **80** 187-97.

Hirt, H. (1997) Multiple roles of MAP kinases in plant signal transduction. *Trends Plant Sci.*, **2** 11-15.

Hoffmann, J.A., Kafatos, F.C., Janeway, C.A. and Ezekowitz, R.A.B. (1999) Phylogenetic perspectives in innate immunity. *Science*, **284** 1313-18.

Jabs, T., Tschöpe, M., Colling, C., Hahlbrock, K. and Scheel, D. (1997) Elicitor-stimulated ion fluxes and O_2(-) from the oxidative burst are essential components in triggering defense gene activation and phytoalexin synthesis in parsley. *Proc. Natl. Acad. Sci. USA*, **94** 4800-805.

Jirage, D., Tootle, T.L., Reuber, L., Frost, L.N., Feys, B.J., Parker, J.E., Ausubel, F.M. and Glazebrook, J. (1999) *Arabidopsis thaliana PAD4* encodes a lipase-like gene that is important for salicylic acid signalling. *Proc. Natl. Acad. Sci. USA*, **96** 13583-88.

Jones, D.A. and Jones, J.D.G. (1997) The role of leucine-rich repeat proteins in plant defences. *Adv. Bot. Res. Adv. Plant Pathol.*, **24** 89-167.

Jorgensen, J.H. (1996) Effect of three suppressors on the expression of powdery mildew resistance genes in barley. *Genome*, **39** 492-98.

Kajava, A.V. (1998) Structural diversity of leucine-rich repeat proteins. *J. Mol. Biol.*, **277** 519-27.

Keller, T., Damude, H.G., Werner, D., Doerner, P., Dixon, R.A. and Lamb, C. (1998) A plant homolog of the neutrophil NADPH oxidase gp91(phox) subunit gene encodes a plasma membrane protein with Ca^{2+} binding motifs. *Plant Cell*, **10** 255-66.

Kenton, P., Mur, L.A.J., Atzorn, R., Wasternack, C. and Draper, J. (1999) (-)-jasmonic acid accumulation in tobacco hypersensitive response lesions. *Mol. Plant-Microbe Interact.*, **12** 74-78.

Köhm, B.A., Goulden, M.G., Gilbert, J.E., Kavanagh, T.A. and Baulcombe, D.C. (1993) A potato virus-X resistance gene mediates an induced, non-specific resistance in protoplasts. *Plant Cell*, **5** 913-20.

Lamb, C. and Dixon, R.A. (1997) The oxidative burst in plant disease resistance. *Annu. Rev. Plant Physiol. Plant Mol. Biol.*, **48** 251-75.

Levine, A., Tenhaken, R., Dixon, R. and Lamb, C. (1994) H_2O_2 from the oxidative burst orchestrates the plant hypersensitive disease resistance response. *Cell*, **79** 583-93.

Ligterink, W., Kroj, T., zur Nieden, U., Hirt, H. and Scheel, D. (1997) Receptor-mediated activation of a MAP kinase in pathogen defense of plants. *Science*, **276** 2054-57.

Machida, Y., Nishihama, R. and Kitakura, S. (1997) Progress in studies of plant homologs of mitogen-activated protein (MAP) kinase and potential upstream components in kinase cascades. *Crit. Rev. Plant Sci.*, **16** 481-96.

Madhani, H.D. and Fink, G.R. (1998) The riddle of MAP kinase signaling specificity. *Trends Genet.*, **14** 151-55.

Martin, G.B., Brommonschenkel, S.H., Chunwongse, J., Frary, A., Ganal, M.W., Spivey, R., Wu, T., Earle, E.D. and Tanksley, S.D. (1993) Map-based cloning of a protein kinase gene conferring disease resistance in tomato. *Science*, **262** 1432-36.

Martinez, C., Montillet, J.L., Bresson, E., Agnel, J.P., Dai, G.H., Daniel, J.F., Geiger, J.P. and Nicole, M. (1998) Apoplastic peroxidase generates superoxide anions in cells of cotton cotyledons undergoing the hypersensitive reaction to *Xanthomonas campestris* pv. *malvacearum* race 18. *Mol. Plant-Microbe Interact.*, **11** 1038-47.

Mauch-Mani, B. and Slusarenko, A.J. (1996) Production of salicylic acid precursors is a major function of phenylalanine ammonia-lyase in the resistance of *Arabidopsis* to *Peronospora parasitica*. *Plant Cell*, **8** 203-12.

McDowell, J.M., Dhandaydham, M., Long, T.A., Aarts, M.G.M., Goff, S., Holub, E.B. and Dangl, J.L. (1998) Intragenic recombination and diversifying selection contribute to the evolution of downy mildew resistance at the *RPP8* locus of *Arabidopsis*. *Plant Cell*, **10** 1861-74.

McDowell, J.M., Dangl, J.L. and Holub, E.B. (2000) Downy mildew (*Peronospora parasitica*) resistance genes in Arabidopsis vary in functional requirements for *NDR1, EDS1, NPR1* and salicylic acid accumulation. *Plant J.* (in press).

Medzhitov, R., PrestonHurlburt, P. and Janeway, C.A. (1997) A human homologue of the *Drosophila* Toll protein signals activation of adaptive immunity. *Nature*, **388** 394-97.

Meyers, B.C., Shen, K.A., Rohani, P., Gaut, B.S. and Michelmore, R.W. (1998) Receptor-like genes in the major resistance locus of lettuce are subject to divergent selection. *Plant Cell*, **10** 1833-46.

Milligan, S.B., Bodeau, J., Yaghoobi, J., Kaloshian, I., Zabel, P. and Williamson, V.M. (1998) The root knot nematode resistance gene, *Mi*, from tomato is a member of the leucine zipper, nucleotide-binding, leucine-rich repeat family of plant genes. *Plant Cell*, **10** 1307-19.

Mitsuhara, I., Malik, K.A., Miura, M. and Ohashi, Y. (1999) Animal cell death suppressors, Bcl-XL and Ced-9, inhibit cell death in tobacco plants. *Curr. Biol.*, **9** 775-78.

Morel, J.B. and Dangl, J.L. (1997) The hypersensitive response and the induction of cell death in plants. *Cell Death Differentiation*, **4** 671-83.

Mur, L., Bi, Y.-M., Darby, R.M., Firek, S. and Draper, J. (1997) Compromising early salicylic acid accumulation delays the hypersensitive response and increases viral dispersal during lesion establishment in TMV-infected tobacco. *Plant J.*, **12** 1113-26.

Nathan, C. (1995) Natural resistance and nitric oxide. *Cell*, **82** 873-76.

Nürnberger, T., Nennsteil, D., Jabs, T., Sacks, W.R., Hahlbrock, K. and Scheel, D. (1994) High affinity binding of a fungal oligopeptide elicitor to parsley plasma membranes triggers multiple defense responses. *Cell*, **78** 449-60.

O'Donnell, P.J., Calvert, C., Atzorn, R., Wasternack, C., Leyser, H.M.O. and Bowles, D.J. (1996) Ethylene as a signal mediating the wound response of tomato plants. *Science*, **274** 1914-17.

Oldroyd, G.E.D. and Staskawicz, B.J. (1998) Genetically engineered broad-spectrum disease resistance in tomato. *Proc. Natl. Acad. Sci. USA*, **95** 10300-305.

Parker, J.E., Holub, E.B., Frost, L.N., Falk, A., Gunn, N.D. and Daniels, M.J. (1996) Characterization of *eds1*, a mutation in *Arabidopsis* suppressing resistance to *Peronospora parasitica* specified by several different *RPP* genes. *Plant Cell*, **8** 2033-46.

Parker, J.E., Coleman, M.J., Szabò, V., Frost, L.N., Schmidt, R., Van der Biezen, E.A., Moores, T., Dean, C., Daniels, M.J. and Jones, J.D.G. (1997) The *Arabidopsis* downy mildew resistance gene, *RPP5*, shares similarity to the Toll and interleukin-1 receptors with *N* and *L6*. *Plant Cell*, **9** 879-94.

Parniske, M., Hammond-Kosack, K.E., Goldstein, C., Thomas, C.M., Jones, D.A., Harrison, K., Wulff, B.B.H. and Jones, J.D.G. (1997) Novel disease resistance specificities result from sequence exchange between tandemly repeated genes at the *Cf-4/9* locus of tomato. *Cell*, **91** 821-32.

Penninckx, I., Eggermont, K., Terras, F.R.G., Thomma, B., De Samblanx, G.W., Buchala, A., Metraux, J.P., Manners, J.M. and Broekaert, W.F. (1996) Pathogen-induced systemic activation of a plant defensin gene in *Arabidopsis* follows a salicylic acid-independent pathway. *Plant Cell*, **8** 2309-23.

Penninckx, I., Thomma, B., Buchala, A., Metraux, J.P. and Broekaert, W.F. (1998) Concomitant activation of jasmonate and ethylene response pathways is required for induction of a plant defensin gene in *Arabidopsis*. *Plant Cell*, **10** 2103-13.

Peterhänsel, C., Freialdenhoven, A., Kurth, J., Kolsch, R. and SchulzeLefert, P. (1997) Interaction analyses of genes required for resistance responses to powdery mildew in barley reveal distinct pathways leading to leaf cell death. *Plant Cell*, **9** 1397-409.

Piedras, P., Hammond-Kosack, K.E., Harrison, K. and Jones, J.D.G. (1998) Rapid, Cf-9- and Avr9-dependent production of active oxygen species in tobacco suspension cultures. *Mol. Plant-Microbe Interact.*, **11** 1155-66.

Pieterse, C.M.J. and van Loon, L.C. (1999) Salicylic acid-independent plant defence pathways. *Trends Plant Sci.*, **4** 52-58.

Pieterse, C.M.J., van Wees, S.C.M., van Pelt, J.A., Knoester, M., Laan, R., Gerrits, N., Weisbeek, P.J. and van Loon, L.C. (1998) A novel signaling pathway controlling induced systemic resistance in *Arabidopsis*. *Plant Cell*, **10** 1571-80.

Rancé, I., Fournier, J. and Esquerré-Tugayé, M.T. (1998) The incompatible interaction between *Phytophthora parasitica* var. *nicotianae* race 0 and tobacco is suppressed in transgenic plants expressing antisense lipoxygenase sequences. *Proc. Natl. Acad. Sci. USA*, **95** 6554-59.

Rathjen, J.P., Chang, J.H., Staskawicz, B.J. and Michelmore, R.W. (1999) Constitutively active Pto induces a Prf-dependent hypersensitive response in the absence of avrPto. *EMBO J.*, **18** 3232-40.

Read, N.D., Shacklock, P.S., Knight, M.R. and Trewavas, A.J. (1993) Imaging calcium dynamics in living plant cells and tissues. *Cell Biol. Internl.*, **17** 111-25.

Reuber, T.L. and Ausubel, F.M. (1996) Isolation of *Arabidopsis* genes that differentiate between resistance responses mediated by the *RPS2* and *RPM1* disease resistance genes. *Plant Cell*, **8** 241-49.

Reuber, T., Plotnikova, J.M., Dewdney, J., Rogers, E.E., Wood, W. and Ausubel, F.M. (1998) Correlation of defence gene induction defects with powdery mildew susceptibility in *Arabidopsis* enhanced disease susceptibility mutants. *Plant J.*, **16** 473-85.

Reymond, P. and Farmer, E.E. (1998) Jasmonate and salicylate as global signals for defense gene expression. *Curr. Opin. Plant Biol.*, **1** 404-11.

Richberg, M.H., Aviv, D.H. and Dangl, J.L. (1998) Dead cells do tell tales. *Curr. Opin. Plant Biol.*, **1** 480-85.

Ritter, C. and Dangl, J.L. (1996) Interference between two specific pathogen recognition events mediated by distinct plant disease resistance genes. *Plant Cell*, **8** 251-57.

Rock, F.L., Hardiman, G., Timans, J.C., Kastelein, R.A. and Bazan, J.F. (1998) A family of human receptors structurally related to *Drosophila* Toll. *Proc. Natl. Acad. Sci. USA*, **95** 588-93.

Rogers, E.E. and Ausubel, F.M. (1997) *Arabidopsis* enhanced disease susceptibility mutants exhibit enhanced susceptibility to several bacterial pathogens and alterations in *PR-1* gene expression. *Plant Cell*, **9** 305-16.

Romeis, T., Piedras, P., Zhang, S.Q., Klessig, D.F., Hirt, H. and Jones, J.D.G. (1999) Rapid Avr9- and Cf-9-dependent activation of MAP kinases in tobacco cell cultures and leaves: convergence of resistance gene, elicitor, wound and salicylate responses. *Plant Cell*, **11** 273-87.

Romeis, T., Piedras, P. and Jones, J.D.G. (2000) Resistance gene-dependent activation of a calcium-dependent protein kinase (CDPK) in the plant defense response. *Plant Cell*, **5** (in press).

Rushton, P.J. and Somssich, I.E. (1998) Transcriptional control of plant genes responsive to pathogens. *Curr. Opin. Plant Biol.*, **1** 311-15.

Salmeron, J.M., Barker, S.J., Carland, F.M., Mehta, A.Y. and Staskawicz, B.J. (1994) Tomato mutants altered in bacterial disease resistance provide evidence for a new locus controlling pathogen recognition. *Plant Cell*, **6** 511-20.

Salmeron, J.M., Oldroyd, G.E.D., Rommens, C.M.T., Scofield, S.R., Kim, H.-S., Lavelle, D.T., Dahlbeck, D. and Staskawicz, B.J. (1996) Tomato *Prf* is a member of the leucine-rich repeat class of plant disease resistance genes and lies embedded within the *Pto* kinase gene cluster. *Cell*, **86** 123-33.

Sanders, D., Brownlee, C. and Harper, J.F. (1999) Communicating with calcium. *Plant Cell*, **11** 691-706.

Sanz, A., Moreno, J.I. and Castresana, C. (1998) PIOX, a new pathogen-induced oxygenase with homology to animal cyclooxygenase. *Plant Cell*, **10** 1523-37.

Schaller, A. and Oecking, C. (1999) Modulation of plasma membrane H^+-ATPase activity differentially activates wound and pathogen defense responses in tomato plants. *Plant Cell*, **11** 263-72.

Schrag, J.D. and Cygler, M. (1997) Lipases and alpha/beta hydrolase fold. *Methods Enzymol.*, **284** 85-107.

Schweizer, P., Buchala, A., Silverman, P., Seskar, M., Raskin, I. and Metraux, J.P. (1997) Jasmonate-inducible genes are activated in rice by pathogen attack without a concomitant increase in endogenous jasmonic acid levels. *Plant Physiol.*, **114** 79-88.

Scofield, S.R., Tobias, C.M., Rathjen, J.P., Chang, J.F., Lavelle, D.T., Michelmore, R.W. and Staskawicz, B.J. (1996) Molecular basis of gene-for-gene specificity in bacterial speck disease of tomato. *Science*, **274** 2063-65.

Segal, A.W. and Abo, A. (1993) The biochemical basis of the NADPH oxidase of phagocytes. *Trends Biol. Sci.*, **18** 43-47.

Seo, S., Okamoto, M., Ishizuka, K., Sano, H. and Ohashi, Y. (1995) Tobacco MAP kinase: a possible mediator in wound signal transduction pathways. *Science*, **270** 1988.

Seo, S., Sano, H. and Ohashi, Y. (1999) Jasmonate-based wound signal transduction requires activation of WIPK, a tobacco mitogen-activated protein kinase. *Plant Cell*, **11** 289-98.

Shah, J., Kachroo, P. and Klessig, D.F. (1999) The *Arabidopsis ssi1* mutation restores pathogenesis-related gene expression in *npr1* plants and renders defensin gene expression salicylic acid dependent. *Plant Cell*, **11** 191-206.

Sheen, J. (1996) Ca^{2+}-dependent protein kinases and stress signal transduction in plants. *Science*, **274** 1900-902.

Shirasu, K., Nakajima, H., Rajasekhar, V.K., Dixon, R.A. and Lamb, C. (1997) Salicylic acid potentiates an agonist-dependent gain control that amplifies pathogen signals in the activation of defense mechanisms. *Plant Cell*, **9** 261-70.

Shirasu, K., Lahaye, T., Tan, M.-W., Zhou, F., Azevedo, C. and Schulze-Lefert, P. (1999) A novel class of eukaryotic zinc-binding proteins is required for disease resistance signaling in barley and development in *C. elegans*. *Cell*, **99** 355-66.

Solomon, M., Belenghi, B., Delledonne, M., Menachem, E. and Levine, A. (1999) The involvement of cysteine proteases and protease inhibitor genes in the regulation of programmed cell death in plants. *Plant Cell*, **11** 431-43.

Song, W.-Y., Wang, G.-L., Chen, L.-L., Kim, H.-S., Pi, L.-Y., Holsten, T., Gardner, J., Wang, B., Zhai, W.-X., Zhu, L.-H., Fauquet, C. and Ronald, P. (1995) A receptor kinase-like protein encoded by the rice disease resistance gene, *Xa21*. *Science*, **270** 1804-806.

Stäb, M.R. and Ebel, J. (1987) Effects of Ca^{2+} on phytoalexin induction by fungal elicitor in soybean cells. *Arch. Biochem. Biophys.*, **257** 416-23.

Staswick, P.E., Yuen, G.Y. and Lehman, C.C. (1998) Jasmonate signaling mutants of *Arabidopsis* are susceptible to the soil fungus, *Pythium irregulare*. *Plant J.*, **15** 747-54.

Stuehr, D.J., Fasehun, O.A., Kwon, N.S., Gross, S.S., Gonzalez, J.A., Levi, R. and Nathan, C. F. (1991) Inhibition of macrophage and endothelial cell nitric oxide synthase by diphenylene iodonium and its analogs. *FASEB J.*, **5** 98-103.

Tang, X., Frederick, R.D., Zhou, J., Halterman, D.A., Jia, Y. and Martin, G.B. (1996) Physical interaction of AvrPto and the Pto kinase defines a recognition event involved in plant disease resistance. *Science*, **274** 2060-63.

Tang, X.Y., Xie, M.T., Kim, Y.J., Zhou, J.M., Klessig, D.F. and Martin, G.B. (1999) Overexpression of Pto activates defense responses and confers broad resistance. *Plant Cell*, **11** 15-29.

Tavernier, E., Wendelhenne, D., Blein, J.-P. and Pugin, A. (1995) Involvement of free calcium in action of cryptogein, a proteinaceous elicitor of hypersensitive reaction in tobacco cells. *Plant Physiol.*, **109** 1025-31.

Thomma, B., Eggermont, K., Penninckx, I., MauchMani, B., Vogelsang, R., Cammue, B.P.A. and Broekaert, W.F. (1998) Separate jasmonate-dependent and salicylate-dependent defense response pathways in *Arabidopsis* are essential for resistance to distinct microbial pathogens. *Proc. Natl. Acad. Sci. USA*, **95** 15107-11.

Thomma, B.P.H.J., Nelissen, I., Eggermont, K., *et al.* (1999) Deficiency in phytoalexin production causes enhanced susceptibility of *Arabidopsis thaliana* to the fungus, *Alternaria brassicicola*. *Plant J.*, **19**(2) 163-71.

Thordal-Christensen, H., Zhang, Z.G., Wei, Y.D. and Collinge, D.B. (1997) Subcellular localization of H_2O_2 in plants: H_2O_2 accumulation in papillae and hypersensitive response during the barley powdery mildew interaction. *Plant J.*, **11** 1187-94.

Torres, M.A., Onouchi, H., Hamada, S., Machida, C., Hammond-Kosack, K.E. and Jones, J.D.G. (1998) Six *Arabidopsis thaliana* homologues of the human respiratory burst oxidase, gp91(phox). *Plant J.*, **14** 365-70.

Van der Biezen, E.A. and Jones, J.D.G. (1998a) Plant disease-resistance proteins and the gene-for-gene concept. *Trends Biochem. Sci.*, **23** 454-56.

Van der Biezen, E.A. and Jones, J.D.G. (1998b) The NB-ARC domain: a novel signalling motif shared by plant resistance gene products and regulators of cell death in animals. *Curr. Biol.*, **8** R226-27.

Viard, M.-P., Martin, F., Pugin, A., Ricci, P. and Blein, J.-P. (1994) Protein phosphorylation is induced in tobacco cells by the elicitor cryptogein. *Plant Physiol.*, **104** 1245-49.

Vidal, S., Ponce de Leon, I., Denecke, J. and Tapio Palva, E. (1997) Salicylic acid and the plant pathogen, *Erwinia carotovora*, induce defense genes via antagonistic pathways. *Plant J.*, **11** 115-23.

Vijayan, P., Shockey, J., Levesque, C.A., Cook, R.J. and Browse, J. (1998) A role for jasmonate in pathogen defense of *Arabidopsis*. *Proc. Natl. Acad. Sci. USA*, **95** 7209-14.

Warren, R.F., Henk, A., Mowery, P., Holub, E. and Innes, R.W. (1998) A mutation within the leucine-rich repeat domain of the *Arabidopsis* disease resistance gene, *RPS5*, partially suppresses multiple bacterial and downy mildew resistance genes. *Plant Cell*, **10** 1439-52.

Warren, R.F., Merritt, P.M., Holub, E. and Innes, R.W. (1999) Identification of three putative signal transduction genes involved in *R* gene-specified disease resistance in *Arabidopsis*. *Genetics*, **152** 401-12.

Whitham, S., Dinesh-Kumar, S.P., Choi, D., Hehl, R., Corr, C. and Baker, B. (1994) The product of the tobacco mosaic virus resistance gene, *N*: similarity to Toll and the interleukin-1 receptor. *Cell*, **78** 1011-15.

Xing, T., Higgins, V.J. and Blumwald, E. (1997) Identification of G proteins mediating fungal elicitor-induced dephosphorylation of host plasma membrane H^+-ATPase. *J. Exp. Bot.*, **48** 229-37.

Xing, T., Higgins, V.J. and Blumwald, E. (1996) Regulation of plant defense response to fungal pathogens: two types of protein kinase in the reversible phosphorylation of the host plasma membrane H^+-ATPase. *Plant Cell*, **8** 555-64.

Xu, H.X. and Heath, M.C. (1998) Role of calcium in signal transduction during the hypersensitive response caused by basidiospore-derived infection of the cowpea rust fungus. *Plant Cell*, **10** 585-97.

Yu, I.C., Parker, J. and Bent, A.F. (1998) Gene-for-gene disease resistance without the hypersensitive response in *Arabidopsis* dnd1 mutant. *Proc. Natl. Acad. Sci. USA*, **95** 7819-24.

Zhang, S.Q. and Klessig, D.F. (1997) Salicylic acid activates a 48-kD MAP kinase in tobacco. *Plant Cell*, **9** 809-24.

Zhang, S.Q. and Klessig, D.F. (1998) The tobacco wounding-activated mitogen-activated protein kinase is encoded by SIPK. *Proc. Natl. Acad. Sci. USA*, **95** 7225-30.

Zhang, S.Q., Du, H. and Klessig, D.F. (1998) Activation of the tobacco SIP kinase by both a cell-wall-derived carbohydrate elicitor and purified proteinaceous elicitins from *Phytophthora* spp. *Plant Cell*, **10** 435-49.

Zhang, Y.L., Fan, W.H., Kinkema, M., Li, X. and Dong, X.N. (1999) Interaction of NPR1 with basic leucine zipper protein transcription factors that bind sequences required for salicylic acid induction of the PR-1 gene. *Proc. Natl. Acad. Sci. USA*, **96** 6523-28.

Zhou, J., Loh, Y.-T., Bressan, R.A. and Martin, G.B. (1995) The tomato gene, *Pti1*, encodes a serine/threonine kinase that is phosphorylated by Pto and is involved in the hypersensitive response. *Cell*, **83** 925-35.

Zhou, J.M., Tang, X.Y. and Martin, G.B. (1997) The Pto kinase conferring resistance to tomato bacterial speck disease interacts with proteins that bind a *cis*-element of pathogenesis-related genes. *EMBO J.*, **16** 3207-18.

Zhou, N., Tootle, T.L., Tsui, F., Klessig, D.F. and Glazebrook, J. (1998) *PAD4* functions upstream from salicylic acid to control defense responses in *Arabidopsis*. *Plant Cell*, **10** 1021-30.

Zhou, N., Tootle, T.L. and Glazebrook, J. (1999) *Arabidopsis PAD3*, a gene required for camalexin synthesis, encodes a putative cytochrome P450 monooxygenase. *Plant cell*, **11**(12) 2419-28.

Zimmermann, S., Nürnberger, T., Frachisse, J.M., Wirtz, W., Guern, J., Hedrich, R. and Scheel, D. (1997) Receptor-mediated activation of a plant Ca^{2+}-permeable ion channel involved in pathogen defense. *Proc. Natl. Acad. Sci. USA*, **94** 2751-55.

7 Programmed cell death in plants in response to pathogen attack

Paul R.J. Birch, Anna O. Avrova, Alia Dellagi,
Christophe Lacomme, Simon Santa Cruz and Gary D. Lyon

7.1 Introduction

In recent years, evidence has emerged to suggest that many developmental and physiological processes in plants, including the hypersensitive response (HR) to pathogen attack, involve programmed cell death (PCD) similar to apoptosis in animals. Morphological features of apoptosis, such as membrane blebbing, chromatin condensation and DNA cleavage, can be observed in the HR. Changes in gene transcription, protein phosphorylation, the oxidative burst, ion fluxes and lipid-based signalling are detected at early stages of PCD in both animals and plants. Furthermore, sequencing projects are revealing plant genes with intriguing similarities to apoptosis-related animal genes. In animals, specific cysteine proteases, termed caspases, are involved in the execution phase of apoptosis. Although cysteine proteases appear to have a role in plant PCD, and at least one of the targets of caspases, poly (ADP-ribose) polymerase, is cleaved during the HR, caspase-encoding genes have not yet been identified in plants. Moreover, a role for the mitochondrion, central in either regulating or effecting apoptosis in animals, has yet to emerge in plant PCD. Evidence exists for more ancient PCD mechanisms than those executed by caspases, and these may ultimately be what is evolutionarily conserved between plants and animals.

The phrase 'programmed cell death' was first used by Lockshin and Williams (1965) to describe a process whereby cells activate suicide pathways in response to external or internal stimuli. Kerr *et al.* (1972) observed that PCD elicited by a variety of mechanisms possessed a distinct morphology and they coined the term 'apoptosis' to describe it. This is now the accepted term for a genetically ordered series of physiological and morphological events that result in the death of an animal cell. Apoptosis occurs during normal development and morphogenesis, to remove those cells whose role is completed, to control cell number, or as a defensive strategy to remove infected, mutated or damaged cells. In essence, it is involved in maintaining the integrity of multicellular organisms, and is distinct from necrosis, which is cell death in response to injurious environmental stimuli and which is not genetically controlled. The gradual acceptance that animal cells have a built-in 'suicide programme' took the best part of 20 years, but the last decade has seen an explosion of discoveries describing the regulation and execution of this process (reviewed in Jacobson, 1997; Jacobson *et al.*, 1997; McKenna *et al.*, 1998; Metzstein *et al.*, 1998; Vaux and Korsmeyer, 1999).

As with animals, PCD is an integral component of a number of developmental processes in plants, including xylogenesis, reproduction, root and leaf development and senescence (reviewed in Greenberg, 1996; Jones and Dangl, 1996; Beers, 1997). Similarly, PCD in plants occurs in response to pathogen attack, most demonstrably in the hypersensitive response that characterises 'incompatible' interactions between a resistant host and an avirulent pathogen. The HR is a localised cell death that serves to prevent the development and spread of a pathogen. It follows recognition of the pathogen either via receptors that bind elicitors generated in the infection process (e.g. Nürnberger *et al.*, 1994), or receptors encoded by resistance (*R*) genes that bind to specific, pathogen-encoded avirulence gene products (Hammond-Kosack and Jones, 1997). In addition to cell death, pathways are activated which lead to the production of pathogenesis-related (PR) proteins, to antimicrobial secondary metabolites termed phytoalexins, and which signal systemic-acquired resistance (Kombrink and Somssich, 1995; Hunt *et al.*, 1996; Greenberg, 1997). It is thus unclear whether the accompanying resistance to a pathogen is due to cell death or whether such cell death is either a 'back-up' or a consequence of the numerous additional defence responses which are activated and which may independently inhibit the pathogen (Greenberg, 1997; Heath, 1998). Indeed, an *Arabidopsis* mutant has recently been identified that is defective in the HR but, nevertheless, possesses highly effective gene-for-gene disease resistance (Yu *et al.*, 1998).

Inevitably, comparisons are being made between cell death in plants and animals, to determine whether there is a central, apoptosis-like mechanism that is conserved between these kingdoms, or whether different mechanisms are exploited in plants and animals. A wealth of information on the morphology, biochemistry, physiology and genetics of plant PCD has been presented in recent years, and a number of reviews have attempted to address the question of whether localised cell death in the HR is analogous to apoptosis (Dangl *et al.*, 1996; Gilchrist, 1997; Greenberg, 1997; Morel and Dangl, 1997; Heath, 1998; Pontier *et al.*, 1998; Richberg *et al.*, 1998).

Before making comparisons, however, it is important to define apoptosis. Initially, the definition was confined to a clear sequence of morphological events that characterise this PCD (Kerr *et al.*, 1972; see also section 7.2). More recently, however, key proteins that execute cell death, cysteine proteases termed caspases, and the complex mitochondrion-based machinery that regulates them, have become synonymous with apoptosis (see section 7.3). Nevertheless, in both yeast and animals, apoptotic cell death morphology has been described in the absence of caspases (see section 7.4), suggesting that the morphology of cell death is more important to the definition of apoptosis, and also that there is more than one pathway to its execution. With this in mind, this chapter aims to describe the pathogen-induced HR in plants and to discuss whether it should be called apoptosis.

7.2 Comparison of the morphologies of apoptosis in animals and PCD in the plant hypersensitive response

Apoptosis is characterised by a distinct set of morphological features (Kerr *et al.*, 1972). Initially, both chromatin and cytoplasm condense and nuclear and cellular outlines appear convoluted or lobed. The cells are seen to shrink. Shortly after this, the nucleus disintegrates and DNA fragmentation occurs. The cells then fragment to form membrane-bound apoptotic bodies containing intact organelles. Apoptotic bodies are finally phagocytosed by neighbouring cells or macrophages.

While there are several signalling routes leading to apoptosis (see section 7.3), chromatin condensation and nuclear DNA fragmentation are common features of the process. A number of cytometric techniques have been central to studying these events. Chromatin condensation may be detected by staining nuclei with propidium iodide (PI) and measuring the level of fluorescence. During apoptosis, there is a reduction in fluorescence due to less PI intercalatory binding as DNA coiling tightens (Darzynkiewicz *et al.*, 1992). High molecular weight DNA cleavage is catalysed by a Ca^{2+}-dependent endonuclease that generates 140–180 base-pair (bp) fragments, which may be visualised as a 'DNA ladder' by gel electrophoresis (Cohen *et al.*, 1994). DNA fragmentation can also be detected by end-labelling of 3′ OH termini with terminal deoxynucleotidyl transferase (TUNEL) (Gorczyca *et al.*, 1993).

There are many differences between plant and animal cells, the most obvious being the presence of chloroplasts and a cell wall in the former. Studies of both of these structures are revealing roles in defence that are clearly specific to plants. In addition, the cell wall precludes phagocytosis, the final stage of apoptosis. If phagocytosis is the defining component of apoptosis, HR in the plant cannot be described as a form of apoptosis. However, many of the key morphological features of apoptosis have been observed in the HR. Plant DNA cleavage has been seen in incompatible reactions with a virus (Mittler *et al.*, 1997), bacteria (Levine *et al.*, 1996; Mittler *et al.*, 1997) and a fungus (Ryerson and Heath, 1996). However, only in the latter case was true oligonucleosomal fragmentation, resulting in DNA laddering, reported. Using TUNEL, the authors showed that an incompatible race of the cowpea rust fungus induced DNA cleavage only in cowpea cells containing haustoria, and this was detectable early in the cell death process. Furthermore, both Wang *et al.* (1996) and Navarre and Wolpert (1999) have reported DNA cleavage in response to fungal pathogen-derived toxins. In the former study, Wang *et al.* (1996) demonstrated not only

DNA laddering but also the formation of membrane-bound structures similar to apoptotic bodies in response to the AAL toxin from *Alternaria alternata* f. sp. *lycopersici*. These events were also seen during developmental PCD and in response to arachidonic acid, an inducer of the HR. Moreover, DNA laddering was enhanced by Ca^{2+} and inhibited by Zn^{2+}, a hallmark of apoptosis. Mittler and Lam (1995) observed that HR-associated nuclease activities in tobacco were also stimulated by Ca^{2+} and inhibited by Zn^{2+}. However, these nucleases were not activated during senescence in plants, suggesting that their involvement is not a universal feature of plant PCD (Mittler and Lam, 1995). Indeed, DNA laddering is not associated with all forms of PCD in animals (Zhang *et al.*, 1998a).

Another well-characterised morphological feature of apoptosis is the loss of membrane phospholipid asymmetry that results in phosphatidylserine (PS) appearing on the outer as well as the inner surface of the plasma membrane. This characteristic of apoptosis may be detected using the PS-binding protein, annexin-V, as a probe for PS on the plasma membrane outer surface. O'Brien *et al.* (1998) reported this phenomenon during PCD in tobacco induced by a number of chemical agents. They also observed that chromatin condensation at the early stages of the death process was reversible on removal of the chemical PCD inducer. Again, this is a feature of apoptosis, in that a number of checkpoints are believed to exist at which the process may be reversed until excessive damage becomes irreparable and unsustainable, leaving death as the only outcome. Such reversibility has been witnessed at early stages of the potato HR to *Phytophthora infestans* (Freytag *et al.*, 1994), demonstrating a flexibility in the control of cell death when the pathogen has been overcome.

7.3 Comparison of the molecular bases of apoptosis in animals and PCD in the plant hypersensitive response

In plants, many events leading to hypersensitive cell death are, at least on the surface, analogous to apoptosis. Receptors are required for the initial recognition of a pathogen. Complex signalling networks are activated, involving changes in protein phosphorylation, generation of reactive oxygen species and modifications to ion fluxes, including an influx of Ca^{2+}. The question is whether, as in animals, the numerous receptor-initiated routes into the HR lead to a conserved death-effector pathway, with similar complex regulation.

Much of the biochemistry and genetics of the effector mechanisms of apoptosis has been well studied, both in the nematode *Caenorhabditis elegans* (Metzstein *et al.*, 1998) and in several mammalian systems. However, relatively little is known about the details of many of the signalling pathways that can induce apoptosis. This section will summarise the current knowledge of these processes in animals, describing the effectors, their regulators and adaptors and finally the signalling pathways that initiate cell death. At each step, analogous studies in plants will be compared.

7.3.1 *Effectors of cell death*

A key feature of the execution phase of apoptosis in animal cells is the tightly-regulated proteolytic cleavage of a limited number of cellular proteins. The main enzymes involved in this process are caspases, aspartate-specific cysteine proteases that are normally present in an inactive form within cells. In addition, other cysteine proteases, such as calpains and cathepsins, serine proteases, such as granzymes, and the proteasome-ubiquitin pathway are all reported to be involved in apoptosis (reviewed in Solary *et al.*, 1998). Substrate proteins for caspases can be either activated or inactivated during cell death. One of the best studied inactivation events is the breakdown of poly (ADP-ribose) polymerase (PARP), an enzyme involved in DNA repair and genome integrity (Jeggo, 1998). Another reported target for cleavage is an inhibitor of caspase-activated DNAse (ICAD), leading to activation of capsase-activated DNAse (CAD) and fragmentation of the cell's DNA (Enari *et al.*, 1998). The identity of the Ca^{2+}-dependent endonuclease(s) responsible for DNA cleavage during apoptosis is a source of debate, but DNAse II (Torriglia *et al.*, 1999) and cyclophilins A, B and C (Montague *et al.*, 1997) are strong candidates.

Cysteine protease involvement in plant PCD has been demonstrated many times during developmentally regulated cell death processes (Jones *et al.*, 1996; Kardailsky and Brewin, 1996; Xu and Chye, 1999), senescence (Drake *et al.*, 1996; Griffiths *et al.*, 1997; Guerrero *et al.*, 1998) and also in the HR (del Pozo and Lam, 1998; D'Silva *et al.*, 1998; Solomon *et al.*, 1999). Nevertheless, the presence in plants of homologues of animal caspases and caspase activity in these processes remains to be conclusively proven. Del Pozo and Lam (1998) reported cleavage of a short oligopeptide at the aspartate residue during infection of resistant tobacco by tobacco mosaic virus (TMV). However, using a longer synthetic peptide, no such activity was seen from cysteine proteases purified from soybean cells undergoing oxidative stress-induced PCD (Solomon *et al.*, 1999). Inhibition of these cysteine proteases with cystatin blocked PCD triggered either by oxidative stress or an avirulent strain of *P. syringae* pv. *glycinea* (Solomon *et al.*, 1999). The authors suggested that PCD was carefully regulated by the activities of cysteine proteases and their inhibitors. D'Silva *et al.* (1998) also suggested a regulatory role for cysteine proteases activated in the HR of cowpea to cowpea rust fungus. In addition, they demonstrated cleavage of PARP for the first time in plants. In an independent study, PARP has been shown to be upregulated during oxidative stress of *Arabidopsis* (Amor *et al.*, 1998).

In animals, at least 10 caspases are present constitutively but as inactive zymogens that are activated by cleavage at aspartate residues by the regulatory caspases 8 and 9. However, caspase activation has also been demonstrated by cathepsin B in animals (Vancompernolle *et al.*, 1998). It is thus interesting to note that, in a variety of plant PCD processes, related, cathepsin-like cysteine proteases are transcriptionally induced (e.g. Drake *et al.*, 1996; Griffiths *et al.*, 1997; Xu and Chye, 1999). Recently, in a cDNA library enriched for sequences upregulated during an early stage of the potato HR to *P. infestans*, a cathepsin-like cysteine protease gene was also detected (Birch *et al.*, 1999). This sequence has subsequently been shown to be tightly regulated and strongly induced at an early time-point in the potato HR to *P. infestans* (Avrova *et al.*, 1999). It will be interesting to discover whether the cathepsin-like proteases encoded by such genes possess caspase-like regulatory functions by activating and inactivating specific proteins during plant PCD.

In addition to the direct evidence of nuclease activation, based on observations of DNA fragmentation, recent evidence of the likely involvement of nucleases in plant PCD comes from the upregulation of two cDNAs in barley developmental processes (Aoyagi *et al.*, 1998). The genes, *ben1* and *zen1*, encode proteins similar to nuclease S1. In addition to these, a cyclophilin-like gene has been shown to be upregulated in the HR, suggesting a functional link between DNA degradation processes in apoptosis and plant PCD (Birch *et al.*, 1999). However, despite the interesting similarities with nucleases involved in animal apoptosis, the precise functions of these genes and their products remain to be tested.

Another active component of the apoptotic execution machinery is the ubiquitin-proteasome. Proteasomes are large multicatalytic proteinase complexes that are responsible for the selective degradation of cellular proteins. They comprise a central catalytic unit (the 20S proteasome) and two terminal regulatory complexes, termed PA700 or PA28. About 40 subunits, with sizes of 20–110 kDa, are assembled to form two types of proteasomal complexes with the same catalytic core and different regulatory modules (Tanaka and Chiba, 1998). Most targeting of unwanted proteins is carried out through co-valent attachment to the highly conserved protein ubiquitin. During apoptosis in animals, ubiquitin and ubiquitin carrier protein are upregulated (Soldatenkov and Dritschilo, 1997). Both are also upregulated in response to stress (Vierstra, 1996); and members of the ubiquitin family may be specifically upregulated during the plant hypersensitive response (Birch *et al.*, 1999).

In addition to these genes, a survey of sequences emerging from the *Arabidopsis* genome project reveals tantalising homologies to both pro- and anti-apoptosis genes (Table 7.1). These sequences provide interesting targets for functional analysis that may allow further parallels to be drawn between animal and plant PCD. Furthermore, studies of genes that are upregulated both in senescence and pathogen-induced PCD reveal potential molecular markers of cell death in plants. Such studies have revealed that transcription of some metallothionein genes is increased both during senescence and the HR

Table 7.1 Arabidopsis genes showing similarity to apoptosis-related proteins

Arabidopsis Accession No.	Similarity to	Putative function
AL031135	Human *nip2* (AA394950)	Cell survival by interacting with Bcl-2
AF075597	RP-8 rat apoptosis protein (AF075597)	Transcription factor containing zinc-finger induced during apoptosis
AL049500	Fly Fas-associated factor (AB013610)	Potentiates Fas-induced apoptosis
AC006418	Human cellular apoptosis susceptibility protein (P55060)	Interferes with selected apoptosis pathways including TNF and ADP-ribosylating toxins
AC002986	Proteosome component (P24495)	Proteolysis by endopeptidase activity
AC003979	Mouse apoptosis protein, MA-3 (D50465)	Possible translation initiation factor induced during apoptosis
AC002986	Fly inhibitor of apoptosis (Q24307)	Transcription factor involved in suppressing apoptosis
Z97343	Bax inhibitor-1 protein (NP003208)	Inhibitor of Bax protein with several predicted membrane spanning domains

Abbreviations: TNF, tumour necrosis factor.

(e.g. Butt *et al.*, 1998; Birch *et al.*, 1999). Metallothioneins are induced during apoptosis in animals, although it is likely that they are induced in tissues immediately surrounding dying cells and have an anti-apoptotic regulatory role (e.g. Abdel-Mageed and Agrawal, 1998).

7.3.2 Adaptors and regulators of cell death

In animals, caspases 8 and 9 may be directly activated by adaptors which bind to them via shared domains. In the case of caspase 8, a death effector domain (DED) in its prodomain binds to a C-terminal DED in the adaptor, FADD (Boldin *et al.*, 1995). The C terminus of FADD also contains a death domain (DD) allowing it to bind to an equivalent region within the cytoplasmic domain of the receptor, CD95 (Fas/Apo-1) (Figure 7.1). Thus, death-inducing signals from outside the cell may be passed from the plasma membrane, via FADD, to caspase 8 (Boldin *et al.*, 1996). Caspase 9 activation follows binding of its caspase recruitment domain (CARD) to the CARD in another adaptor protein, Apaf1 (Figure 7.2). The domains, DD, CARD and DED are all structurally similar, suggesting a common evolutionary origin, and are, for the most part, associated only with proteins with a role in cell death. Although such domains have yet to be identified in plants, intriguing structural similarities have been shown between the adaptor Apaf-1 and a number of *R* genes (van Der Biezen and Jones, 1998; Inohara *et al.*, 1999) (see also section 7.3.3).

The adaptors that activate caspases in animals may themselves be regulated. A protein called FLIP may bind to FADD, preventing activation of caspase 8 (Irmler *et al.*, 1997). However, many of the key regulatory processes controlling

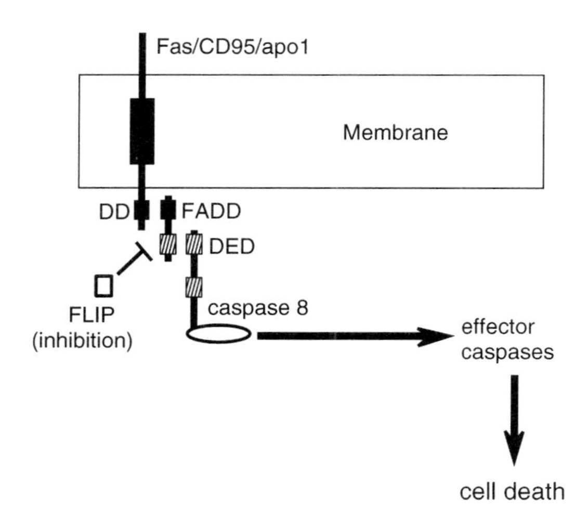

Figure 7.1 Direct receptor-mediated activation of caspase 8 in animals. Death-inducing signals activate the tumour necrosis factor receptor family member, CD95 (Fas/Apo1), which in turn binds the adaptor-protein, FADD, via common death domains (DD). Caspase 8 is then directly activated by interaction with FADD via a shared death effector domain (DED). Caspase 8 cleaves other effector caspases at aspartate residues, activating them and causing cell death. A regulator protein, FLIP, inhibits activation of caspase 8 by binding to FADD.

apoptosis are centred around the mitochondrion. Mitochondrial damage is an important amplification step in the death pathway in animals and is the site of action for pro- and anti-apoptotic members of the Bcl-2 family (Figure 7.2). Bcl-2 itself binds directly to Apaf-1 and prevents its action. It is also involved in reducing cytochrome *c* release by regulating permeability of the mitochondrial membrane; cytochrome *c* combines with Apaf-1, causing conformational changes that allow it to process caspase 9, which then directly activates effector caspases (reviewed in Reed, 1997). The combination of cytochrome *c*, Apaf1 and caspase 9 is termed the apoptosome. Anti-apoptosis proteins (Bcl-2, BclX, BclW, A1 and mcl-1) interact with proapoptosis proteins (e.g. Bax, Bid and Bad) via the Bcl-homology (BH) domain 3 to inhibit their pro-death functions (reviewed in Golstein, 1997; Green and Reed, 1998).

Evidence for a similar, complex regulatory control of cell death in plants has yet to emerge. Nevertheless, a homologue of Bcl-2, localising to mitochondria, chloroplasts and nuclei, has been detected in tobacco leaf cells using an antibody (Dion *et al.*, 1997). Moreover, sequences matching genes coding for Bcl-2 binding proteins are emerging from genomic sequencing projects (Table 7.1). For example, a mammalian Bax inhibitor-1 (BI-1) protein was recently identified by functional screening in yeast (Xu and Reed, 1998). This protein, which contains several membrane-spanning domains and localises to intracellular membranes, shows significant similarity (29% identity, 45%

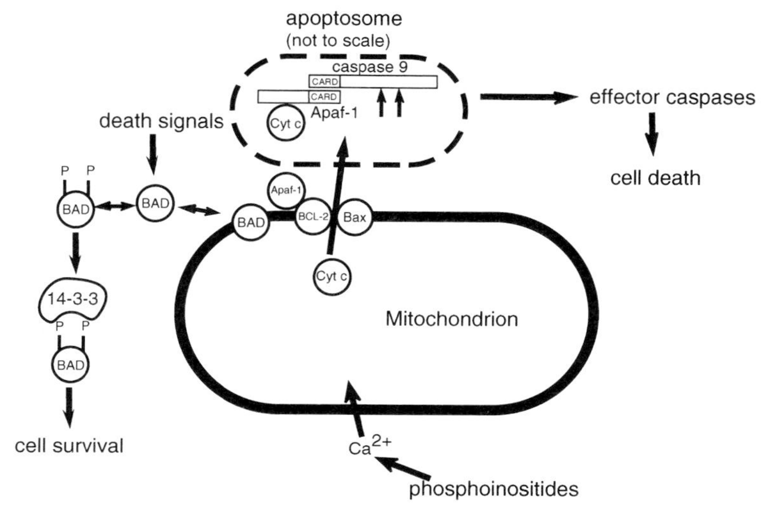

Figure 7.2 Regulation of cell death by mitochondria in animals. The adaptor protein, Apaf1, is activated by cyochrome *c*, causing conformational changes which allow it to interact with caspase 9 via shared caspase recruitment domains (CARDs) in a molecular combination termed the apoptosome. Apaf1 then cleaves caspase 9 (upward arrows in the apoptosome) at aspartate residues, activating it so that it, in turn, may activate effector caspases, resulting in an irreversible pathway leading to cell death. Cytochrome *c* (cyt *c*) release from the mitochondrion is promoted by an influx of Ca^{2+} and occurs via regulation of the voltage-dependent anion channel (VDAC), which is closed by Bcl-2 and opened by Bax (Green and Reed, 1998; Shimizu *et al.*, 1999). Bcl-2 may also inhibit apoptosis by direct interaction with Apaf1. The pro-apoptotic protein, BAD, promotes cell death by interaction with Bcl-2. BAD may be phosphorylated and inactivated by binding to a 14-3-3 protein, inducing a pathway leading to cell survival (Wang *et al.*, 1999).

similarity at the amino acid level) to the putative protein product of an *Arabidopsis* gene (Table 7.1) (Xu and Reed, 1998).

Lacomme and Santa Cruz (1999) introduced the gene coding for the pro-apoptotic Bax protein into tobacco using a TMV expression system, and found that it triggered a cell death response with similarities to the HR. Bax deletion mutants revealed that the BH1 and BH3 domains were required for rapid cell death. Furthermore, removal of the C-terminal transmembrane (TM) domain prevented cell death. When the TM domain was expressed as a C-terminal fusion with the jellyfish green fluorescent protein, it was shown to localise to the mitochondrion. Bax-induced cell death correlated with an accumulation of the defence protein PR1, which is upregulated in the HR, suggesting that an endogenous cell death pathway had been activated. However, in an independent study, constitutive expression of the anti-apoptotic Bcl-XL in transgenic tobacco did not inhibit the HR (Mittler *et al.*, 1996), showing that complete conservation of the Bcl-associated cell death machinery is unlikely.

Cell death is believed to be a default pathway in animals, which is constantly held in check by negative regulators, such as Bcl-2. However, a number of other mechanisms may also be involved in preventing PCD, in defining the limits of its spread and in reversing the process. For example, a family of evolutionarily ancient inhibitor of apoptosis proteins (IAPs), which include *survivin* from mammalian cells, directly inhibit caspases and regulate the NF-κB transcription factor (reviewed in LaCasse *et al.*, 1998). Intriguingly, similarities have been noted, particularly in the C-terminal zinc finger domain, between *Arabidopsis* sequences and genes encoding IAPs (Table 7.1; accession number AC002986).

Direct evidence of mechanisms to prevent or suppress PCD, including the HR, in plants is provided by lesion mimic mutants, in which HR-like lesions spontaneously develop in the absence of pathogens (reviewed in Dangl *et al.*, 1996). The *Arabidopsis lsd1* mutant phenotype confers increased disease resistance and cell death. The *lsd1* gene encodes a putative transcription factor, which either represses a pro-cell-death pathway or activates an anti-cell-death pathway (Dietrich *et al.*, 1997). Other negative controllers of PCD in plants include an *Arabidopsis* gene induced by oxidative stress, *oxy5*, which both protects bacterial cells from death caused by oxidative stress and cancer cells from tumour necrosis factor (TNF)-induced apoptosis (Janicke *et al.*, 1998).

7.3.3 The initial recognition of a pathogen

The initial perception of stimuli leading to apoptosis in animals is through a number of death receptors belonging to the TNF receptor superfamily (Ashkenazi and Dixit, 1998). They can activate caspases within seconds of ligand binding followed by rapid apoptotic demise of the cell. Such receptors are associated mainly with removal of cancerous cells or in response to pathogens. Nevertheless, signalling from these receptors may also lead to either cell proliferation or survival, as opposed to cell death. Plants also possess receptors (encoded by *R* genes) specifically involved in response to pathogens. However, in addition to signalling the HR they simultaneously initiate numerous defence signalling pathways not associated with developmental PCD processes (Heath, 1998). Furthermore, they are activated through specific interactions with avirulence proteins generated only by certain races of a pathogen in a 'gene-for-gene' manner (Hammond-Kosack and Jones, 1997). So far, there is no evidence that *R* gene products signal cell proliferation or survival in addition to cell death.

In recent years, considerable effort has been directed toward the isolation and characterisation of a number of *R* genes and their products. The *R* genes isolated so far encode proteins that fall into several classes. *R* genes often encode domains including nucleotide binding sites (NBSs), leucine-rich repeats (LRRs), transmembrane domains (TMs) and serine/threonine protein kinases (PK), to give proteins with NBS-LRR, LRR-TM-PK, LRR-TM or PK structures. The majority of R proteins characterised so far are classified as NBS-LRR and are believed to be functionally confined to disease resistance (reviewed in Hammond-Kosack and Jones, 1997).

The NBS-LRR class of *R* genes may be further subdivided according to the presence or absence of an N-terminal TIR domain. The TIR domain shares amino acid sequence and predicted structural similarities to the cytoplasmic signalling domains of the *Drosophila* Toll and mammalian interleukin-1 (IL-1) receptors. This subgroup includes resistance genes, *N* (Witham *et al.*, 1994), *L6* (Lawrence *et al.*, 1995), *M* (Anderson *et al.*, 1997), *RPP5* (Parker *et al.*, 1997), and three recently described genes of the *Arabidopsis RPP1* complex resistance locus (Botella *et al.*, 1998). Recently, the Toll domain has been shown to directly mediate defence signalling by lipopolysaccharide, a major component of the cell walls of invading bacterial pathogens (Yang *et al.*, 1998). In *Drosophila*, Toll interacts with a serine-threonine protein kinase, Pelle, which signals degradation of the protein, Cactus, and activation of the NF-κB-like transcription factor, Dorsal. Similarly, protein kinases, IRAK1 and IRAK2, which signal from the IL-1 receptor in animals, are analogues of Pelle. Intriguingly, Pelle resembles not only IRAK1 and IRAK2, but also the R protein, Pto, a tomato cytoplasmic serine threonine protein kinase involved in the HR to *Pseudomonas syringae* containing *AvrPto* (Tang *et al.*, 1996). Protein intermediaries, Tube in the case of Pelle, and myeloid differentiation antigen 88 (MyD88) in the case of IRAK1 and IRAK2, are involved in the interaction between these protein kinases and the TIR domains of their respective receptors. Pelle, IRAK1, IRAK2, Tube and MyD88 all possess death domains (Muzio *et al.*, 1997). Recently, isolation of putative plant analogues of the Cactus-like transcription factor inhibitor IκB (Ryals *et al.*, 1997) and the transcription factor NF-κB (Cao *et al.*, 1997) has provided further evidence that some plant and animal signalling processes are conserved.

In addition to the TIR domain, the NBS domain shows homology to regions found in the pro-apoptotic regulator, Apaf-1 (van der Biezen and Jones, 1998). Apaf-1 and these R proteins also share a similar structural organisation (Figure 7.3). Thus, the common nucleotide binding (NBS or NB-ARC) domain shared by these proteins links an N-terminal effector domain (CARD in Apaf1 and TIR in the R proteins) to a C-terminal domain likely to be involved in protein-protein interactions (WD-40 in Apaf1 and LRR in R proteins), and both are involved in cell death. Recently, a search for additional *Apaf1*-like sequences in humans revealed a novel gene, *Nod1*, encoding a protein with N-terminal CARD, followed by NB-ARC domains, and a C-terminal LRR domain, further increasing the similarity between R proteins and an apoptosis adaptor protein (Figure 7.3) (Inohara *et al.*, 1999). Like Apaf-1, the Nod1 protein activates caspase 9. However, in addition, it activates a pathway leading to the induction of NF-κB (Inohara *et al.*, 1999).

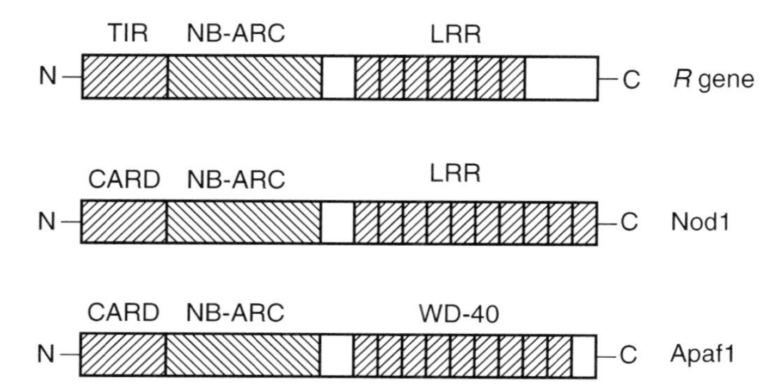

Figure 7.3 Domain structures of some NBS-LRR plant resistance (R) proteins and the pro-apoptotic adaptor proteins, Apaf-1 and Nod1. Plant resistance (R) proteins encoded by genes such as the *N* gene of tobacco, possess a nucleotide binding domain (NB-ARC) which links an N-terminal effector domain (TIR) to a C-terminal leucine-rich repeat (LRR), likely to be involved in protein-protein interactions (van der Biezen and Jones, 1998). Similarly, the pro-apoptotic adaptor proteins, Apaf-1 and Nod1, each possess an NB-ARC domain which links an N-terminal effector domain (the caspase recruitment domain or CARD) to a C-terminal protein-protein interaction domain (LRR in Nod1, Inohara *et al.*, 1999; and WD-40 repeats in Apaf-1, see van der Biezen and Jones, 1998).

7.3.4 Signalling and signal transduction

In plants, as in animals, signalling and signal transduction following the activation of a receptor is probably complex and likely to involve a number of parallel and/or interacting pathways. Other than cell death, a number of defence-related genes are upregulated and local and systemic disease resistances are activated. The cell death response itself may be reversible up to a certain point and does not progress beyond the location of the pathogen, implying that it is tightly, negatively controlled. As an example of the potential complexity of signal transduction pathways, the mammalian genome encodes approximately 4000 intracellular kinases and phosphatases and 10,000 transcription factors (Crabtree, 1999). If this is a measure of the complexity that can be expected in plants, potentially amplified by 'cross-talk' between pathways, then our understanding of the signalling of plant PCD is at a very early stage. Nevertheless, key components of signalling and signal transduction in the plant HR are common to apoptosis.

7.3.4.1 Protein phosphorylation

In apoptosis, both protein phosphorylation, in the form of mitogen-activated protein kinase (MAPK) cascades, and dephosphorylation, play important roles in signalling the execution, spread and control of cell death (e.g. Jarpe *et al.*, 1998). Similarly, one of the most likely early events following recognition of a plant pathogen is a change in protein phosphorylation (Scheel, 1998). Use of non-specific fungal elicitors in cell cultures causes rapid and transient changes in protein phosphorylation (Felix *et al.*, 1994; Viard *et al.*, 1994; Chandra and Low, 1995). Kinase inhibitors, such as K252a, inhibit HR-related cellular events, such as the oxidative burst and defence gene activation (Levine *et al.*, 1994). Conversely, protein phosphatase (PP) inhibitors have been shown to generate elicitor-like responses (Felix *et al.*, 1994; Mackintosh *et al.*, 1994). Nevertheless, PP activity is apparently required for developmental PCD processes in plants (e.g. Kuo *et al.*, 1996), and a regulatory subunit of PP2A is activated by caspase 3 in animals, after which it regulates MAPK cascades (Santoro *et al.*, 1998). In tobacco, blocking PP activity inhibited HR mediated both by the *N* gene (Dunigan and Madlener, 1995) and by expression of the pro-apoptotic Bax protein (Lacome and Santa Cruz, 1999).

The *R* gene product, Pto, is a cytoplasmic serine/threonine protein kinase, which has been shown to interact with other proteins, including putative transcription factors thought to activate genes encoding pathogenesis-related (PR) proteins (Zhou *et al.*, 1997). Increasing evidence supports a role for MAPK in plant signal transduction leading to defence responses (Ligterink *et al.*, 1997) and, indeed, a number of transcription factors believed to activate defence genes appear to be phosphorylated (e.g. Dröge-Laser *et al.*, 1997).

Salicylic acid, an important signalling molecule in local and systemic plant defences, has been shown to activate a MAPK in tobacco, called SIPK (Zhang and Klessig, 1997). SIPK is also activated by TMV in resistant plants (Zhang and Klessig, 1998) and by *Phytophthora*-derived elicitins. The elicitins cause hypersensitive cell death and prolonged SIPK activation (Zhang *et al.*, 1998b).

An additional MAPK from tobacco, WIPK, is wound inducible. This also is induced locally and systemically by TMV in resistant plants (Zhang and Klessig, 1998). In this case, the post-translational activation of WIPK follows accumulation of the corresponding transcript and protein. Recently, both SIPK and WIPK have been shown to be activated in transgenic tobacco containing the *R* gene, *Cf-9*. The activation was in an avirulence gene- and *Cf-9*-dependent manner (Romeis *et al.*, 1999). Taken together, these observations indicate that a complex range of stimuli, from wounding to *R* gene activation, can activate common signalling pathways.

7.3.4.2 Reactive oxygen species

Another early event that is common to both apoptosis and the HR is the oxidative burst. Both superoxide (Auh and Murphy, 1995) and/or hydrogen peroxide (Levine *et al.*, 1994) may be detected in the apoplast of infected cells, and may participate in plant defence by direct toxicity to the pathogen, by promoting cell wall strengthening, and by inter- and intra-cell signalling (reviewed in Lamb and Dixon, 1997; Jabs, 1999). Recently, a gene, *Atao1*, encoding a protein which generates H_2O_2, has been cloned from *Arabidopsis* (Moller and McPherson, 1998). This gene is expressed in developmental tissues destined to undergo PCD. The membrane-bound enzyme, NADPH oxidase, is believed to mediate reactive oxygen species (ROS) production both in animal (Segal and Abo, 1993) and plant (Auh and Murphy, 1995) cells. Recently, a plant analogue of a subunit of the human respiratory burst NADPH oxidase has been isolated (Keller *et al.*, 1998), and encodes a plasma membrane protein with calcium binding motifs. It is thus interesting to note that calcium influx, as well as elevated cytosolic calcium levels, are required for elicitor induction of the oxidative burst (Chandra *et al.*, 1997; Jabs *et al.*, 1998).

In addition to ROS, nitrogen oxide has been implicated in PCD in plants (Delledonne *et al.*, 1998; Durner *et al.*, 1998) and animals (reviewed in Brüne *et al.*, 1998). In soybean, its inhibition blocks the HR to avirulent bacteria and resistance is reduced. Whereas H_2O_2 is required for systemic lesion formation and systemic acquired resistance (SAR), local lesions require both H_2O_2 and NO (Alvarez *et al.*, 1998; Delledonne *et al.*, 1998). In animals, exogenous NO can trigger apoptosis (Huwiler *et al.*, 1999), or prevent apoptosis by inhibiting caspase-3 (Rössig *et al.*, 1999).

Infection with a bacterial strain mutated in the *hrp* locus, and thus defective in initiating the HR, nevertheless elicited an oxidative burst similar to the wild-type interaction (Glazener *et al.*, 1996). Therefore the oxidative burst does not appear to be necessary for such cell death. It has been suggested that the initial oxidative burst is not specific to HR but that a later, more sustained generation of ROS is. Both are required for SAR (Alvarez *et al.*, 1998). ROS also have a dual role in animals: a signalling function in the induction phase of apoptosis; followed by a second phase involved in the effector stage of apoptosis (Jabs, 1999). This second, sustained generation of ROS is a consequence of a decrease in the coupling efficiency of electron chain transport, leading to the production and release of superoxide from the mitochondrion (Green and Reed, 1998). This can be a result of alterations to the mitochondrial permeability transition (PT), caused by regulating the voltage-dependent anion channel (VDAC) (Shimizu *et al.*, 1999). VDAC is held closed by Bcl-2 and opened by Bax and physiological levels of Ca^{2+} (Green and Reed, 1998; Shimizu *et al.*, 1999). Another potential source of ROS, which is specific to plants, is the chloroplast, and recent studies suggest that it may play a role in ROS production during the HR in photosynthetic tissues (Allen *et al.*, 1999; Navarre and Wolpert, 1999). It is thus interesting to note a Bcl-2 antibody localised not only to plant mitochondria but also to chloroplasts (Dion *et al.*, 1997), raising the possibility of similar regulation of membrane permeability in each.

7.3.4.3 Ion fluxes

In apoptosis, Ca^{2+} influx is a universal feature, mediating release of cytochrome *c* from the mitochondrion and DNA degradation (Krebs, 1998). Elevation of Ca^{2+} levels in plants has been demonstrated in response to microbial elicitors (e.g. Levine *et al.*, 1996), and patch-clamp experiments have suggested that elicitors may affect the activity of plasma membrane calcium channels (Gelli and Blumwald, 1997; Zimmerman *et al.*, 1997). Xu and Heath (1998) have shown that cytoplasmic Ca^{2+} is increased prior to cowpea cell penetration by an avirulent race of the cowpea rust fungus. No such response was observed in susceptible interactions. They concluded that Ca^{2+} is involved in signal transduction leading to the HR. Jabs *et al.* (1998) attempted to order the early signalling events of plant PCD and found that elevated Ca^{2+} was required to stimulate the oxidative burst.

7.3.4.4 Lipid-based signalling

Over recent years, considerable attention has been directed towards lipid-based signalling systems, especially the sphingolipids, and their involvement in regulating cell growth, differentiation and apoptosis in animals (Merrill *et al.*, 1997; Peña *et al.*, 1997). Sphingolipid polymers are abundant in the membranes of all eukaryotic cells, although sphingomyelin is present in animal but not plant cells (Heinz, 1996). The first biosynthetic step in sphingolipid production is the generation of sphinganine by serine palmitoyltransferase (SPT). This is then converted either to sphingosine-1-phosphate (SPP) or to ceramide, both of which are potent second messengers (Merrill *et al.*, 1997). SPP and ceramide have opposite balancing effects, the former promoting cell survival and proliferation and the latter promoting apoptosis (Spiegel, 1999). Addition of either sphingosine or ceramide to animal cells results in apoptosis via regulation of MAPK or stress-activated protein kinase (SAPK or c-jun kinase), respectively (Jarvis *et al.*, 1997). Ceramide is rapidly generated from sphingomyelin by the activation of a membrane-bound phospholipase C-like enzyme, sphingomyelinase, either directly by stresses, such as H_2O_2, or via the TNF family or IL-1 receptors (reviewed in Peña *et al.*, 1997). Sphingosine and SPP are involved in regulating calcium levels in animal cells.

Interestingly, many studies of sphingolipid signalling in animals have involved the use of sphinganine analogue mycotoxins derived from plant pathogenic fungi, in particular, fumonisins, derived from *Fusarium moniliforme* and AAL toxins, derived from *Alternaria alternata* fsp. *lycopersici* (Merrill *et al.*, 1997). Both have been shown to induce apoptosis-like cell death in plants and animals at similar concentrations (Wang *et al.*, 1996; Gilchrist, 1997). The toxins competitively inhibit ceramide synthase, resulting in an accumulation of sphinganine. Recently, cerebrosides A and C have been shown to elicit HR and phytoalexin accumulation in rice, demonstrating that sphingolipid signalling can occur in plants (Koga *et al.*, 1998). In addition, the upregulation of the potato serine palmitoyltransferase gene has been demonstrated early in the HR to *P. infestans* (Birch *et al.*, 1999). Upregulation of serine palmitoyltransferase in animal cells results in apoptotic death (Shimabukuro *et al.*, 1998). Thus, these preliminary studies suggest that, as with animals, sphingolipid signalling may play a role in plant PCD.

7.4 Conclusion: is the plant hypersensitive response apoptosis?

That PCD occurs in plants as well as animals, and for much the same reasons, can be of little doubt. Developmental and reproductive processes, senescence, and the responses to damage or pathogen invasion often involve or result in cell suicide. A number of striking morphological similarities exist between apoptosis and plant processes such as the pathogen-induced HR. Pathogen recognition in plants may be mediated by receptors with domains that are conserved in animals, and signalling/signal transduction in each involve changes in protein phosphorylation, ion fluxes, ROS and may involve lipid-based second messengers. Intriguing similarities have been observed between plant DNA sequences and apoptosis-related animal genes, and cysteine proteases appear to be involved in all plant cell death processes, notably cleaving at least one of the highly specific targets of caspases, PARP, during the HR. Have we reached the stage where we can think of the HR as apoptosis?

From an evolutionary viewpoint, the case for calling the HR apoptosis would be strengthened if evidence linking these PCD processes in a common ancestor was found. The common ancestor of both plants and animals was presumably single-celled. There is evidence for PCD occurring in single-celled organisms, but no evidence of the presence of caspases (reviewed in Vaux and Korsmeyer, 1999). Yeast also lacks analogues of apoptosis-related genes, yet some components of apoptosis retain a cell death function when introduced into yeast. This is the case for the pro-apoptotic *Bax* gene. In addition, many of the morphological characteristics of apoptosis are evoked in yeast by Bax, including loss of PS asymmetry in the plasma membrane, chromatin condensation and DNA fragmentation (Ligr *et al.*, 1998). Conversely, introduction of the main apoptotic effector, caspase-3, into yeast did not cause death, although growth was slowed, suggesting that some but not all caspase targets are present (Wright *et al.*, 1998). So, how does Bax cause death in the absence of caspases? One way may be to simply provoke release of ROS from the mitochondrion by altering membrane permeability (see section 7.3.4.2). In addition, however, Bax has been shown to mediate cell death through interaction with promyelocytic leukaemia (PML) nuclear bodies in animals, and this occurred in the presence of a caspase inhibitor but nevertheless retained the cytoplasmic features of apoptosis (Quignon *et al.*, 1998). Both this report, and the demonstration of a Bax-induced apoptotic morphology in yeast cells (Ligr *et al.*, 1998), therefore, show that caspases are not essential for PCD.

The ceramide signalling pathway is a highly evolutionarily conserved signalling system that provides further evidence of a caspase-independent route(s) to cell death. In animals, stress-activated ceramide signalling leads to apoptosis. However, in yeast, stress-induced ceramide signalling also leads to cell death, but in the absence of caspases. This suggests that the older ceramide pathway has become linked to the apoptotic pathway at a later stage in evolution (reviewed in Mathias *et al.*, 1998).

In conclusion, we are very much at an early stage in understanding PCD in the plant hypersensitive response. The many intriguing similarities to PCD in animals will need to be rigorously tested biochemically and physiologically to demonstrate that they are conserved, and are derived from a common ancestral origin. Whether PCD in plants, such as that found in the HR to pathogen attack, should be called apoptosis, depends on whether the participation of caspases is important to the definition of this form of cell death. Such enzymes have not yet been found in plants, although cysteine proteases with specific activities have been reported in the HR (D'Silva *et al.*, 1998). There may, as is becoming apparent in animals, be more than one PCD pathway, and the next few years is likely to see intense research activity to discover if this is so.

Acknowledgements

The Scottish Crop Research Institute is grant-aided from the Scottish Office Agriculture Environment and Fisheries Department. A.O.A. was funded by a NATO grant and A.D. acknowledges the support of the European Union (project ERBIC-15-CT-960908).

References

Abdel-Mageed, A.B. and Agrawal, K.C. (1998) Activation of nuclear factor kappa B: potential role in metallothionein-mediated mitogenic response. *Cancer Res.*, **58** 2335-38.

Allen, L.J., MacGregor, K.B., Koop, R.S., Bruce, D.H., Karner, J. and Brown, A.W. (1999) The relationship between photosynthesis and a mastoparan-induced hypersensitive response in isolated mesophyll cells. *Plant Physiol.*, **19** 1233-42.

Alvarez, M.E., Pennel, R.I., Meijer, P.-J., Ishikawa, A., Dixon, R.A. and Lamb, C. (1998) Reactive oxygen intermediates mediate a systemic network in the establishment of plant immunity. *Cell*, **92** 773-84.

Anderson, P.A., Lawrence, G.J., Morrish, B.C., Ayliffe, M.A., Finnegan, E.J. and Ellis, J.G. (1997) Inactivation of the flax rust resistance gene, *M*, associated with loss of a repeated unit within the leucine-rich repeat coding region. *Plant Cell*, **9** 641-51.

Amor, Y., Babiychuk, E., Inze, D. and Levine, A. (1998) The involvement of poly(ADP-ribose) polymerase in the oxidative stress responses in plants. *FEBS Lett.*, **440**, 1-7.

Aoyagi, S., Sugiyama, M. and Fukuda, H. (1998) BEN1 and ZEN1 cDNAs encoding S1-type DNases that are associated with programmed cell death in plants. *FEBS Lett.*, **429** 134-38.

Ashkenazi, A. and Dixit, V.M. (1998) Death receptors: signalling and modulation. *Science*, **281** 1305-308.

Auh, C.-K. and Murphy, T.M. (1995) Plasma membrane redox enzyme is involved in the synthesis of O_2^- and H_2O_2 by *Phytophthora* elicitor-stimulated rose cells. *Plant Physiol.*, **107** 1241-47.

Avrova, A.O., Stewart, H.E., De Jong, W., Heilbronn, J., Lyon, G.D. and Birch, P.R.J. (1999) A cysteine protease gene is expressed early in resistant potato interactions with *Phytophthora infestans*. *Mol. Plant-Microbe Interact.*, **12** 1114-19.

Beers, E.P. (1997) Programmed cell death during plant growth and development. *Cell Death Differ.*, **4** 649-61.

Birch, P.R.J., Avrova, A.O., Duncan, J.M., Lyon, G.D. and Toth, R.L. (1999) Isolation of potato genes which are induced during an early stage of the hypersensitive response to *Phytophthora infestans*. *Mol. Plant-Microbe Interact.*, **12** 356-61.

Boldin, M.P., Varfolomeev, E.E., Pancer, Z., Mett, I.L., Camonis, J.H. and Wallach, D. (1995) A novel protein that interacts with the death domain of Fas/APO1 contains a sequence motif related to the death domain. *J. Biol. Chem.*, **270** 7795-98.

Boldin, M.P., Goncharov, T.M., Goltsev, Y.V. and Wallach, D. (1996) Involvement of MACH, a novel MORT1/FADD-interacting protease, in Fas/APO-1- and TNF receptor-induced cell death. *Cell*, **85** 803-15.

Botella, M.A., Parker, J.E., Frost, L.N., Bittner-Eddy, P.D., Beynon, J.L., Daniels, M.J., Holub, E.B. and Jones, J.D. (1998) Three genes of the *Arabidopsis RPP1* complex resistance locus recognise distinct *Peronospora parasitica* avirulence determinants. *Plant Cell*, **10** 1847-60.

Brüne, B., von Knethen, A. and Sandau, K.B. (1998) Nitric oxide and its role in apoptosis. *Eur. J. Pharmacol.*, **351** 261-72.

Butt, A., Mousley, C., Morris, K., Beynon, J., Can, C., Holub, E., Greenberg, J.T. and Buchanan-Wollaston, V. (1998) Differential expression of a senescence-enhanced metallothionein gene in *Arabidopsis* in response to isolates of *Peronospora parasitica* and *Pseudomonas syringae*. *Plant J.*, **16** 209-21.

Cao, H., Glazebrook, J., Clarke, J.D., Volko, S. and Dong, X. (1997) The *Arabidopsis NPR1* gene that controls systemic acquired resistance and encodes a novel protein containing ankyrin repeats. *Cell*, **88** 57-63.

Chandra, S. and Low, P.S. (1995) Role of phosphorylation in elicitation of the oxidative burst in cultured soybean cells. *Proc. Natl. Acad. Sci. USA*, **92** 4120-23.

Chandra, S., Stennis, M. and Low, P.S. (1997) Measurement of Ca^{2+} fluxes during elicitation of the oxidative burst in aequorin-transformed tobacco cells. *J. Biol. Chem.*, **272** 28274-80.

Cohen, G.M., Sun, X., Fearnhead, H., MacFarlane, M., Brown, D.G., Snowden, R.T. and Dinsdale, D. (1994) Formation of large molecular weight fragments of DNA is a key committed step of apoptosis in thymocytes. *J. Immunol.*, **153** 507-16.

Crabtree, G.R. (1999) Generic signals and specific outcomes: signalling through Ca^{2+}, calcineurin and NF-AT. *Cell*, **96** 611-14.

Dangl, J.L., Dietrich, R.A. and Richberg, M.H. (1996) Death don't have no mercy: cell death programs in plant-microbe interactions. *Plant Cell*, **8** 1793-807.

Darzynkiewicz, Z., Bruno, S., del Bino, G., Gorczyca, W., Hotz, M.A., Lassota, P. and Traganos, F. (1992) Features of apoptotic cells measured by flow cytometry. *Cytometry*, **13** 795-808.

Delledonne, M., Xia, Y., Dixon, R.A. and Lamb, C.J. (1998) Nitric oxide functions as a signal in plant disease resistance. *Nature*, **394** 585-88.

del Pozo, O. and Lam, E. (1998) Caspases and programmed cell death in the hypersensitive response of plants to pathogens. *Curr. Biol.*, **8** 1129-32.

Dietrich, R.A., Richberg, M.H., Schmidt, R., Dean, C. and Dangl, J.L. (1997) A novel zinc finger protein is encoded by the *Arabidopsis LSD1* gene and functions as a negative regulator of cell death. *Cell*, **88** 685-94.

Dion, M., Chamberland, H., St-Michel, C., Plante, M., Darveau, A., Lafontaine, J.G. and Brisson, L.F. (1997) Detection of a homologue of Bcl-2 in plant cells. *Biochem. Cell Biol.*, **75** 457-61.

Drake, R., John, I., Farrell, A., Cooper, W., Schuch, W. and Grierson, D. (1996) Isolation and analysis of cDNA encoding tomato cysteine proteases expressed during leaf senescence. *Plant Mol. Biol.*, **30** 755-67.

Dröge-Laser, W., Kaiser, A., Lindsay, W.P., Halkier, B.A., Loake, G.J., Doerner, P., Dixon, R.A. and Lamb, C. (1997) Rapid stimulation of a soybean protein-serine kinase which phosphorylates a novel bZIP DNA-binding protein, G/HBF-1, during the induction of early transcription-dependent defenses. *EMBO J.*, **16** 726-38.

D'Silva, I., Poirier, G.C. and Heath, M.C. (1998) Activation of a cysteine protease in cowpea plants during the hypersensitive response—a form of programmed cell death. *Exp. Cell Res.*, **245** 389-99.

Dunigan, D.D. and Madlener, J.C. (1995) Serine/threonine protein phosphatase is required for tobacco mosaic virus-mediated programmed cell death. *Virology*, **207** 460-66.

Durner, J., Wendehenne, D. and Klessig, D.F. (1998) Defense gene induction in tobacco by nitric oxide, cyclic GMP and cyclic ADP-ribose. *Proc. Natl. Acad. Sci. USA*, **5** 10328-33.

Enari, M., Sakahira, H., Yokoyama, H., Okawa, K., Iwamatsu, A. and Nagata, S. (1998) A caspase-activated DNase that degrades DNA during apoptosis and its inhibitor ICAD. *Nature*, **391** 43-50.

Felix, G., Regenass, M., Spanu, P. and Boller, T. (1994) The protein phosphatase inhibitor, calyculin A, mimics elicitor action in plant cells and induces rapid hyperphosphorylation of specific proteins as revealed by pulse labelling with [^{33}P] phosphate. *Proc. Natl. Acad. Sci. USA*, **91** 952-56.

Freytag, S., Arabatzis, N., Hahlbrock, K. and Schmelzer, E. (1994) Reversible cytoplasmic rearrangements precede wall apposition, hypersensitive cell death and defense-related gene activation in potato/*Phytophthora infestans* interactions. *Planta*, **194** 123-35.

Gelli, A. and Blumwald, E. (1997) Hyperpolarization-activated Ca^{2+}-permeable channels in the plasma membrane of tomato cells. *J. Membr. Biol.*, **155** 35-45.

Gilchrist, D.G. (1997) Mycotoxins reveal connections between plants and animals in apoptosis and ceramide signalling. *Cell Death Differ.*, **4** 689-98.

Glazener, J.A., Orlandi, E.W. and Baker, C.J. (1996) The active oxygen response of cell suspensions to incompatible bacteria is not sufficient to cause hypersensitive cell death. *Plant Physiol.*, **110** 759-63.

Golstein, P. (1997) Controlling cell death. *Science*, **275** 1081-82.

Gorczyca, W., Gong, J. and Darzynkiewicz, Z. (1993) Detection of DNA strand breaks in individual apoptotic cells by the *in situ* terminal deoxynucleotidyl transferase and nick translation assays. *Cancer Res.*, **53** 1445-51.

Green, D.R. and Reed, J.C. (1998) Mitochondria and apoptosis. *Science*, **281** 1309-12.

Greenberg, J.T. (1996) Programmed cell death: a way of life for plants. *Proc. Natl. Acad. Sci. USA*, **93** 12094-97.

Greenberg, J.T. (1997) Programmed cell death in plant-pathogen interactions. *Annu. Rev. Plant Physiol. Plant Mol. Biol.*, **48** 525-45.

Griffiths, C.M., Hosken, S.E., Oliver, D., Chojecki, J. and Thomas, H. (1997) Sequencing, expression pattern and RFLP mapping of a senescence-enhanced cDNA from *Zea mays* with high homology to oryzain gamma and aleurain. *Plant Mol. Biol.*, **34** 815-21.

Guerrero, C., de la Calle, M., Reid, M.S. and Valpuesta, V. (1998) Analysis of the expression of two thiolprotease genes from day lily (*Hemerocallis* spp.) during flower senescence. *Plant Mol. Biol.*, **36** 565-71.

Hammond-Kosack, K.E. and Jones, J.D.G. (1997) Plant disease resistance genes. *Annu. Rev. Plant Physiol. Plant Mol. Biol.*, **48** 575-607.

Heath, M.C. (1998) Apoptosis, programmed cell death and the hypersensitive response. *Eur. J. Plant Pathol.*, **104** 117-24.

Heinz, E. (1996) Plant glycolipids: structure, isolation and function, in *Advances in Lipid Methodology III*. (ed. W.W. Christie), The Oily Press, UK.

Hershko, A. and Ciechanover, A. (1998) The ubiquitin system. *Annu. Rev. Biochem.*, **67** 425-79.

Hunt, M.D., Neuenschwander, U.H., Delaney, T.P., Weymann, K.B., Friedrich, L.B., Lawton, K.A., Steiner, H.Y. and Ryals, J.A. (1996) Recent advances in systemic acquired resistance research—a review. *Gene*, **79** 89-95.

Huwiler, A., Pfeilschifter, J. and van den Bosch, H. (1999) Nitric oxide donors induce stress signalling via ceramide formation in rat renal mesangial cells. *J. Biol. Chem.*, **274** 7190-95.

Inohara, N., Koseki, T., del Peso, L., Hu, Y., Yee, C., Chen, S., Carrio, R., Merino, J., Liu, D., Ni, J. and Nunez, G. (1999) Nod1, an apaf-1-like activator of caspase-9 and nuclear factor-kappaB. *J. Biol. Chem.*, **274** 14560-67.

Irmler, M., Thome, M., Hahne, M., Schneider, P., Hofmann, K., Steiner, V., Bodmer, J.L., Schroter, M., Burns, K., Mattmann, C., Rimoldi, D., French, L.E. and Tschopp, J. (1997) Inhibition of death receptor signals by cellular FLIP. *Nature*, **388** 190-95.

Jabs, T. (1999) Reactive oxygen intermediates as mediators of programmed cell death in plants and animals. *Biochem. Pharmacol.*, **57** 231-45.

Jabs, T., Tschope, M., Colling, C., Hahlbrock, K. and Scheel, D. (1998) Elicitor-stimulated ion fluxes and O_2^- from the oxidative burst are essential components in triggering defense gene activation and phytoalexin synthesis in parsley. *Proc. Natl. Acad. Sci. USA*, **94** 4800-4805.

Jacobson, M.D. (1997) Apoptosis: Bcl-2-related proteins get connected. *Curr. Biol.*, **7** 277-81.

Jacobson, M.D., Weil, M. and Raff, M.C. (1997) Programmed cell death in animal development. *Cell*, **88** 347-54.

Janicke, R.U., Porter, A.G. and Kush, A. (1998) A novel *Arabidopsis thaliana* protein protects tumor cells from tumor necrosis factor-induced apoptosis. *Biochim. Biophys. Acta*, **402** 70-78.

Jarpe, M.B., Widmann, C., Knall, C., Schlesinger, T.K., Gibson, S., Yujiri, T., Fanger, G.R., Gelfand, E.W. and Johnson, G.L. (1998) Anti-apoptotic *versus* pro-apoptotic signal transduction: check-points and stop signs along the road to death. *Oncogene*, **17** 1475-82.

Jarvis, W.D., Fornari, F.A., Jr., Auer, K.L., Freemerman, A.J., Szabo, E., Birrer, M.J., Johnson, C.R., Barbour, S.E., Dent, P. and Grant, S. (1997) Coordinate regulation of stress- and mitogen-activated protein kinases in the apoptotic actions of ceramide and sphingosine. *Mol. Pharmacol.*, **52** 935-47.

Jeggo, P.A. (1998) PARP—Another guardian angel? *Curr. Biol.*, **8** 49-51.

Jones, A.L. and Dangl, J.L. (1996) Logjam at the Styx: programmed cell death in plants. *Trends Plant Sci.*, **1** 114-19.

Jones, C.G., Tucker, G.A. and Lycett, G.W. (1996) Pattern of expression and characteristics of a cysteine proteinase cDNA from germinating seeds of pea (*Pisum sativum* L.). *Biochim. Biophys. Acta*, **129** 613-15.

Kardailsky, I.V. and Brewin, N.J. (1996) Expression of cysteine protease genes in pea nodule development and senescence. *Mol. Plant-Microbe Interact.*, **9** 689-95.

Keller, T., Damude, H.G., Werner, D., Doerner, P., Dixon, R.A. and Lamb, C. (1998) A plant homolog of the neutrophil NADPH oxidase $gp91^{phox}$ subunit gene encodes a plasma membrane protein with Ca^{2+} binding motifs. *Plant Cell*, **10** 255-66.

Kerr, J.F., Wyllie, A.H. and Currie, A.R. (1972) Apoptosis: a basic biological phenomenon with wide-ranging implications in tissue kinetics. *Br. J. Cancer*, **26** 239-57.

Koga, J., Yamauchi, T., Shimura, M., Ogawa, N., Oshima, K., Umemura, K., Kikuchi, M. and Ogasawara, N. (1998) Cerebrosides A and C, sphingolipid elicitors of hypersensitive cell death and phytoalexin accumulation in rice plants. *J. Biol. Chem.*, **273** 1985-91.

Kombrink, E. and Somssich, I.E. (1995) Defense responses of plants to pathogens. *Adv. Botanic. Res.*, **21** 2-33.

Krebs, J. (1998) The role of calcium in apoptosis. *Biometals*, **11** 375-82.

Kuo, A., Cappelluti, S., Cervantes-Cervantes, M., Rodriguez, M. and Bush, D.S. (1996) Okadaic acid, a protein phosphatase inhibitor, blocks calcium changes, gene expression and cell death induced by gibberellin in wheat aleurone cells. *Plant Cell*, **8** 259-69.

LaCasse, E.C., Baird, S., Korneluk, R.G. and MacKenzie, A.E. (1998) The inhibitors of apoptosis (IAPs) and their emerging role in cancer. *Oncogene*, **17** 3247-59.

Lacomme, C. and Santa Cruz, S. (1999) Bax-induced cell death in tobacco is similar to the hypersensitive response. *Proc. Natl. Acad. Sci. USA*, **96** 7956-61.

Lamb, C. and Dixon, R.A. (1997) The oxidative burst in disease resistance. *Annu. Rev. Plant Physiol. Plant Mol. Biol.*, **48** 251-75.

Lawrence, G.J., Finnegan, E.J., Ayliffe, M.A. and Ellis, J.G. (1995) The *L6* gene for flax rust resistance is related to the *Arabidopsis* bacterial resistance gene, *RPS2*, and the tobacco viral resistance gene, *N*. *Plant Cell*, **7** 1195-206.

Levine, A., Tenhaken, R., Dixon, R. and Lamb, C. (1994) H_2O_2 from the oxidative burst orchestrates the plant hypersensitive response. *Cell*, **79** 583-93.

Levine, A., Pennell, R.I., Alvarez, M.E., Palmer, R. and Lamb, C. (1996) Calcium-mediated apoptosis in a plant hypersensitive disease resistance response. *Curr. Biol.*, **6** 427-37.

Ligr, M., Madeo, F., Frohlich, E., Hilt, W., Frohlich, K.U. and Wolf, D.H. (1998) Mammalian Bax triggers apoptotic changes in yeast. *FEBS Lett.*, **438** 61-65.

Ligterink, W., Kroj, T., zur Nieden, U., Hirt, H. and Scheel, D. (1997) Receptor-mediated activation of a MAP kinase in pathogen defense of plants. *Science*, **276** 2054-57.

Lockshin, R. and Williams, C. (1965) Programmed cell death. II. Endocrine potentiation of the breakdown of the intersegmental muscles of silkworms. *J. Insect Physiol.*, **11** 803-809.

Mackintosh, C., Lyon, G.D. and Mackintosh, R.W. (1994) Protein phosphatase inhibitors activate antifungal defense responses of soybean cotyledons and cell cultures. *Plant J.*, **5** 137-47.

Mathias, S., Pena, L.A. and Kolesnick, R.N. (1998) Signal transduction of stress via ceramide. *Biochem. J.*, **335** 465-80.

McKenna, S.L., McGowan, A.J. and Cotter, T.G. (1998) Molecular mechanisms of programmed cell death. *Adv. Biochem. Eng. Biotechnol.*, **62** 1-31.

Merrill, A.H., Jr., Schmelz, E.-M., Dillehay, D.L., Spiegel, S., Shayman, J.A., Schroeder, J.J., Riley, R.T., Voss, K.A. and Wang, E. (1997) Sphingolipids—the enigmatic lipid class: biochemistry, physiology and pathophysiology. *Toxicol. Appl. Pharmacol.*, **142** 208-25.

Metzstein, M.M., Stanfield, G.M. and Horvitz, H.R. (1998) Genetics of programmed cell death in *C. elegans*: past, present and future. *Trends Genet.*, **14** 410-16.

Mittler, R. and Lam, E. (1995) Identification, characterization and purification of a tobacco endonuclease activity induced upon hypersensitive response cell death. *Plant Cell*, **7** 1951-62.

Mittler, R., Shulaev, V., Seskar, M. and Lam, E. (1996) Inhibition of programmed cell death in tobacco plants during a pathogen-induced hypersensitive response at low oxygen pressure. *Plant Cell*, **8** 1991-2001.

Mittler, R., Simon, L. and Lam, E. (1997) Pathogen-induced programmed cell death in tobacco. *J. Cell Sci.*, **110** 1333-44.

Moller, S.G. and McPherson, M.J. (1998) Developmental expression and biochemical analysis of the *Arabidopsis atao1* gene encoding an H_2O_2-generating diamine oxidase. *Plant J.*, **13** 781-91.

Montague, J.W., Hughes Jr., F.M. and Cidlowski, J.A. (1997) Native recombinant cyclophilins A, B and C degrade DNA independently of peptidylprolyl *cis-trans*-isomerase activity: potential roles of cyclophilins in apoptosis. *J. Biol. Chem.*, **272** 6677-84.

Morel, J.B. and Dangl, J.L. (1997) The hypersensitive response and the induction of cell death in plants. *Cell Death Differ.*, **4** 671-83.

Muzio, M., Ni, J., Feng, P. and Dixit, V.M. (1997) IRAK (pelle) family member, IRAK-2, and MyD88 as proximal members of IL-1 signalling. *Science*, **278** 1612-15.

Navarre, D.A. and Wolpert, T.J. (1999) Victorin induction of an apoptotic/senescence-like response in oats. *Plant Cell*, **11** 237-49.

Nürnberger, T., Nennstiel, D., Jabs, T., Sacks, W.R., Hahlbrock, K. and Scheel, D. (1994) High affinity binding of a fungal oligopeptide elicitor to parsley plasma membranes triggers multiple defense responses. *Cell*, **78** 449-60.

O'Brien, I.E.W., Baguley, B.C., Murray, B.G., Morris, B.A.M. and Ferguson, I.B. (1998) Early stages of the apoptotic pathway in plant cells are reversible. *Plant J.*, **13** 803-14.

Parker, J.E., Coleman, M.J., Szabo, V., Frost, L.N., Schmidt, R., van der Biezen, E.A., Moores, T., Dean, C., Daniels, M.J. and Jones, J.D. (1997) The *Arabidopsis* downy mildew resistance gene, *RPP5*, shares similarity to the Toll and interleukin-1 receptors with *N* and *L6*. *Plant Cell*, **9** 879-94.

Peña, L.A., Fuks, Z. and Kolesnick, R. (1997) Stress-induced apoptosis and the sphingomyelin pathway. *Biochem. Pharmacol.*, **53** 615-21.

Pontier, D., Balague, C. and Roby, D. (1998) The hypersensitive response: a programmed cell death associated with plant resistance. *C. R. Acad. Sci. III*, **321** 721-34.

Quignon, F., De Bels, F., Koken, M., Feunteun, J., Ameisen, J.C. and de The, H. (1998) PML induces a novel caspase-independent death process. *Nature Genet.*, **20** 259-65.

Reed, J.C. (1997) Cytochrome *c*: can't live with it—can't live without it. *Cell*, **91** 559-62.

Richberg, M.H., Aviv, D.H. and Dangl, J.L. (1998) Dead cells do tell tales. *Curr. Opin. Plant Biol.*, **1** 480-85.

Romeis, T., Piedras, P., Zhang, S., Klessig, D.F., Hirt, H. and Jones, J.D. (1999) Rapid Avr9- and Cf-9-dependent activation of MAP kinases in tobacco cell cultures and leaves: convergence of resistance gene, elicitor, wound and salicylate responses. *Plant Cell*, **11** 273-87.

Rössig, L., Fichtlscherer, B., Breitschopf, K., Haendeler, J., Zeiher, A.M., Mülsch, A. and Dimmeler, S. (1999) Nitric oxide inhibits caspase-3 by S-nitrosation *in vivo*. *J. Biol. Chem.*, **274** 6823-26.

Ryals, J., Weymann, K., Lawton, K., Friedrich, L., Ellis, D., Steiner, H.-Y., Johnson, J., Delaney, T.P., Jesse, T., Vos, P. and Uknes, S. (1997) The *Arabidopsis NIM1* protein shows homology to the transcription factor inhibitor I*κB*. *Plant Cell*, **9** 425-39.

Ryerson, D.E. and Heath, M.C. (1996) Cleavage of DNA into oligonucleosomal fragments during cell death induced by fungal infection or by abiotic treatments. *Plant Cell*, **8** 393-402.

Santoro, M.F., Annand, R.R., Robertson, M.M., Peng, Y.-W., Brady, M.J., Mankovich, J.A., Hackett, M.C., Ghayur, T., Walter, G., Wong, W.W. and Giegel, D.A. (1998) Regulation of protein phosphatase 2A activity by caspase-3 during apoptosis. *J. Biol. Chem.*, **273** 13119-28.

Scheel, D. (1998) Resistance response physiology and signal transduction. *Curr. Opin. Plant Biol.*, **1** 305-10.

Segal, A.W. and Abo, A. (1993) The biochemical basis of the NADPH oxidase of phagocytes. *Trends Biochem. Sci.*, **18** 43-47.

Shimabukuro, M., Higa, M., Zhou, Y.T., Wang, M.Y., Newgard, C.B. and Unger, R.H. (1998) Lipoapoptosis in beta-cells of obese prediabetic fa/fa rats: role of serine palmitoyltranferase overexpression. *J. Biol. Chem.*, **273** 32487-90.

Shimizu, S., Narita, M. and Tsujimoto, Y. (1999) Bcl-2 family proteins regulate the release of apoptogenic cytochrome *c* by the mitochondrial channel VDAC. *Nature*, **399** 483-87.

Solary, E., Eymin, B., Droin, N. and Haugg, M. (1998) Proteases, proteolysis and apoptosis. *Cell. Biol. Toxicol.*, **14** 121-32.

Soldatenkov, V.A. and Dritschilo, A. (1997) Apoptosis of Ewing's sarcoma cells is accompanied by accumulation of ubiquitinated proteins. *Cancer Res.*, **18** 3881-85.

Solomon, M., Belenghi, B., Delledonne, M., Menachem, E. and Levine, A. (1999) The involvement of cysteine proteases and protease inhibitor genes in the regulation of programmed cell death in plants. *Plant Cell*, **11** 431-43.

Spiegel, S. (1999) Sphingosine-1-phosphate: a prototype of a new class of second messengers. *J. Leukoc. Biol.*, **65** 341-44.

Tanaka, K. and Chiba, T. (1998) The proteasome: a protein-destroying machine. *Genes Cells*, **3** 499-510.

Tang, X.Y., Frederick, R.D., Zhou, J.M., Halterman, D.A., Jia, Y.L. and Martin, G.B. (1996) Initiation of plant disease resistance by physical interaction of AVRPTO and PTO kinase. *Science*, **274** 2060-63.

Torriglia, A., Negri, C., Chaudun, E., Prosperi, E., Courtois, Y., Counis, M.F. and Scovassi, A.I. (1999) Differential involvement of DNases in HeLa cell apoptosis induced by etoposide and long-term culture. *Cell Death Differ.*, **6** 234-44.

Vancompernolle, K., Van Herreweghe, F., Pynaert, G., Van de Craen, M., de Vos, K., Totty, N. and Sterling, A. (1998) Atractyloside-induced release of cathepsin B, a protease with caspase-processing activity. *FEBS Lett.*, **438** 150-58.

van der Biezen, E.A. and Jones, J.D.G. (1998) The NB-ARC domain: a novel signalling motif shared by plant resistance gene products and regulators of cell death in animals. *Curr. Biol.*, **8** 226-67.

Vaux, D.L. and Korsmeyer, S.J. (1999) Cell death in development. *Cell*, **96** 245-54.

Viard, M.-P., Martin, F., Pugin, A., Ricci, P. and Blein, J.-P. (1994) Protein phosphorylation is induced in tobacco cells by the elicitor, cryptogein. *Plant Physiol.*, **104** 1245-49.

Vierstra, R.D. (1996) Proteolysis in plants: mechanisms and functions. *Plant Mol. Biol.*, **32** 275-302.

Wang, H., Li, J., Bostock, R.M. and Gilchrist, D.G. (1996) Apoptosis: a functional paradigm for programmed plant cell death induced by a host-selective phytotoxin and invoked during development. *Plant Cell*, **8** 375-91.

Wang, H.-G., Pathan, N., Ethell, I.M., Krajewski, S., Yamaguchi, Y., Shibasaki, F., McKeon, F., Bobo, T., Franke, F.F. and Reed, J.C. (1999) Ca^{2+}-induced apoptosis through calcineurin dephosphorylation of BAD. *Science*, **284** 339-43.

Witham, S., Dinesh-Kumar, S.P., Choi, D., Hehl, R., Corr, C. and Baker, B. (1994) The production of the tobacco mosaic virus resistance gene, *N*: similarity to toll and the interleukin-1 receptor. *Cell*, **78** 1101-15.

Wright, M.E., Han, D.K., Carter, L., Fields, S., Schwartz, S.M. and Hockenbery, D.M. (1998) Caspase-3 inhibits growth in *Saccharomyces cerevisiae* without causing cell death. *FEBS Lett.*, **446** 9-14.

Xu, F.X. and Chye, M.L. (1999) Expression of cysteine proteinase during developmental events associated with programmed cell death in brinjal. *Plant J.*, **17** 321-27.

Xu, H. and Heath, M.C. (1998) Role of calcium in signal transduction during the hypersensitive response caused by basidiospore-derived infection of the cowpea rust fungus. *Plant Cell*, **10** 585-98.

Xu, Q. and Reed, J.C. (1998) Bax inhibitor-1, a mammalian apoptosis suppressor identified by functional screening in yeast. *Mol. Cell*, **1** 337-46.

Yang, R.B., Mark, M.R., Gray, A., Huanh, A., Xie, M.H., Zhang, M., Goddard, A., Wood, W.I., Gurney, A.L. and Godowski, P.J. (1998) Toll-like receptor-2 mediates lipopolysaccharide-induced cellular signalling. *Nature*, **395** 284-88.

Yu, I.C., Parker, J. and Bent, A.F. (1998) Gene-for-gene disease resistance without the hypersensitive response in *Arabidopsis dnd1* mutant. *Proc. Natl. Acad. Sci. USA*, **95** 7819-24.

Zhang, J.H., Liu, X.S., Scherer, D.C., Vankaer, L., Wang X.D. and Xu, M. (1998a) Resistance to DNA fragmentation and chromatin condensation in mice lacking the DNA fragmentation factor 45. *Proc. Natl. Acad. Sci. USA*, **95** 12480-85.

Zhang, S. and Klessig, D.F. (1997) Salicylic acid activates a 48-kDa. MAP kinase in tobacco. *Plant Cell*, **9** 809-24.

Zhang, S. and Klessig, D.F. (1998) Resistance gene *N*-mediated *de novo* synthesis and activation of a tobacco mitogen-activated protein kinase by tobacco mosaic virus infection. *Proc. Natl. Acad. Sci. USA*, **95** 7433-38.

Zhang, S., Du, H. and Klessig, D.F. (1998b) Activation of the tobacco SIP kinase by both a cell-wall-derived carbohydrate elicitor and purified proteinaceous elicitins from *Phytophthora* spp. *Plant Cell*, **10** 435-49.

Zhou, J., Tang, X. and Martin, G.B. (1997) The pto kinase conferring resistance to tomato bacterial speck disease interacts with proteins that bind a *cis*-element of pathogenesis-related genes. *EMBO J.*, **11** 3207-18.

Zimmerman, S., Nürnberger, T., Frachisse, J.-M., Wirtz, W., Guern, J., Hedrich, R. and Scheel, D. (1997) Receptor-mediated activation of a plant Ca^{2+}-permeable ion channel involved in pathogen defense. *Proc. Natl. Acad. Sci. USA*, **94** 2751-55.

8 Systemic acquired resistance

Claire Barker

8.1 Introduction

Systemic acquired resistance (SAR) (Ross, 1961) is a form of inducible resistance which is activated, upon primary infection, throughout the whole plant in a resistance gene independent manner (Kuc, 1982; Lawton *et al.*, 1995). Often paralleled historically to animal immunity (Chester, 1933), this long-lasting, broad spectrum resistance mechanism manifests itself as a reduction in lesion size and number after subsequent infection by avirulent pathogens or the activation of hypersensitive response (HR)-like defences in response to virulent challenge (Ryals *et al.*, 1994, 1996). The nature of the resistance induced has led to the concept that SAR acts to prime host cells for a more rapid future deployment of defences (Kessmann *et al.*, 1994).

Biologically reproducible examples of SAR have been established in many species, including cucurbits, bean, tomato and *Arabidopsis*, upon induction by bacterial, fungal and viral pathogens (Kuc, 1982; Uknes *et al.*, 1992; Kessmann *et al.*, 1994). The spectrum of pathogens against which systemic resistance is effective remains constant for each plant species, irrespective of the nature of the inducing pathogen. However, this spectrum varies between species; thus, SAR can be considered to provide a characteristic 'fingerprint of protection' that has proved useful in discriminating SAR from other resistance mechanisms (Ryals *et al.*, 1996). SAR is not effective against all pathogens; notable exceptions include the lack of protection of tobacco against challenge by *Botrytis cinerea* and *Pseudomonas syringae* pv. *tomato* DC3000 (Pst DC3000) (Friedrich *et al.*, 1996) and the inability to immunise cucurbits with, and against, powdery mildew (Kuc, 1982).

To facilitate the identification of SAR at the molecular level, pathogenesis related (PR) gene expression and protein accumulation have been used extensively as markers of the resistant state (Ryals *et al.*, 1996). Initially, Ward *et al.* (1991) determined that the expression of messenger ribonucleic acids (mRNAs) encoding nine different PR protein isoforms could be positively correlated, both temporally and quantitatively, with the onset of resistance in tobacco. The induced genes comprised a subset of the known tobacco PR genes, with those encoding an acidic PR-1, a class II glucanase and a class III chitinase being most strongly expressed (Ward *et al.*, 1991). A similar subset of acidic isoforms has since been correlated with resistance in cucumber and *Arabidopsis*. In

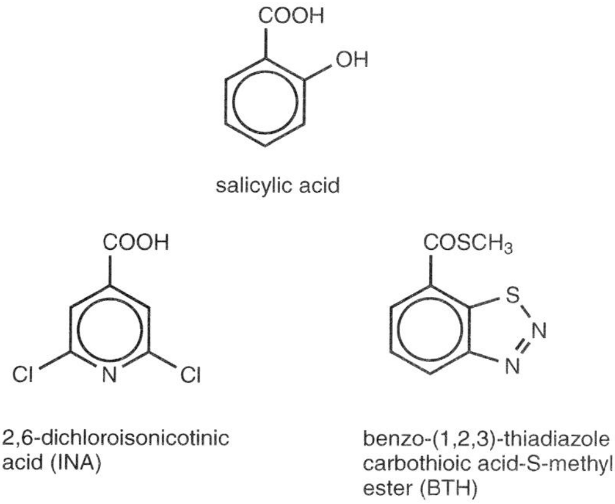

COOH
OH

salicylic acid

COOH
Cl N Cl

2,6-dichloroisonicotinic acid (INA)

$COSCH_3$
S
N
N

benzo-(1,2,3)-thiadiazole carbothioic acid-S-methyl ester (BTH)

Figure 8.1 Structural formulae of the three proven chemical activators of systemic acquired resistance.

Arabidopsis, PR-1, PR-2 and *PR-5* transcripts principally accumulate during SAR and, as with tobacco, *PR-1* transcripts accumulate most extensively (Uknes *et al.*, 1992, 1993). Consequently, PR-1 expression at the mRNA or protein level is often used as the predominant marker for SAR.

In addition to biological induction, certain chemicals also have SAR activating properties. These include salicylic acid (SA) (White, 1979), 2,6 dichloroisonicotinic acid (INA) (Uknes *et al.*, 1992; Vernooij *et al.*, 1995) and benzo (1,2,3) thiadiazole-7-carbothioic acid S-methyl ester (BTH) (Friedrich *et al.*, 1996; Görlach *et al.*, 1996; Lawton *et al.*, 1996). All three induce the same biochemical markers and spectrum of resistance as in the biologically induced state and are proposed to act at similar positions in the SAR signalling pathway. Interestingly, these three activators show structural similarity not only to one another (Figure 8.1) but also to the natural stress messenger molecules, trigonellin and nicotinamide (Schneider *et al.*, 1996).

8.2 Biochemical analysis of systemic signalling

8.2.1 The role of salicylic acid in achieving resistance

SA and H_2O_2 are the only two molecules known, so far, to have a proven role in the onset and maintenance of SAR. SA and its conjugated glycoside derivative (SAG) have repeatedly been shown to accumulate to high levels, in both local and systemic tissue, after primary infection of several plant species (Malamy *et al.*, 1990; Lawton *et al.*, 1995). Accumulation has also been observed in the phloem (Métraux *et al.*, 1990; Smith-Becker *et al.*, 1998). The role of such accumulation has been investigated through the generation of transgenic

Arabidopsis and tobacco lines, which are incapable of accumulating SA due to the constitutive expression of the *nahG* transgene. This bacterial gene encodes the enzyme, salicylate hydroxylase, which converts SA into biologically inactive catechol. As these lines neither accumulated SA, expressed *PR-1* systemically, nor exhibited enhanced resistance after inducing infection, it was concluded that SA is essential for SAR (Gaffney *et al.*, 1993; Ryals *et al.*, 1995). However, grafting experiments using wild-type and *nahG* rootstocks and scions demonstrated that, although SA accumulation is necessary for the manifestation of enhanced resistance in systemic tissue, it is not required at the site of systemic signal release (Ryals *et al.*, 1995). Therefore, SA is not the systemically transported SAR signal. Similar conclusions were reached by Pallas *et al.* (1996), after conducting grafting experiments using transgenic tobacco lines in which sense suppression of the endogenous phenylalanine ammonium lyase (*PAL*) gene was used to prevent SA accumulation.

An unexpected secondary observation concerning the depletion of SA in *nahG* lines was that SA also appeared to alter the efficacy of local responses during the HR. A breakdown of genetic resistance was observed for several *R-Avr* gene interactions where SA accumulation was prevented. Therefore, it appears that SA is instrumental in regulating both local and systemic defence responses (Ryals *et al.*, 1995).

8.2.2 SA responsive signal transduction and gene expression

In an attempt to identify components that directly interact with SA, two SA binding proteins have been isolated: SABP (Chen *et al.*, 1993) and SABP2 (Du and Klessig, 1997a). The former is a highly abundant protein which binds SA with a rather low affinity ($K_D = 14\,\mu M$). Confirmed as being a catalase isoform, binding of SA can inhibit the activity of the enzyme by 50–80% (Chen *et al.*, 1993; Du and Klessig, 1997a). Catalase is considered to be the major cellular sink for H_2O_2 in the cell (Willekens *et al.*, 1997) and SA inhibits the enzyme by acting as an electron donor to the slow beta-peroxidative cycle, rather than by directly chelating the haem group (Durner and Klessig, 1996). INA and BTH are also effective inhibitors of catalase and ascorbate peroxidase in tobacco (Conrath *et al.*, 1995; Wendehenne *et al.*, 1998). However, the analysis of catalase deficient tobacco lines has shown that H_2O_2 accumulation occurs upstream of SA accumulation, rather than SA binding of catalase increasing cellular H_2O_2 levels (Du and Klessig, 1997b; Chamnongpol *et al.*, 1998). Thus, the most probable role for SA binding to haem-containing, active oxygen species (AOS) generating enzymes is in the feedback propagation of the H_2O_2 signal, amplifying the cellular response.

SABP2 ($K_D = 90\,nM$) has a far greater affinity for SA than SABP, and the binding and displacement kinetics of SA to SABP2 are far more rapid. As this protein is in low abundance in the cell, yet binds with high enough affinity to be active even at the relatively low levels of SA found in systemically induced tissue, SABP2 has potential receptor-like properties. It remains to be seen whether SABP2 is, indeed, a receptor, or acts in SA transport, non-defence-related pathways or has some biologically significant activity (Du and Klessig, 1997a).

Downsteam of SA perception, recent evidence has raised the possibility that calmodulin may be involved in the signalling cascade leading to defence gene expression. Constitutive expression of two soybean calmodulin isoforms in transgenic tobacco plants triggered spontaneous lesion formation and enhanced resistance to a variety of virulent and avirulent pathogens. Resistance was associated with the constitutive, high level expression of all nine SAR marker genes, irrespective of whether lesions had formed. As this expression was independent of SA accumulation, as shown by generating *nahG* double transgenic lines, the authors proposed that the expression of resistance occurred via an SA-independent pathway (Heo *et al.*, 1999). However, as the genes induced were identical to the SAR marker genes previously identified in tobacco by Ward *et al.* (1991), it is also possible that the calmodulin isoforms expressed act downstream of SA accumulation to induce defences.

The definitive role of SA in achieving the resistant state remains to be determined. However, many genes have been identified which are directly responsive to SA, in addition to the characterised SAR marker genes (Ward *et al.*, 1991). These include genes encoding mannitol dehydrogenase, anionic peroxidase and the ethylene response protein, EREBP1 (Horvath *et al.*, 1998; Katz *et al.*, 1998; Thulke and Conrath, 1998). The presence of SAR activators may also indirectly affect gene expression by potentiating responses to other stimuli, evident for genes such as *PAL, hydroxyproline rich glycoprotein* and *4-coumarate:coA ligase* (Kästner *et al.*, 1998; Katz *et al.*, 1998; Thulke and Conrath, 1998). Prior treatment with activators such as SA, INA and BTH can render cells more responsive to even very low elicitor concentrations, analogous to the priming seen during SAR. Moreover, augmentation of responses has been observed to be time dependent (Kästner *et al.*, 1998; Katz *et al.*, 1998). This suggests that it is not merely the presence of the SAR activator that is required for potentiation but rather some prior cellular response to it, which may involve the expression of directly responsive genes and the synthesis of new cellular components.

8.2.3 The role of hydrogen peroxide in establishing systemic immunity

Hydrogen peroxide has, on numerous occasions, been reported to induce SAR and is required upstream of SA accumulation (Neuenschwander *et al.*, 1995; Sharma *et al.*, 1996). However, as no rise in H_2O_2 was observed in non-inoculated leaves of tobacco during the onset and maintenance of resistance, H_2O_2 was not originally considered to act as an integral factor in the establishment of resistance in systemic tissue (Neuenschwander *et al.*, 1995).

Recent evidence challenges this view and suggests that the earlier inability to detect H_2O_2 in systemic tissue may have been due to its transient production in very few, highly localised cells (Alvarez *et al.*, 1998). Alvarez and co-workers initially delineated a time window of response in systemic tissue by detecting the expression of the antioxidant *glutathione-S-transferase* (*gst*) gene. Two waves of systemic expression were observed upon inoculation of *Arabidopsis* with avirulent Pst DC3000 (*avrRpt2*), 4–6 and 55–70 h post-inoculation. In contrast, a single wave of expression occurred locally after 1–6 h.

As *gst* was considered to reflect the possible production of AOS, diaminobenzidine (DAB) was used to histochemically detect endogenous H_2O_2. Production occurred just once in local and systemic tissue, 2 h and 3–4 h post-inoculation, respectively. Inhibition of the H_2O_2 generating plasma membrane NADPH oxidase complex with exogenous diphenylene iodinium (DPI) proved H_2O_2 production to be essential for triggering both the first and second waves of systemic *gst* induction and nuclear apoptosis. DPI applied at the site of inoculation prevented H_2O_2 production, *gst* transcription and cell death in both primary and systemic tissue. The development of resistance to virulent pathogens was also precluded and similar results were obtained when catalase was co-infiltrated along with the inducing Pst pathogen, to degrade any H_2O_2 produced. Consequently, H_2O_2 production at the site of inoculation is a necessary, early event in the establishment of systemic signalling activity (Alvarez *et al.*, 1998).

However, if DPI was injected into systemic tissue 2.5 h after inducing inoculation, i.e. just prior to the systemic production of H_2O_2, resistance was again abolished. Thus, a micro oxidative burst in systemic tissue immediately following systemic signal perception initiates resistance activation. This single systemic microburst directs both the initial and later waves of *gst* induction and the accompanying production of apoptotic features. Alvarez *et al.* (1998), therefore, proposed that multiple reiterations of micro AOS generation are required to produce and sustain systemic immunity. H_2O_2 has previously been shown to act as a local signal for *gst* induction (Levine *et al.*, 1994). Whether it has sufficient range to act as a systemic signal remains to be determined, due to its rapid rate of degradation.

8.3 Genetic analysis of the SAR pathway

8.3.1 Lesion mimic mutants

In an attempt to dissect the SAR signalling pathway at the genetic level, a number of *Arabidopsis* mutants have been generated which display aberrant SAR gene and resistance activation (Table 8.1). Accelerated cell death (*acd1*, *acd2*) (Greenburg *et al.*, 1994) and the lesions simulating disease (*lsd*) mutant

Table 8.1 A summary of the characteristics of the known SAR mutants and their possible role in the systemic signalling pathway

Mutant	Dominant/ recessive	Lesions formed	SAR genes expressed	ISR genes expressed	Possible role	References
acd2	Recessive	Yes	Lesion +ve only	Lesion +ve only	Cross-talk?	Greenburg *et al.*, 1994
cpr1	Recessive	No	Submaximal	Submaximal	Cross-talk?	Bowling *et al.*, 1994
cpr5-1	Recessive	Yes	Yes	Yes	Pathogen sensing?	Bowling *et al.*, 1997
cpr6-1	Dominant	No	Submaximal	Yes	Cross-talk?	Clarke *et al.*, 1998
lsd1	Recessive	Yes	Yes	?	Control of lesion spread	Dietrich *et al.*, 1997
lsd2, 4	Dominant	Yes	Lesion +ve only	?	Lesion regulation?	Dietrich *et al.*, 1994
lsd3, 5	Recessive	Yes	Lesion +ve only	?	Lesion regulation?	Dietrich *et al.*, 1994
lsd6, 7	Dominant	Yes	Lesion +ve only	?	Feedback control of SA?	Dietrich *et al.*, 1994
npr1	Recessive	No	No	?	Partitioning of defence gene expression	Cao *et al.*, 1994
ssi1	Dominant	Yes	Submaximal	Yes	Regulation of SA/NPR1 independent gene expression	Shah *et al.*, 1999

Abbreviations: SAR, systemic acquired resistance; ISR, induced systemic resistance; *acd*, accelerated cell death; *cpr*, constitutive expresser of pathogenesis related (PR) proteins; *lsd*, lesions simulating disease; *npr*, non-expresser of PR proteins; *ssi*, suppressor of salicylic acid (SA) insensitivity.

classes (Dietrich *et al.*, 1994) are examples of this type. Both mutant classes are associated with the spontaneous formation of HR-like lesions, elevated SA and SAG levels, constitutively high expression of SAR marker genes and enhanced resistance to normally virulent pathogens (Dietrich *et al.*, 1994; Greenburg *et al.*, 1994; Weymann *et al.*, 1995).

All of the *lsd* lines, except *lsd1*, formed determinate lesions, which were conditional to day length and/or humidity. Under suppressive environmental conditions, which prevented lesion formation, no enhanced resistance was observed. However, upon shift to permissive conditions, the formation of lesions was accompanied by SA accumulation and the onset of resistance to virulent pathogens (Dietrich *et al.*, 1994; Weymann *et al.*, 1995). Elevation of SA was responsible for *PR* gene expression and resistance in *lsd2* and *lsd4*, as shown by introducing *nahG* into the mutant background, although lesion formation was SA independent. Therefore, *lsd2* and *lsd4* are considered to reside at a position upstream of SA accumulation, but before a branch point initiating lesion formation, in the SAR signalling pathway (Weymann *et al.*, 1995; Hunt *et al.*, 1996).

In contrast, suppression of SA accumulation in *lsd6* and *lsd7* prevented *PR* gene expression, the onset of resistance and lesion formation. The co-segregating, distorted leaf phenotype of *lsd6* was also suppressed in the presence of *nahG*. As INA and SA application could restore resistance and the phenotype of *lsd6*, it was concluded that this gene may act in a feedback loop involved in the regulation of SA accumulation and lesion formation. However, no lesion restoration was evident in the *nahG lsd7* line after SA or INA treatment (Weymann *et al.*, 1995). The roles of the wild-type gene products of *lsd 2-7* are unknown. Nevertheless, as both dominant and recessive *lsd* mutations have arisen, the encoded products must be involved in both positive and negative regulatory control of signalling pathways.

8.3.2 LSD1, a regulator of lesion spread

During initial characterisation, it became apparent that *lsd1* showed remarkably different properties to any of the other lines, particularly since this mutant was resistant in both the lesion positive and lesion negative state. The regulation of lesion formation in this line was highly aberrant, with lesions induced by very low inoculum loads, 10^5 colony-forming units (cfu)/ml of bacteria, which would have no discernible effect on the wild-type. Lesions also formed in response to the non-pathogen, *P. s.* pv. *phaseolicola*, and by mutant pathogens which were unable to elicit HR in the wild-type. However, no lesions formed in response to wounding (Dietrich *et al.*, 1994).

Curiously, application of SAR activating chemicals caused lesions, leading to the implication that the maintenance of SAR may modulate any subsequent lesion development. Dietrich *et al.* (1994) concluded that *lsd1* represented a class of *lsd* mutant defective in the feedback, or propagation, of a signal, resulting in the inability of the mutant to control the advancing lesion and producing a spreading lesion habit, which eventually causes lethality.

Jabs *et al.* (1996) demonstrated that lesion spread in *lsd1* is controlled by apoplastic superoxide, as spot inoculation of xanthine/xanthine oxidase (which generates superoxide) into the intercellular spaces of *lsd1* leaves caused spreading lesions. Similar application of glucose/glucose oxidase (to generate H_2O_2) did not trigger lesion formation, and exogenous application of 30 mM H_2O_2 caused cell death at the point of application but lesions remained confined. Therefore, the *lsd1* mutation appears to affect a key component in the control of signalling between cells, mediated by superoxide (Jabs *et al.*, 1996).

The *Lsd1* gene has been mapped and cloned. The wild-type ~1.2 kb transcript is rare, but constitutively expressed in all tissues and two homologous EST sequences identified indicate that the transcript can be alternatively spliced. Transcription is not affected by INA application and the encoded protein contains zinc finger motifs, with the greatest homology observed with the plant GATA1 family of transcription factors. The function of LSD1 cannot be predicted on the basis of homology alone; however, the negative regulatory nature inferred by the *lsd1* mutation suggests that it may act to dampen cellular responses to superoxide until a critical threshold is reached. As HR mediated cell death in *lsd1* mutants (a null allele) is unaffected, it appears that only extracellular signalling is compromised by the mutation (Dietrich *et al.*, 1997).

Significantly, SA application has been shown to stimulate superoxide generation via salicylhydroxamic acid (SHAM) sensitive enzymes, possibly extracellular peroxidases, such as guaiacol peroxidase (Kawano *et al.*, 1998). Thus, it is possible that the initiation of lesions in *lsd1* by SA application is due to superoxide generated by these enzymes. Once the threshold of tolerance to superoxide, maintained via *lsd1* and related mechanisms, is surpassed, then pro-cell-death signals are generated in adjacent cells. As a time lag of 12–16 h is observed between the generation of superoxide and the initiation of cell death in *lsd1* mutants, this time lag may allow for the induction of pro-cell-death-intermediates (Dietrich *et al.*, 1997). Whether LSD1 responds to superoxide directly, or via superoxide-derived signals, now requires clarification.

8.3.3 Second site mutations: the phx *loci*

To identify additional cell death and resistance regulatory genes, *lsd5* has been further mutagenised and second site mutations have been isolated, which abolish the lesion forming phenotype of the single mutant. The screen for lesion suppression involved growing mutagenised *lsd5* seedlings under short day conditions, a conditional lethal environment for *lsd5*, and screening for survivors. Nine individual loci were identified in this way, the loci being designated *phx* (after the mythological bird, phoenix, which rises from the ashes) (Morel and Dangl, 1999).

Of the nine loci, *phx 2, 3, 6* and *11-1* displayed the strongest suppression of *lsd5* phenotype, totally preventing lesion formation and *PR-1* expression. However, *PR-1* could be induced by exogenous SA application, hence the signalling pathway downstream of SA accumulation remains intact in these mutants. When disease resistance was examined in the *phx/lsd5* lines, only the strongest *phx* suppressers displayed altered disease phenotypes upon infection. *phx2/lsd5* and *phx3/lsd5* showed a distinct breakdown of gene-for-gene resistance when inoculated with avirulent Pst DC3000 (*avrRpm1*), allowing bacterial growth to occur to a level comparable to the compatible interaction. Chlorotic symptoms formed after 5 days, despite the presence of *lsd5*, which would normally condition enhanced resistance to pathogens. The double mutant line, *phx2/lsd5*, was also more susceptible to the compatible *Peronospora parasicita* isolate, Emwa1.

The role of the genes affected by the *phx* mutations remains to be determined. However, the four strongest suppressors are very different in their genetic derivation: *phx2* is dominant and linked to *lsd5; phx3* is dominant and unlinked; *phx6* is recessive and unlinked; *phx11-1* is recessive and linked to *lsd5* (Morel and Dangl, 1999). Consequently, it will be interesting to discover the roles these loci play in the cell death and disease resistance signalling pathways.

8.3.4 NPR1 in SA regulated gene expression

So far, only one mutant line has been isolated in which the activation of SAR genes and resistance is prevented, even in the presence of SAR activators, such as SA, INA and BTH. This mutant, therefore, defines a genetic lesion in a component essential for signalling events downstream of SA accumulation and has been designated *npr1* (non-expresser of PR proteins) (Cao *et al.*, 1994).

npr1 mutants were originally identified in a screen of the mutagenised progeny of primary *BGL2:GUS* transformants for those incapable of expressing the reporter gene in the presence of SA and INA. The mutant line isolated, *npr1*, expressed neither the GUS reporter gene, nor endogenous SAR genes, in response to infection, SA or INA treatment. Accordingly, these mutants were also incapable of displaying enhanced resistance to virulent pathogens after SA or INA treatment, indicative of an inability to activate SAR (Cao *et al.*, 1994).

Numerous mutations allelic to *npr1* have since been isolated using several different screening criteria: six *nim1* (non-inducible immunity) alleles were isolated due to their susceptibility to virulent *P. parasicita* infection even after INA application (Ryals *et al.*, 1997); *sai1* (salicylic acid insensitive) was identified via its inability to activate an SA responsive promoter linked to a selectable marker gene (Shah *et al.*, 1997); and *eds*-17 (enhanced disease susceptibility) was selected for its production of symptoms upon infection with inoculum titres below that which would cause symptoms on wild-type plants (Volko *et al.*, 1998). The repeated occurrence of *npr1* alleles indicates that this locus may play an important role in the regulation of SAR gene expression.

As the *npr1* mutation prevents SA induced gene expression, it might also be expected to affect HR-mediated resistance, as this response equally requires the accumulation of high levels of SA (Ryals *et al.*, 1995). However, at present, this relationship is unresolved. Cao *et al.* (1994) reported that *npr1* mutants were unaffected in their response to avirulent pathogens, consistent with the response of *eds-17* (Volko *et al.*, 1998). Yet, bacterial growth rates in *sai1* mutants were reported to be 20 times higher than in the wild-type after inoculation with avirulent *Pseudomonas* strains (Shah *et al.*, 1997). Consequently, comparison of alleles under uniform conditions is required to confirm this point.

Nevertheless, it is clear that NPR1 acts locally to reduce the severity of symptoms following infection by virulent pathogens. This conclusion was reached following the formation of larger, less defined yellow lesions and an extended area of pathogen colonisation upon infection of *npr1-1* plants with a virulent Pst strain (Cao *et al.*, 1994). Severe symptoms were evident even if inoculum load was reduced tenfold (Cao *et al.*, 1997), consistent with the enhanced disease susceptibility phenotype identified by Volko *et al.* (1998) in the isolation of *eds-17*. Thus, NPR1 acts as a downstream element in a signalling cascade which is activated independently of HR-mediated resistance and which appears to be potentiated upon induction of SAR. The role of NPR1 may also involve feedback control of SA biosynthesis, as both *sai1* and *npr1-1* plants appear to hyper-accumulate SA upon infection (Cao *et al.*, 1997; Shah *et al.*, 1997).

The *NPR1* gene resides on the short arm of chromosome 1 and has been isolated by map-based cloning (Cao *et al.*, 1997; Ryals *et al.*, 1997). The gene comprises four exons, which produce an approximately 2-kb mRNA. The predicted NPR1 translation product includes four regions with homology to the ankyrin consensus sequence, identified in a wide range of proteins, such as mammalian transcription factors and cellular regulators. These repeats are generally found in at least four copies and are thought to promote protein-protein interactions and cooperative folding (Cao *et al.*, 1997). In addition, Ryals *et al.* (1997) noted that the protein bore a striking resemblance to members of the $I_{\kappa}B_{\alpha}$ class of transcription factors. $I_{\kappa}B_{\alpha}$ and NPR1 were found to share an overall sequence similarity of 80%, with two N-terminal serine residues, a pair of lysines and an acidic C-terminus conserved between the two. Thus, it has been proposed that NPR1 may act in a manner analogous to its animal counterparts, which are involved in a highly conserved pathway regulating defensive and immunological reactions in animals and flies (Ryals *et al.*, 1997).

The function of $I_{\kappa}B_{\alpha}$ in animals is to inhibit the transcription factor, NF-κB. In the non-induced cell, $I_{\kappa}B_{\alpha}$ binds to NF-κB to form a complex, which prevents the transcription factor from entering the nucleus. Reception of a suitable stimulus initiates the phosphorylation of two particular serine residues of the $I_\kappa B_\alpha$ protein, which enables it to be ubiquitinated at the lysine pair. These residues appear to be conserved in the NPR1 protein. Upon addition of the ubiquitin signal, the whole complex is transported for selective proteolytic degradation, to release NF-κB whilst destroying $I_\kappa B_\alpha$. The free NF-κB is then able to enter the nucleus and positively promote gene transcription (Ryals *et al.*, 1997).

For this mechanism to be active in plants, the NF-κB transcription factor corresponding to NPR1 must be a repressor (rather than an activator) of defence gene expression, in order to account for the phenotype of the *npr1* mutant alleles. In favour of this hypothesis, overexpression of NPR1 in *npr1* mutants (Cao *et al.*, 1997) and wild-type plants (Cao *et al.*, 1998) to produce a moderate 1.5–3 fold increase in NPR1 protein levels led to significant resistance to the virulent pathogens, *P. s.* pv. *maculicola* (Psm) ES4326 and *P. parasicita* NOCO. Resistance was associated with higher level, but not more rapid, *PR-1, PR-2* and *PR-5* expression, which could be positively correlated with the extent of NPR1 protein accumulation. Importantly, overexpression of NPR1 did not lead to constitutive defence activation and a stimulus was still required to signal through the defence gene-inducing pathway (Cao *et al.*, 1998). Moreover, as *NPR1* itself is upregulated twofold by SA and INA (Cao *et al.*, 1997), a mechanism can be proposed whereby the systemic signal, upon triggering SA accumulation, leads to enhanced NPR1 protein accumulation. The abundance of NPR1 protein more effectively prevents negative regulation of defence genes, thus upon secondary stimulation the defence response is potentiated. As such, NPR1 may act as a partitioning factor, favouring positive SAR gene inducing signals over those that may act to repress the response.

In support of such a ubiquitin-dependent pathway, it is interesting that defence activation is constitutively enhanced if the ubiquitin proteolytic degradation pathway is disturbed. Becker *et al.* (1993) altered this activity by expressing a non-functional ubiquitin that prevents proteolysis in some tobacco lines, and by co-suppressing the endogenous gene in others. Like the *lsd* mutant lines, ubiquitin transgenics displayed necrotic lesions in a conditional manner when placed under stress. Additionally, high expression of the variant ubiquitin protein led to stunted growth and crippled leaves, reminiscent of the phenotype of *lsd6*. Lesions were induced in response to TMV infection, even though the transformed lines were genetically susceptible to the pathogen, and in each case lesion formation paralleled *PR-1* gene expression.

At the time, Becker *et al.* (1993) speculated that a particular element of the signalling pathway may be constitutively expressed at a low level, but maintained in an inactive state by ubiquitination. However, these results can now be interpreted in terms of NPR1 activity, whereby disturbance of the proteolytic degradation of NPR1 prevents a negative regulator from entering the nucleus, favouring defence gene induction over other genes that may be expressed in response to simultaneous stimuli.

8.4 SA/NPR1 independent resistance pathways

8.4.1 Alternative resistance pathways: induced systemic resistance

During the characterisation of the various alleles of *npr1*, it became apparent that, although the major regulation of *PR-1* expression occurs via SA/NPR1 dependent pathways, some *PR-1* expression was retained in *npr1* mutants. As *PR-1* expression occurred after inoculation with a pathogen, but not in response to SA or INA treatment, this expression is likely to be regulated via an independent signalling pathway (Ryals *et al.*, 1997; Shah *et al.*, 1997). Therefore, other signalling pathways must be active, in addition to those responsive to SA/NPR1, in order to produce the resistant state.

The phenomenon of induced systemic resistance (ISR) has recently been reviewed by van Loon *et al.* (1998). Like SAR, this resistance mechanism involves the induction of systemic resistance, affording enhanced protection to normally virulent pathogens (Pieterse *et al.*, 1996). However, rather than induction by phytopathogens, ISR is activated in a plant ecotype and pathogen strain specific manner by plant growth promoting rhizobacteria (PGPR) (van Wees *et al.*, 1997). ISR is not associated with SAR marker gene expression and is independent of SA accumulation (Pieterse *et al.*, 1996; van Wees *et al.*, 1997). It also appears to provide protection against pathogens, such as *B. cinerea* and *Alternaria brassicicola*, for which SAR is ineffective and vice versa (Thomma *et al.*, 1998). Nevertheless, evidence exists for overlap between the mechanisms regulating resistance in SAR and ISR.

As *PR-1, PR-2* and *PR-5* are taken as the marker genes of SAR in *Arabidopsis* (Uknes *et al.*, 1992, 1993), so *PR-3, PR-4, Thi2.1* and *PDF1.2* are the equivalent local and systemic markers of ISR activation (Penninckx *et al.*, 1996; Thomma *et al.*, 1998). *PDF1.2* is a pathogen inducible defensin gene, which is not responsive to SA or INA application but which is induced by paraquat (which forms superoxide), rose bengal (which forms singlet oxygen) and *A. brassicicola* infection. The major regulatory molecules in *PDF1.2* expression appear to be ethylene and jasmonic acid (JA), and a regulatory pathway has been proposed whereby the specific interaction between plant and biocontrol bacterium leads to the production of AOS. Once generated, the AOS triggers ethylene and JA response pathways in parallel, the activation of both of which is necessary for defensin gene expression (Penninckx *et al.*, 1996, 1998).

Although NPR1 and SA accumulation are not major factors in the induction of defensin expression, they may play a modulating role, as the level of ISR protection is reduced in *nahG* and *npr1* plants (Van Wees *et al.*, 1997; Pieterse *et al.*, 1998; Thomma *et al.*, 1998). Several lines of evidence indicate that there may be cross-talk between the SA and JA response pathways, and elevated levels of defensin are observed in *nahG* transgenic lines. Additionally, the *acd2* mutant (Greenburg *et al.*, 1994) was found to produce high levels of defensin and accumulate JA in addition to SA. In young, necrotic leaves of *acd2* plants, defensin contributed ~5% of the total soluble protein and JA levels were elevated threefold. These values were increased to ~10% and ninefold in older, necrotic leaves (Penninckx *et al.*, 1996). Consequently, the *acd2* mutation may either lie upstream of a branch point between these two defence responses or may represent an element that regulates cross-talk between the two pathways.

8.4.2 *The* cpr *mutant class*

One particular class of SAR regulatory mutants which appear to be affected in their regulation of both SAR and ISR are the *cpr* (constitutive expresser of *PR* protein) mutants, identified in a screen developed by Bowling *et al.* (1994). The screen involved mutagenisation of *Arabidopsis* transformants to identify those that constitutively expressed a *BGL2:GUS* transgene. Three *cpr* mutants have been described so far, *cpr1, cpr5-1* and *cpr6-1*, all of which show constitutive expression of *PR-1, PR-2, PR-5* and *PDF1.2*, heightened resistance to virulent pathogens and elevated endogenous SA and SAG levels. *cpr5-1* additionally displays spontaneous lesion formation, whereas *cpr1* and *cpr6-1* are not associated with a lesion mimic phenotype (Bowling *et al.*, 1994, 1997; Penninckx *et al.*, 1996; Clarke *et al.*, 1998).

The first of the class to be described, *cpr1*, is a recessive mutation that lies upstream of SA accumulation. Although *cpr1* shows constitutively high *PDF1.2* expression, clearly it is not maximal in this mutant as *A. brassicola* infection causes further induction of the defensin gene (Penninckx *et al.*, 1996). Likewise, SAR was not considered to be maximally activated, as INA application could further enhance SAR gene expression and pathogen resistance (Bowling *et al.*, 1994). Thus, in *cpr1*, both the ISR and SAR pathways are operative, yet neither is activated to full capacity.

The *cpr1* mutant has a distinctive phenotype of reduced plant size and dense production of trichomes on the adaxial leaf surface. The leaves are also dark and narrow, reminiscent of the phenotype of *lsd6* (Dietrich *et al.*, 1994) and, as with *lsd6*, the co-segregating phenotype is SA dependent (Bowling *et al.*, 1994). Interestingly, the phenotype of *cpr5-1* displays the opposite characteristics, with a reduced trichome number evident. The opposing effects of *cpr5-1* and *cpr1* on trichome number may reflect negative and positive regulatory functions of these two genes on the same developmental process.

Defence activation in *cpr5-1* mutants is considered maximally induced, as mutants were as resistant to virulent *P. parasitica* and Psm isolates as INA treated wild-types. *PR-1* expression in this line was fully SA/NPR1 dependent, whereas some expression of the *BGL2:GUS* transgene was retained in *cpr5-1 nahG* lines, indicative of SA independent induction. The expression of *PDF1.2* was unaffected by the presence of *nahG* or *NPR1* in the genetic background (Bowling *et al.*, 1997). In *cpr5-1* mutants, therefore, SA/NPR1 dependent and independent pathways are active and, in contrast to *cpr1*, the SA/NPR1 dependent branch is fully induced.

The *cpr6-1* mutation produced a distinctive co-segregating phenotype of delayed flowering, early cotyledon senescence and a loss of apical dominance. Interestingly, in this mutant neither the SAR marker genes nor *PDF1.2* and *Thi2.1* were subject to regulation by SA or NPR1. Although *cpr6-1* mutants were more resistant to Psm than wild-type plants, the level of protection afforded was considerably less than that of INA treated wild-type plants. INA treatment of *cpr6-1* enhanced both SAR marker gene expression and resistance to Psm, yet transiently decreased *PDF1.2* and *Thi2.1* expression. Thus, activation of the INA dependent SAR signalling pathway appears to partition resources towards transcription of the SAR marker genes.

As INA mediated resistance is NPR1 dependent, the authors proposed that NPR1 may act as a modifier of CPR6 responsive gene expression, favouring the transcriptional activation of particular subsets of defence genes. In support of this theory, despite the maintenance of defence gene expression and *P. parasitica* resistance in *cpr6-1 npr1* double mutants, resistance to Psm was abolished by *npr1* (Clarke *et al.*, 1998). A similar breakdown in resistance to Psm, but not *P. parasitica*, was previously observed for *cpr5-1 npr1* mutants (Bowling *et al.*, 1997). As *P. parasitica* is a biotrophic oomycete and Psm is a necrotrophic bacterium, NPR1 may facilitate the expression of particular defences against specific types of pathogen (Clarke *et al.*, 1998).

8.4.3 Regulation of SA/NPR1 independent resistance

A potential regulator of NPR1 independent gene expression has recently been described by Shah *et al.* (1999). *ssi1* (suppressor of SA insensitivity) was identified by mutagenisation of *npr1-5* seed and screening for individuals that show constitutive, high level expression of *PR-1*, despite the presence of the *npr1* mutation. This mutant shows constitutively elevated *PR-1, PR-2, PR-5* and *PDF1.2* expression, spontaneous lesion formation, elevated SA and SAG levels and a dwarf phenotype.

ssi1 is a dominant, *NPR1* unlinked mutation, in which the accumulation of SA is responsible for the dwarf phenotype and, unusually, the expression of *PDF1.2*. As the suppression of *PDF1.2* in *ssi1 nahG* lines can be reversed by BTH application, in this mutant the defensin gene appears to act like an SAR marker gene. However, co-regulation is unlikely as, although *PR-1* expression was further enhanced by Pst infection, *PDF1.2* was not. Instead, a transient decrease in transcript accumulation was observed, similar to INA treated *cpr6-1* (Clarke *et al.*, 1998; Shah *et al.*, 1999).

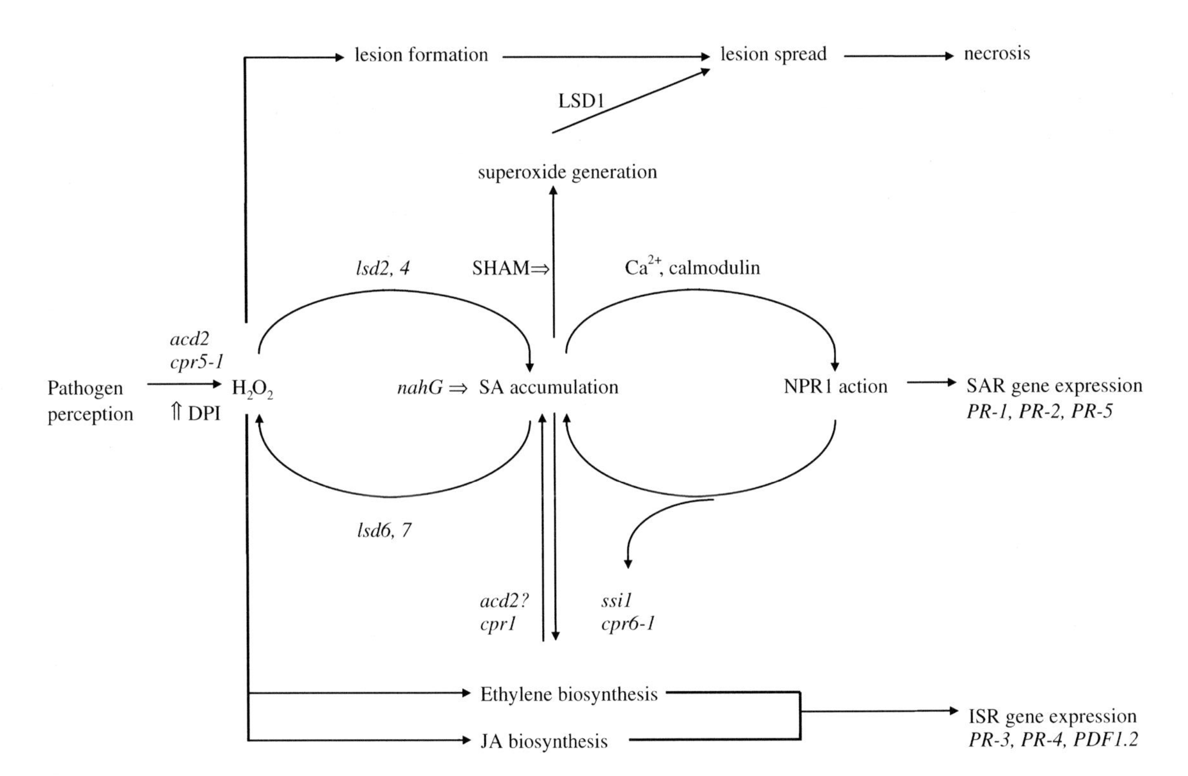

Figure 8.2 Schematic summary pathway of the events leading to the expression of systemic resistance. Abbreviations: DPI, diphenylene iodinium; JA, jasmonic acid; SHAM, salicylhydroxamic acid; ⇒, inhibited by; *lsd*, lesion simulating disease; *acd*, accelerated acid; SAR, systemic acquired resistance; ISR, induced systemic resistance; *ssi*, suppressor of SA insensitivity; NPR, non-expresser of PR proteins.

Analysis of defensin expression in homozygous and heterozygous *ssi1* mutants, in *npr1* and *NPR1* backgrounds, has again suggested that NPR1 regulates defensin expression in a negative manner. In contrast, as heterozygotes (*SSI1 ssi1*) express higher levels of defensin than homozygous mutants (*ssi1 ssi1*), SSI1 may have a positive regulatory influence on defensin expression.

In summary, a situation can be envisaged whereby SAR and ISR activation leads to the expression of particular sets of defence genes which condition resistance against a defined, exclusive spectrum of pathogens. Cross-talk must exist between the two pathways, with NPR1 acting as an oscillating switch, which commits the cell to one defence pathway (SAR) at the expense of others (ISR) (Figure 8.2). However, as SA and NPR1 are required in part for ISR (van Wees *et al.*, 1997; Pieterse *et al.*, 1998; Thomma *et al.*, 1998) and *PDF1.2* behaves, under certain circumstances, like an SAR gene (Shah *et al.*, 1999), these two mechanisms must have a subtle and highly coordinated degree of interface.

8.5 Concluding remarks

As knowledge concerning the regulation of SAR is enhanced, so the prospect of manipulating this endogenous resistance pathway for agricultural benefit draws ever closer. BTH has previously been shown to act as an effective crop protectant when applied to wheat and rice in field trials (Görlach *et al.*, 1996), and has since been made commercially available. Moreover, it has recently been suggested that the efficacy of certain fungicides may be due to synergism between fungicide action and the induction of SAR (Molina *et al.*, 1998), making systemic resistance activation a potentially important facet of crop protection. However, as the present wealth of information regarding SAR concerns predominantly dicotyledonous species, yet the agriculturally important crop species are monocotyledonous, a void of information relating to resistance in this latter class of plants must be filled before manipulation of the SAR pathway can be reliably utilised in agriculture.

Interestingly, it appears that *NPR1* homologues exist in numerous species, including *Brassica napus*, potato and wheat (Cao *et al.*, 1998). It is, therefore, possible that *NPR1* represents a highly conserved aspect of the resistance signal transduction pathway. If so, this gene would be an attractive target for the genetic manipulation of SAR. Overexpression of NPR1 is known to confer resistance in *Arabidopsis* (Cao *et al.*, 1998), and as defences remain inducible, the developmental abnormalities and metabolic stress associated with constitutive defence activation, as evidenced in mutants such as *lsd6* (Dietrich *et al.*, 1994) and *cpr1* (Bowling *et al.*, 1994), are avoided. Unfortunately, NPR1 does not appear to regulate all plant defence responses and its action provides protection against a subset of pathogens. Therefore, future schemes to elevate plant resistance may favour the dual activation of both SAR- and ISR-like defences concomitantly, to attain a broader spectrum of resistance. Such dual activation may be achieved by the use of jasmonates together with SAR activating agents (Thomma *et al.*, 1998). Further studies examining the regulatory mechanism governing SAR, ISR and their relationship to one another, may yield important information pertinent to the manipulation of plant defences in this manner.

Acknowledgements

Saara Lång is gratefully acknowledged for critical reading of this manuscript.

References

Alvarez, M.E., Pennell, R.I., Meijer, P.-J., Ishikawa, A., Dixon, R.A. and Lamb, C. (1998) Reactive oxygen intermediates mediate a systemic signal network in the establishment of plant immunity. *Cell*, **92** 773-84.

Becker, F., Buschfeld, E., Schnell, J. and Bachmir, A. (1993) Altered response to viral infection by tobacco plants perturbed in the ubiquitin system. *Plant J.*, **3** 875-81.

Bowling, S.A., Guo, A., Cao, H., Gordon, A.S., Klessig, D.F. and Dong, X. (1994) A mutation in *Arabidopsis* that leads to constitutive expression of systemic acquired resistance. *Plant Cell*, **6** 1845-57.

Bowling, S.A., Clarke, J.D., Liu, Y., Klessig, D.F. and Dong, X. (1997) The *cpr5* mutant of *Arabidopsis* expresses both *NPR1*-dependent and *NPR1*-independent resistance. *Plant Cell*, **9** 1573-84.

Cao, H., Bowling, S.A., Gordon, A.S. and Dong, X. (1994) Characterisation of an *Arabidopsis* mutant that is nonresponsive to inducers of systemic acquired resistance. *Plant Cell*, **6** 1583-92.

Cao, H., Glazebrook, J., Clarke, J.D., Volko, S. and Dong, X. (1997) The *Arabidopsis NPR1* gene that controls systemic acquired resistance encodes a novel protein containing ankyrin repeats. *Cell*, **88** 57-63.

Cao, H., Li, X. and Dong, X. (1998) Generation of broad-spectrum disease resistance by overexpression of an essential regulatory gene in systemic acquired resistance. *Proc. Natl. Acad. Sci. USA*, **95** 6531-36.

Chamnongpol, S., Willekens, H., Moeder, W., Langebartels, C., Sandermann, H., van Montagu, M., Inzé, D. and van Camp, W. (1998) Defense activation and enhanced pathogen tolerance induced by H_2O_2 in transgenic tobacco. *Proc. Natl. Acad. Sci. USA*, **95** 5818-23.

Chen, Z., Silva, H. and Klessig, D.F. (1993) Active oxygen species in the induction of plant systemic acquired resistance by salicylic acid. *Science*, **262** 1883-85.

Chester, K.S. (1933) The problem of acquired physiological immunity in plants. *Q. Rev. Biol.*, **8** 275-324.

Clarke, J.D., Liu, Y., Klessig, D.F. and Dong, X. (1998) Uncoupling *PR* gene expression from NPR1 and bacterial resistance: characterization of the dominant *Arabidopsis cpr6-1* mutant. *Plant Cell*, **10** 557-69.

Conrath, U., Chen, Z., Ricigliano, J.R. and Klessig, D.F. (1995) Two inducers of plant defense responses, 2,6-dichloroisonicotinic acid and salicylic acid, inhibit catalase activity in tobacco. *Proc. Natl. Acad. Sci. USA*, **92** 7143-47.

Dietrich, R.A., Delaney, T.P., Uknes, S.J., Ward, E.R., Ryals, J.A. and Dangl, J.L. (1994) *Arabidopsis* mutants simulating disease resistance response. *Cell*, **77** 565-77.

Dietrich, R.A., Richberg, M.H., Schmidt, R., Dean, C. and Dangl, J.L. (1997) A novel zinc finger protein is encoded by the *Arabidopsis LSD1* gene and functions as a negative regulator of plant cell death. *Cell*, **88** 685-94.

Du, H. and Klessig, D.F. (1997a) Identification of a soluble, high affinity salicylic acid-binding protein in tobacco. *Plant Physiol.*, **113** 1319-27.

Du, H. and Klessig, D.F. (1997b) Role for salicylic acid in the activation of defense responses in catalase-deficient transgenic tobacco. *Mol. Plant-Microbe Interact.*, **10** 922-25.

Durner, J. and Klessig, D.F. (1996) Salicylic acid is a modulator of tobacco and mammalian catalases. *J. Biol. Chem.*, **271** 28492-501.

Friedrich, L., Lawton, K., Ruess, W., Masner, P., Specker, N., Gut Rella, M., Meier, B., Dichner, S., Staub, T., Uknes, S., Métraux, J.-P., Kessmann, H. and Ryals, J. (1996) A benzothiadiazole derivative induces systemic acquired resistance in tobacco. *Plant J.*, **10** 61-71.

Gaffney, T., Friedrich, L., Vernooij, B., Negretto, D., Nye, G., Uknes, S., Ward, E., Kessmann, H. and Ryals, J. (1993) Requirement of salicylic acid for the induction of systemic acquired resistance. *Science*, **261** 754-56.

Görlach, J., Volrath, S., Knauf-Beiter, G., Hengy, G., Beckhove, U., Kogel, K.-H., Oostendorp, M., Staub, T., Ward, E., Kessmann, H. and Ryals, J. (1996) Benzothiadiazole, a novel class of inducers of systemic acquired resistance, activates gene expression and disease resistance in wheat. *Plant Cell*, **8** 629-43.

Greenburg, J.T., Guo, A., Klessig, D.F. and Ausubel, F.M. (1994) Programmed cell death in plants: a pathogen-triggered response activated coordinately with multiple defense function. *Cell*, **77** 551-63.

Heo, W.D., Lee, S.H., Kim, M.C., Kim, J.C., Chung, W.S., Chun, H.J., Lee, K.J., Park, C.Y., Pary, H.C., Choi, J.Y. and Cho, M.J. (1999) Involvement of specific calmodulin isoforms in salicylic acid-independent activation of plant disease resistance responses. *Proc. Natl. Acad. Sci. USA*, **96** 766-71.

Horvath, D.M., Huang, D.J. and Chau, N.-H. (1998) Four classes of salicylate-induced tobacco genes. *Mol. Plant-Microbe Interact.*, **11** 895-905.

Hunt, M.D., Neuenschwander, U.H., Delaney, T.P., Weymann, K.B., Friedrich, L.B., Lawton, K.A., Steiner, H.-Y. and Ryals, J.A. (1996) Recent advances in systemic acquired resistance research—a review. *Gene*, **179** 89-95.

Jabs, T., Dietrich, R.A. and Dangl, J.L. (1996) Initiation of runaway cell death in an *Arabidopsis* mutant by extracellular superoxide. *Science*, **273** 1853-56.

Kästner, B., Tenhaken, R. and Kauss, H. (1998) Chitinase in cucumber hypocotyls is induced by germinating fungal spores and by fungal elicitor in synergism with inducers of acquired resistance. *Plant J.*, **13** 447-54.

Katz, V.A., Thulke, O.U. and Conrath, U. (1998) A benzothiadiazole primes parsley cells for augmented elicitation of defense responses. *Plant Physiol.*, **117** 1333-39.

Kawano, T., Sahashi, N., Takahashi, K., Uozumi, N. and Muto, S. (1998) Salicylic acid induces extracellular superoxide generation followed by an increase in cytosolic calcium ion in tobacco suspension culture: the earliest events in salicylic acid signal transduction. *Plant Cell Physiol.*, **39** 721-30.

Kessmann, H., Staub, T., Hofmann, C., Maetzke, T., Herzog, J., Ward, E., Uknes, S. and Ryals, J. (1994) Induction of systemic acquired disease resistance in plants by chemicals. *Annu. Rev. Phytopathol.*, **32** 439-59.

Kuc, J. (1982) Induced immunity to plant disease. *BioScience*, **32** 854-60.

Lawton, K., Weymann, K., Friedrich, L., Vernooij, B., Uknes, S. and Ryals, J. (1995) Systemic acquired resistance in *Arabidopsis* requires salicylic acid but not ethylene. *Mol. Plant-Microbe Interact.*, **8** 863-70.

Lawton, K., Friedrich, L., Hunt, M., Weymann, K., Delaney, T., Kessmann, H., Staub, T. and Ryals, J. (1996) Benzothiadiazole induces disease resistance in *Arabidopsis* by activation of the systemic acquired resistance signal transduction pathway. *Plant J.*, **10** 71-82.

Levine, A., Tenhaken, R., Dixon, R. and Lamb, C. (1994) H_2O_2 from the oxidative burst orchestrates the plant hypersensitive disease resistance response. *Cell*, **79** 583-93.

Malamy, J., Carr, J.P., Klessig, D.F. and Raskin, I. (1990) Salicylic acid: a likely endogenous signal in the resistance response of tobacco to viral infection. *Science*, **250** 1002-1004.

Métraux, J.P., Signer, H., Ryals, J., Ward, E., Wyss-Benz, M., Gaudin, J., Raschdorf, K., Schmid, E., Blum, W. and Inverardi, B. (1990) Increase in salicylic acid at the onset of systemic acquired resistance in cucumber. *Science*, **250** 1004-1006.

Molina, A., Hunt, M.D. and Ryals, J.A. (1998) Impaired fungicide activity in plants blocked in disease resistance signal transduction. *Plant Cell*, **10** 1903-14.

Morel, J.-B. and Dangl, J.L. (1999) Suppressors of the *Arabidopsis lsd5* cell death mutation identify genes involved in regulating disease resistance responses. *Genetics*, **151** 305-19.

Neuenschwander, U., Vernooij, B., Friedrich, L., Uknes, S., Kessmann, H. and Ryals, J. (1995) Is hydrogen peroxide a second messenger of salicylic acid in systemic resistance. *Plant J.*, **8** 227-33.

Pallas, J.A., Paiva, N.L., Lamb, C. and Dixon, R.A. (1996) Tobacco plants epigenetically suppressed in phenylalanine ammonium lyase expression do not develop systemic acquired resistance in response to infection by tobacco mosaic virus. *Plant J.*, **10** 281-93.

Penninckx, I.A.M.A., Eggermont, K., Terras, F.R.G., Thomma, B.P.H.J., de Samblanx, G.W., Buchala, A., Métraux, J.-P., Manners, J.M. and Broekaert, W.F. (1996) Pathogen-induced systemic activation of a plant defensin gene in *Arabidopsis* follows a salicylic acid-independent pathway. *Plant Cell*, **8** 2309-23.

Penninckx, I.A.M.A., Thomma, B.P.H.J., Buchala, A., Métraux, J.-P. and Broekaert, W.F. (1998) Concomitant activation of jasmonate and ethylene response pathways is required for induction of a plant defensin gene in *Arabidopsis. Plant Cell*, **10** 2103-13.

Pieterse, C.M.J., van Wees, S.C.M., Hoffland, E., van Pelt, J.A. and van Loon, L.C. (1996) Systemic resistance in *Arabidopsis* induced by biocontrol bacteria is independent of salicylic acid accumulation and pathogenesis-related gene expression. *Plant Cell*, **8** 1225-37.

Pieterse, C.M.J., van Wees, S.C.M., van Pelt, J.A., Knoester, M., Laan, R., Gerrits, H., Weisbeek, P.J. and van Loon, L.C. (1998) A novel signaling pathway controlling induced systemic resistance in *Arabidopsis. Plant Cell*, **10** 1571-80.

Ross, A.F. (1961) Systemic acquired resistance induced to plant virus infection in hypersensitive hosts. *Virology*, **14** 329-39.

Ryals, J., Uknes, S. and Ward, E. (1994) Systemic acquired resistance. *Plant Physiol.*, **104** 1109-12.

Ryals, J., Lawton, K.A., Delaney, T.P., Friedrich, L., Kessmann, H., Neuenschwander, U., Uknes, S., Vernooij, B. and Weymann, K. (1995) Signal transduction in systemic acquired resistance. *Proc. Natl. Acad. Sci. USA*, **92** 4202-205.

Ryals, J.A., Neuenschwander, U.H., Willits, M.G., Molina, A., Steiner, H.-Y. and Hunt, M.D. (1996) Systemic acquired resistance. *Plant Cell*, **8** 1809-19.

Ryals, J., Weymann, K., Lawton, K., Friedrich, L., Ellis, D., Steiner, H.-Y., Johnson, J., Delaney, T.P., Jesse, T., Vos, P. and Uknes, S. (1997) The *Arabidopsis* NIM1 protein shows homology to the mammalian transcription factor inhibitor I B. *Plant Cell*, **9** 425-39.

Schneider, M., Schweizer, P., Meuwly, P. and Métraux, J.-P. (1996) Systemic acquired resistance in plants. *Int. Rev. Cytol.*, **168** 303-34.

Shah, J., Tsui, F. and Klessig, D.F. (1997) Characterisation of a salicylic acid-insensitive mutant (*sai1*) of *Arabidopsis thaliana*, identified in a selective screen utilizing the SA-inducible expression of the *tms2* gene. *Mol. Plant-Microbe Interact.*, **10** 69-78.

Shah, J., Kachroo, P. and Klessig, D.F. (1999) The *Arabidopsis ssi1* mutation restores pathogenesis-related gene expression in *npr1* plants and renders defensin gene expression salicylic acid dependent. *Plant Cell*, **11** 191-206.

Sharma, Y.K., León, J., Raskin, I. and Davis, K.R. (1996) Ozone-induced responses in *Arabidopsis thaliana*: the role of salicylic acid in the accumulation of defense-related transcripts and induced resistance. *Proc. Natl. Acad. Sci. USA*, **93** 5099-104.

Smith-Becker, J., Marios, E., Huguet, E.J., Midland, S.L., Sims, J.J. and Keen, N.T. (1998) Accumulation of salicylic acid and 4-hydroxybenzoic acid in phloem fluids of cucumber during systemic acquired resistance is preceded by a transient increase in phenylalanine ammonium-lyase activity in petioles and stems. *Plant Physiol.*, **116** 231-38.

Thomma, B.P.H.J., Eggermont, K., Penninckx, I.A.M.A., Mauch-Mani, B., Vogelsang, R., Cammue, B.P.A. and Broekaert, W.F. (1998) Separate jasmonate-dependent and salicylate-dependent defense-response pathways in *Arabidopsis* are essential for resistance to distinct microbial pathogens. *Proc. Natl. Acad. Sci. USA*, **95** 15107-11.

Thulke, O. and Conrath, U. (1998) Salicylic acid has a dual role in the activation of defence-related genes in parsley. *Plant J.*, **14** 35-42.

Uknes, S., Mauch-Mani, B., Moyer, M., Potter, S., Williams, S., Dincher, S., Chandler, D., Slusarenko, A., Ward, E. and Ryals. J. (1992) Acquired resistance in *Arabidopsis. Plant Cell*, **4** 645-56.

Uknes, S., Winter, A.M., Delaney, T., Vernooij, B., Morse, A., Friedrich, L., Nye, G., Potter, S., Ward, E. and Ryals, J. (1993) Biological induction of systemic acquired resistance in *Arabidopsis. Mol. Plant-Microbe Interact.*, **6** 692-98.

van Loon, L.C., Bakker, P.A.H.M. and Pieterse, C.M.J. (1998) Systemic resistance induced by rhizosphere bacteria. *Annu. Rev. Phytopathol.*, **36** 453-83.

van Wees, S.C.M., Pieterse, C.M.J., Trijssenaar, A., van't Westende, Y.A.M., Hartog, F. and van Loon, L.C. (1997) Differential induction of systemic resistance in *Arabidopsis* by biocontrol bacteria. *Mol. Plant-Microbe Interact.*, **10** 716-24.

Vernooij, B., Friedrich, L., Ahl Goy, P., Staub, T., Kessmann, H. and Ryals, J. (1995) 2,6-dichloroisonicotinic acid induced resistance to pathogens without the accumulation of salicylic acid. *Mol. Plant-Microbe Interact.*, **8** 228-34.

Volko, S.M., Boller, T. and Ausubel, F.M. (1998) Isolation of new *Arabidopsis* mutants with enhanced disease susceptibility to *Pseudomonas syringae* by direct screening. *Genetics*, **149** 537-48.

Ward, E.R., Uknes, S.J., Williams, S.C., Dichner, S.S., Wiederhold, D.L., Alexander, D.C., Ahl-Goy, P., Métraux, J.-P. and Ryals, J. (1991) Coordinate gene activity in response to agents that induce systemic acquired resistance. *Plant Cell*, **3** 1085-94.

Wendehenne, D., Durner, J., Chen, Z. and Klessig, D.F. (1998) Benzothiadiazole, an inducer of plant defenses, inhibits catalase and ascorbate peroxidase. *Phytochemistry*, **47** 651-57.

Weymann, K., Hunt, M., Uknes, S., Neuenschwander, U., Lawton, K., Steiner, H.-Y. and Ryals, J. (1995) Suppression and restoration of lesion formation in *Arabidopsis lsd* mutants. *Plant Cell*, **7** 2013-22.

White, R.F. (1979) Acetyl salicylic acid (aspirin) induces resistance to tobacco mosaic virus in tobacco. *Virology*, **99** 410-12.

Willekens, H., Chamnongpol, S., Davey, M., Schrauder, M., Langebartels, C., van Montagu, M., Inzé, D. and van Camp, W. (1997) Catalase is a sink for H_2O_2 and is indispensable for stress defence in C3 plants. *EMBO J.*, **16** 4806-16.

9 Transgenic approaches to disease resistant plants as exemplified by viruses

John A. Walsh

9.1 Introduction

The advent of plant transformation has provided the opportunity to capitalise on existing conventional forms of plant resistance and to develop new and novel ones. In a review chapter on viral 'cross protection', Hamilton (1980) suggested that if complementary deoxyribonucleic acid (cDNA) of mild virus strains could be transferred to plants, these might provide protection against other strains of the virus. This was taken a step further by Sanford and Johnston (1985), who suggested the possibility of inducing resistance (to viruses) by transforming a susceptible plant with genes derived from the pathogen itself. This suggested form of resistance was termed 'parasite-derived resistance' and was subsequently called 'pathogen-derived resistance'. The first demonstration of this concept was described the following year by Powell Abel *et al.* (1986). Since then, there have been many more examples and transgenic crops with pathogen-derived resistance are now grown commercially in a number of countries. The first virus-resistant transgenic crop plant commercially released in the USA was a squash F_1 hybrid cultivar that was highly resistant to zucchini yellow mosaic virus (ZYMV) and watermelon mosaic virus 2 (Fuchs and Gonsalves, 1995). Subsequently, a transgenic squash line with resistance to the same two viruses plus cucumber mosaic virus (CMV) (Tricoli *et al.*, 1995) has been deregulated and a transgenic papaya line with resistance to papaya ringspot type p virus (PRSV) (Gonsalves, 1998) has been commercialised. Tobacco made from virus-resistant transgenic plants is reported to be commercially available in China (Plafker, 1994; Birch, 1997). A number of excellent reviews on pathogen-derived resistance have appeared in recent years (Beachy, 1988; Wilson, 1993; Baulcombe, 1994a; Grumet, 1994; Lomonossoff, 1995; Malpica *et al.*, 1998). Some of these reviews also cover other transgenic forms of resistance not dependent on pathogen-derived sequences. The ability to identify and clone natural plant resistance genes to pathogens has now provided further opportunities to confer plants with new sources of resistance, facilitating interspecific transfer of genes.

The number of plant species for which transformation and regeneration systems have been developed has increased dramatically in recent years (Pierpoint *et al.*, 1995), opening up the possibility of engineering transgenic resistance to diseases of most crops.

This chapter describes different transgenic approaches to producing disease resistant plants. Resistance to viruses is used as the model, as most progress has been made in this area and a wide range of approaches has been developed. A number of the strategies described are also applicable to bacterial and fungal pathogens of plants.

9.2 Pathogen-derived resistance to plant viruses

The majority of examples of successful pathogen-derived resistance involve viruses with positive-strand RNA genomes, including members of the tobamo-, cucumo-, potex-, furo-, poty-, luteo-, carla-, ilar-, tenui-, tobra-, tombus-, nepo-, tymo- and alfalfa mosaic virus groups. To some extent, this is due to the importance and relative abundance of viruses in these groups. However, the concept has also been successfully exploited for the ambisense RNA tospovirus group and the single-stranded (ss) DNA-containing geminivirus group. Attempts have been reported (Passelègue and Kerlan, 1996) but there have been no examples, so far, of successful pathogen-derived resistance to double-stranded (ds) DNA-containing plant viruses.

The idea that plants expressing viral sequences could be resistant to subsequent infection arose from the phenomenon of 'cross-protection', where a mild, symptomless or attenuated strain of a virus is used to infect and protect plants from subsequent infection by severely pathogenic strains of the same virus (Hamilton, 1980; Ponz and Bruening, 1986). The underlying molecular mechanisms of cross-protection are still not understood. Postulated mechanisms for the different types of pathogen-derived resistance to viruses described here are discussed. The type of resistance obtained and its effectiveness is influenced by the promoter used, the number of copies of transgenes and the position of transgenes in the plant genome. These and a number of other factors need to be considered on a case-by-case basis when trying to establish consensus models for the various forms of pathogen-derived resistance. Advantages, limitations and potential problems of the different approaches are also discussed.

9.2.1 Coat-protein-mediated resistance

The first report concerning the expression of the coat protein of a virus in a plant was published in 1985, when Bevan *et al.* (1985) described the transgenic expression of tobacco mosaic virus (TMV) coat protein in tobacco. However, progeny of the original transformants were as susceptible to TMV as the non-transgenic controls. The first example of pathogen-derived resistance also involved the transformation of tobacco plants to express the coat protein of TMV. Seedlings that expressed the coat protein gene showed a delay in symptom development and 10–60% failed to develop symptoms (Powell Abel *et al.*, 1986). Following this first example, there have been a large number of reports of coat-protein-mediated resistance being effective against many RNA viruses, including: several tobamoviruses (Nelson *et al.*, 1987; Nejidat and Beachy, 1990; Sanders *et al.*, 1992); alfalfa mosaic virus (AlMV) (Loesch-Fries *et al.*, 1987); tobacco streak virus (Van Dun *et al.*, 1988a); beet necrotic yellow vein virus (Kallerhoff *et al.*, 1990); CMV (Gonsalves *et al.*, 1992; Gielen *et al.*, 1996); rice stripe virus (Hayakawa *et al.*, 1992); tobacco rattle virus (TRV) (Van Dun *et al.*, 1987; Ploeg *et al.*, 1993); potato virus X (PVX) (Hemenway *et al.*, 1988; Fehér *et al.*, 1992); potato virus S (MacKenzie *et al.*, 1991); tomato spotted wilt virus (TSWV) (MacKenzie and Ellis, 1992; Pang *et al.*, 1994); potato leafroll virus (PLRV) (Barker *et al.*, 1992); arabis mosaic virus (Bertioli *et al.*, 1992); cymbidium ringspot virus (CyRSV) (Rubino *et al.*, 1993); grapevine chrome mosaic (Brault *et al.*, 1993); grapevine fanleaf virus (Bardonnet *et al.*, 1994); potato aucuba mosaic virus (Leclerc and AbouHaidar, 1995); and several potyviruses (Stark and Beachy, 1989; Namba *et al.*, 1992; Regner *et al.*, 1992; Dinant *et al.*, 1993; Fang and Grumet, 1993; Farinelli and Malnoë, 1993; Murray *et al.*, 1993; Cassidy and Nelson, 1995; Scorza *et al.*, 1995; Lehmann *et al.*, 1996; Gonsalves, 1998). Coat-protein-mediated resistance has also been shown to be effective against the DNA geminivirus, tomato yellow leaf curl (Kunik *et al.*, 1994).

Despite the extensive literature, there is no single consensus model for the mechanism of coat-protein-mediated resistance. It is thought that the resistance results from an early event(s) in the infection process, possibly prior to uncoating of the would-be infecting virus. Evidence for this comes from partial breakdown of resistance to TMV and AlMV following inoculation of viral RNA (Nelson *et al.*, 1987; Van Dun *et al.*, 1987). However, as these authors observed a degree of protection against viral RNA, this was taken as suggesting other steps in infection may also be inhibited or slowed (Beachy, 1988). For TMV, AlMV and PVX, there appears to be a correlation between the amount of coat protein expressed in plants and the efficacy of coat-protein-mediated resistance (Loesch-Fries *et al.*, 1987; Hemenway *et al.*, 1988; Powell *et al.*, 1990); the more protein expressed, the greater the resistance.

Further evidence that the coat protein itself rather than the mRNA encoding it is necessary for some examples of coat-protein-mediated resistance has been derived from studies on the effect of temperature on protein levels and resistance. For TMV and CMV, elevated temperatures led to decreased levels of coat protein and of resistance (Nejidat and Beachy, 1989; Okuno *et al.*, 1993). Powell *et al.* (1990) also provided evidence for the requirement of protein expression for transgenic resistance to TMV. They showed that plants transformed with the TMV coat protein gene lacking an initiation codon showed similar levels of coat protein mRNA to transformants containing the intact gene, but lacked resistance. Register and Beachy (1988) suggested that coat-protein-mediated resistance against TMV was effected by inhibition of virion disassembly in initially infected cells. Alternatively, the viral RNA could be re-encapsidated by coat protein molecules expressed by the transgene (Beachy *et al.*, 1996). The studies of Clark *et al.* (1995) support the hypothesis that if 1–4 coat protein molecules can be replaced (by re-encapsidation) following release, infection can be blocked. If more than 5–10 molecules are released, this probably results in ribosome binding and infection.

Further evidence for the inhibition of virion disassembly has been obtained from experiments with 'pseudovirions' and partially stripped TMV. A chimeric construct containing β-glucuronidase (GUS) mRNA and the TMV origin of assembly were coated with TMV coat protein. The resulting pseudovirions were inoculated to protoplasts of transgenic plants exhibiting coat-protein-mediated resistance and to protoplasts of non-transgenic plants (Osbourn *et al.*, 1989). GUS activity in protoplasts from transgenic plants was only about 1% of that observed in protoplasts from non-transformed plants, which is consistent with an inhibition of pseudovirion disassembly (Osbourn *et al.*, 1989). When protoplasts from transgenic plants were inoculated with naked, uncoated constructs, GUS activity was 30 to 50-fold greater than in similar protoplasts inoculated with the pseudovirions (Osbourn *et al.*, 1989). The suggested mechanisms of coat-protein-mediated resistance have been discussed in detail previously (Beachy *et al.*, 1990, 1996; Reimann-Philipp and Beachy, 1993; Hackland *et al.*, 1994; Baulcombe, 1996a).

It is thought that one of the limitations of coat-protein-mediated resistance is that it only operates well against closely-related virus strains. Even point mutations in coat proteins have been shown to result in breakdown of coat-protein-mediated resistance in some instances (Wilson, 1993). However, using a range of tobamoviruses, resistance was detectable when their coat proteins amino acid sequences were more than 60% homologous (Nejidat and Beachy, 1990). Contrary to this, Sanders *et al.* (1992) showed that tomato plants expressing the coat protein gene of TMV showed either a low level, or no resistance to several TMV strains, one of which was 88% homologous to the transgene at the amino acid level. Coat-protein-mediated resistance to potato virus Y (PVY) induced by the coat protein of one strain of the virus has been shown to be effective against other strains (Van den Heuvel *et al.*, 1994; Hefferon *et al.*, 1997). Transforming potato plants with lettuce mosaic virus coat protein has induced immunity to PVY also (Hassairi *et al.*, 1998). Expressing chimeric coat protein constructs derived from two different potyviruses in *Nicotiana benthamiana* conferred resistance to two potyviruses that were unrelated to those from which the transgenes were derived (Hammond and Kamo, 1995). A low but detectable level of coat-protein-mediated resistance against unrelated viruses has been claimed for transgenic tobacco plants expressing TMV, AlMV, CMV, soybean mosaic virus or ZYMV coat protein (Anderson *et al.*, 1989; Stark and Beachy, 1989; Fang and Grumet, 1993; Nakajima *et al.*, 1993).

As mentioned previously for coat-protein-mediated resistance to TMV and CMV, high temperatures can result in decreased resistance (Nejidat and Beachy, 1989; Okuno *et al.*, 1993). Also, high concentrations of virus inoculum can overcome coat-protein-mediated resistance. Coat-protein-mediated resistance identified by mechanical inoculation of virus to plants can break down in the field when plants are subject to vector transmission of the virus. Transgenic tobacco plants expressing the coat protein of TRV were resistant to mechanical inoculation of the virus, but the resistance was not effective against infection via viruliferous nematodes (Ploeg *et al.*, 1993). Coat-protein-mediated resistance to PRSV has been shown to be stable in papaya in the field for 2 yrs, indicating that this form of resistance can provide effective protection against viral diseases over a significant proportion of the crop cycle of a perennial species (Lius *et al.*, 1997).

An assumed risk of coat-protein-mediated resistance is transencapsidation, also known as heterologous encapsidation. This phenomenon involves the encapsidation of infecting viral genomes by transgenically expressed coat protein, and has been observed for: TMV (Osbourn *et al.*, 1990; Holt and Beachy, 1991); barley yellow dwarf virus (Wen and Lister, 1991); potato virus Y (Farinelli *et al.*, 1992); a number of other potyviruses (Hammond, 1993); CMV/AlMV (Candelier-Harvey and Hull, 1993); and ZYMV/plum pox virus (PPV) (Lecoq *et al.*, 1993). In this last instance, a non-aphid transmissible strain of ZYMV was rendered aphid transmissible from transgenic plants expressing the coat protein of PPV (Lecoq *et al.*, 1993). Transencapsidation no doubt occurs in nature and has been observed in mixed infections of viral cultures; observations of such phenomena may give indications of potential risks. An example of this is the aphid transmission of potato spindle tuber viroid (which is not normally aphid transmissible) encapsidated in potato leafroll virus particles (Syller *et al.*, 1997). All the above examples of transencapsidation have been observed in artificial/experimental circumstances, often involving mechanical inoculation of transgenic plants with large amounts of virus, probably vastly in excess of anything they would experience in the field and under conditions of high selection pressure to recover viruses with new properties. It is difficult to extrapolate from these experiments how frequently transencapsidation would occur in field circumstances. Even if transencapsidation does occur, long-term stable genetic or epidemiological effects have not yet been demonstrated. For example, non-aphid-transmissible ZYMV transmitted from transgenic plants expressing PPV coat protein is not aphid transmissible from subsequently infected plants not expressing PPV coat protein. In a recently reported field experiment conducted over two consecutive years, transgenic melon and squash expressing coat protein genes of aphid transmissible isolates of CMV, ZYMV and watermelon mosaic virus 2 were infected with an aphid non-transmissible isolate of CMV. Aphid-vectored spread of the aphid non-transmissible CMV isolate to CMV-susceptible non-transgenic plants growing alongside the transgenic plants did not occur (Fuchs *et al.*, 1998). Surprisingly, this was the first report on field experiments designed to determine the potential of transgenic plants expressing coat protein genes for triggering changes in virus vector specificity.

The specificity and significance of heterologous encapsidation of virus and virus-like RNAs has been reviewed by Falk *et al.* (1995).

9.2.2 Replicase-mediated resistance

Following coat-protein-mediated resistance, the next viral gene shown to be capable of conferring resistance was the viral replicase. This again was first demonstrated for TMV (Golemboski *et al.*, 1990). Replicase-mediated resistance was subsequently found to be effective against tobra-, cucumo-, potex-, tombus- and potyviruses. In addition to TMV, the RNA viruses it has been demonstrated for include: CMV (Anderson *et al.*, 1992; Zaitlin *et al.*, 1994); PVX (Braun and Hemenway, 1992; Longstaff *et al.*, 1993; Mueller *et al.*, 1995); PVY (Audy *et al.*, 1994); cowpea mosaic virus (CoMV) (Sijen *et al.*, 1995); pea early-browning virus (PEBV) (MacFarlane and Davies, 1992); pepper mild mottle virus (PMMV) (Tenllado *et al.*, 1995); CyRSV (Rubino and Russo, 1995); PPV (Guo and García, 1997); and pea seed-borne mosaic virus (PSbMV) (Jones *et al.*, 1998a). It has also been found to be effective against the ss DNA-containing geminivirus, African cassava mosaic virus (Hong and Stanley, 1996).

Replicase-mediated resistance has been reviewed previously (Lomonossoff and Davies, 1992; Carr and Zaitlin, 1993; Baulcombe, 1994b, 1996a). Replicase-mediated resistance is often stronger than coat-protein-mediated resistance as it is effective even when plants are challenged with high levels of virus and infectious viral RNA. In experiments on PVX, Braun and Hemenway (1992) judged replicase-mediated resistance to be more effective than coat-protein-mediated resistance. The resistance is quite specific to the virus from which the replicase is derived and does not usually extend to other viruses of the same group, not even closely-related ones. Only one case of broad-spectrum resistance has been described (to tobamoviruses; Donson *et al.*, 1993). In this case, however, the replicase gene was interrupted by a bacterial insertion element (Donson *et al.*, 1993). Full-length replicase genes conferred resistance in the cases of PVX, CoMV and CyRSV (Braun and Hemenway, 1992; Rubino and Russo, 1995; Sijen *et al.*, 1995), but not for AlMV (Van Dun *et al.*, 1988b) and brome mosaic virus (BMV) (Mori *et al.*, 1992). Deliberately defective constructs have been made and found to confer resistance (e.g. for CMV, Gal-On *et al.*, 1998); this could possibly be explained by RNA-mediated resistance (see section 9.2.6). Guo and García (1997) induced resistance to PPV by expressing intact NIb replicase genes of the virus in *N. benthamiana*; the resistance was largely overcome at high inoculum doses. They also induced a total recovery phenotype by expressing mutated NIb gene sequences. Intact replicase genes can confer high levels of resistance (Braun and Hemenway, 1992).

There is no single model that fully explains all the phenomena observed when viral replicase genes are expressed in plants. Baulcombe (1996a) reviewed these phenomena and concluded that a number, including CoMV (Sijen *et al.*, 1995), PVX (Mueller *et al.*, 1995) and PMMV (Tenllado *et al.*, 1995), involved RNA-mediated, homology-dependent resistance. In other examples, including PEBV (MacFarlane and Davies, 1992) and CyRSV (Rubino and Russo, 1995), it was thought likely that the transgene conferred homology-dependent resistance, although the data were not conclusive. For the resistance conferred by the modified TMV replicase (Donson *et al.*, 1993), Baulcombe (1996a) considered that gene silencing could be ruled out completely. Jones *et al.* (1998a) concluded that replicase-mediated resistance to PSbMV was mediated through post-transcriptional gene silencing and that it was induced following virus infection.

Replicase-mediated resistance scores over coat-protein-mediated resistance in that it appears to be stronger and, of course, cannot lead to transencapsidation. However, being highly strain-specific and only effective against virus strains that are closely related to the source of the transgene, it does not usually give such broad-spectrum resistance as coat-protein-mediated resistance.

9.2.3 Movement-protein-mediated resistance

Plant viruses encode proteins that facilitate movement from cell to cell. Viral movement proteins and their modes of action have been reviewed by Wolf and Lucas (1994). Lapidot *et al.* (1993) demonstrated that tobacco plants that were transgenic for a TMV movement protein that had been deliberately rendered dysfunctional by deleting three amino acids from near its N-terminus, showed much delayed virus spread, replication and lesion development. Tobacco plants expressing a non-functional movement protein of BMV had a degree of resistance to the unrelated TMV (Malyshenko *et al.*, 1993), and plants transgenic for a mutant movement protein of white clover mosaic virus (WClMV) had broad-spectrum resistance against systemic infection by plant viruses with a triple gene block (Beck *et al.*, 1994). Plants transgenic for a defective mutant of the TMV movement protein were resistant to a wide range of plant viruses, including members of the nepo-, alfamo-, caulimo-, cucumo- and tobra-virus groups, in addition to tobamoviruses (Cooper *et al.*, 1995).

Movement proteins of viruses have been shown to accumulate near plasmodesmata (see, for example, Atkins *et al.*, 1991) and directly or indirectly change the exclusion properties of plasmodesmata. Under normal physiological conditions, plasmodesmata allow the transport of low molecular weight compounds (less than 1 kDa and about 2.0 nm in diameter) and allow viruses to move from cell to cell. According to Citovsky and Zambryski (1993), the size of virus particles, e.g. TMV (18 × 300 nm) or its free-folded RNA (10 nm in diameter), indicates that the plasmodesmata gate size must increase to allow passage of virus and/or viral RNA. Micro-injection of TMV movement protein into cells increased permeability up to 9 nm (Wolf *et al.*, 1991). Several movement proteins have been characterized as single-stranded nucleic acid binding proteins. Citovsky *et al.* (1992) proposed a model in which TMV movement protein interacts with TMV RNA to make a 2 nm thin transport complex. It is postulated to have two distinct domains, one that binds to plasmodesmata and allows its size exclusion limit to increase and another that binds unspecifically to single-strand nucleic acids.

Carrington *et al.* (1996) recently reviewed the mechanisms of virus movement for different viral groups in plants. Resistance conferred by transgenic expression of a dysfunctional TMV movement protein is likely to be due to competition for plasmodesmatal binding sites between the mutant movement protein and the wild-type movement protein of the inoculated virus (Lapidot *et al.*, 1993). The broad-spectrum resistance of plants expressing dysfunctional movement proteins highlighted above, indicates that movement proteins of several different viruses may interact with the same plasmodesmatal components (Baulcombe, 1996a). In some virus groups, three movement proteins are encoded by a series of overlapping reading frames, referred to as the triple gene block (Beck *et al.*, 1991). Transgenic expression of a mutant of one of the triple gene block proteins of WClMV conferred resistance against a narrower range of viruses than did TMV movement-protein-mediated resistance (Beck *et al.*, 1994), indicating that the movement proteins of these two viruses do not interact with plasmodesmata in exactly the same way. The P30 movement protein of TMV has been found to have the capacity to bind single-stranded nucleic acids (Citovsky *et al.*, 1990); consequently, it has been suggested that dysfunctional movement proteins might have modified nucleic acid binding properties. The dysfunctional protein might bind non-productively to the viral RNA, and thus compete with the invading functional protein (Lomonossoff, 1995). The possible mechanisms of movement-protein-mediated resistance have been reviewed previously (Baulcombe, 1996a; Beachy *et al.*, 1996).

Movement-protein-mediated resistance appears particularly promising in that it seems to be capable of inducing very broad-spectrum resistance. It is considered that greater knowledge of the role of viral movement proteins will make it possible to create mutations in the movement proteins that will provide both broad-spectrum and durable resistance (Beachy *et al.*, 1996).

9.2.4 Other viral proteins

There have been examples of resistance induced by transforming plants with viral proteases. A small proportion of plants (3 lines out of 50) transformed with the NIa viral protein were found to be fully protected against PVY for as long as 60 days after inoculation (Vardi *et al.*, 1993). NIa and CI proteins of tobacco vein mottling virus (TVMV) have been expressed in transgenic plants. The NIa induced good resistance that was effective against inoculums of 100 μg/ml, but was only effective against the homologous virus, whereas the CI conferred no resistance (Maiti *et al.*, 1993). The CI protein gene of PPV did not confer resistance when transformed into *N. benthamiana*; however, one transgenic line expressing a CI gene that had been mutated in the nucleotide binding motif region was resistant (Wittner *et al.*, 1998). Plants of some potato lines transformed with the P1 gene of PVY showed a high level of resistance to the homologous strain (PVY^{o}) of the virus but not a heterologous strain (PVY^{N}) (Pehu *et al.*, 1995). Tobacco plant lines transformed with the P1 or P3 gene of TVMV were protected against the homologous TVMV strain and showed variable proportions of two resistance phenotypes: asymptomatic plants; or symptomatic plants that recovered from infection (Moreno *et al.*, 1998). The resistance conferred by both genes was effective against high inoculum doses and the former was not effective against a heterologous strain of TVMV, whereas the latter was.

The NIa is the major protease of potyviruses and is responsible for cleaving the single large polyprotein, characteristic of potyviruses, into several mature proteins (Hellmann *et al.*, 1988). Other potyvirus proteins have also been shown to be proteases (Carrington *et al.*, 1989). The mechanism of protease-mediated resistance is unclear, but it is comparatively stronger than that conferred by coat proteins. The CI protein of potyviruses has RNA helicase activity, is involved in virus cell-to-cell movement and contains a conserved nucleotide binding motif. The mutant CI that conferred resistance to PPV was predicted to be defective, but the authors were unsure of the process involved in the resistance (Wittner *et al.*, 1998). They suggested it could be RNA-mediated, as the CI protein was undetectable in transgenic plants and RNA accumulation was very low. The low levels of P1 and P3 transgene expression in resistant tobacco plants suggested that these resistances were RNA-mediated (Moreno *et al.*, 1998). In contrast to most reports of virus-activated gene silencing, some plant lines expressing P3 and showing predominantly the recovery phenotype, showed silencing of the transgene that was activated at a certain developmental stage of the plant, independently of virus infection (Moreno *et al.*, 1998).

There have been three reports of virus resistance in transgenic plants that were expressing replicating viral RNAs. The first report involved tobacco plants expressing a mild strain of TMV that accumulated virus particles but were symptom-free and resistant to a related severe strain of TMV (Yamaya *et al.*, 1988a,b). The results suggested that the transgenic expression of the viral genome mimicked the protection observed naturally when a plant is infected by a mild virus strain. However, it was not possible to say whether the resistance factor was protein or RNA. Protoplasts of plants expressing RNA1 and RNA2 of BMV showed BMV-specific resistance that was dependent on replication of the

BMV RNAs (Kaido *et al.*, 1995). There was a lower level of transgenic RNAs in the resistant lines than in the susceptible lines that expressed a similar but non-replicable construct. It is likely that this resistance was RNA-mediated. Resistance was induced to CMV in tobacco transformed to produce RNA1 and RNA2 of CMV (Suzuki *et al.*, 1996). This resistance was similar to that conferred by the replicable BMV RNAs outlined above, in that it required replication of the viral RNAs. However, resistance was not associated with a lower level of transgenic RNA and was not specific to CMV. Resistance was more effective against virion rather than RNA inocula (Suzuki *et al.*, 1996), suggesting that this resistance mechanism was protein rather than RNA-mediated.

9.2.5 Defective-interfering RNA or DNA sequences

This form of pathogen-derived resistance is due to direct inhibition of the viral infection cycle by the transgene itself or its RNA transcript. Satellite RNA-mediated resistance (Gerlach *et al.*, 1987; Harrison *et al.*, 1987; Taliansky *et al.*, 1998), resistance to geminiviruses conferred by a transgenic defective-interfering DNA (Stanley *et al.*, 1990; Frischmuth and Stanley, 1994) and resistance to CyRSV due to transgenic expression of a replicable defective-interfering RNA (Kollàr *et al.*, 1993) are all examples of this form of transgenic resistance. Transgenic expression of the 3′ non-coding region of turnip yellow mosaic virus (TYMV) partially protected against TYMV infection (Zaccomer *et al.*, 1993).

Defective-interfering sequences are deletion mutants of viral genomes that have generally lost all essential viral genes needed for movement, replication and encapsidation. Such RNAs or DNAs are dependent upon a helper virus for *trans*-acting factors necessary for replication, and often multiply and accumulate at the expense of the helper virus from which they are derived. Baulcombe (1996a) suggested that this resistance could operate if the transgenic nucleic acids act as decoy molecules, competing with the viral genome to redirect host- or viral-encoded proteins into interactions that would be non-productive for replication or spread of the virus in the infected plant. In other words, the transgene induces a competitive inhibition. Baulcombe (1996a) also suggested that the resistance from the transgenically expressed 3′ non-coding region of TYMV RNA was conferred through direct competitive inhibition of the viral genome.

The limitation of this approach is that not many viruses have satellites or naturally occurring defective-interfering sequences. In the latter case, this has been overcome by the use of artificial defective-interfering RNAs for TMV (Raffo and Dawson, 1991) and BMV (Huntley and Hall, 1996). A further disadvantage of the satellite approach is the ease with which satellite RNAs can mutate to more severe forms and the fact that new satellite-virus combinations can result in more severe symptoms (Devic *et al.*, 1990; Sleat and Palukaitis, 1992).

9.2.6 *RNA-mediated resistance*

The first indication that expression of protein by a transgene is not always necessary for resistance came from the finding that the level of RNA accumulating from a virus-derived transgene and the levels of transgene protein of PVY and potato leafroll virus did not always correlate with the level of resistance (Kawchuk *et al.*, 1990; Lawson *et al.*, 1990). Subsequently, it was found that resistance was conferred by modified viral transgenes that encoded untranslatable RNAs of TSWV, tobacco etch virus (TEV) and PVY (De Haan *et al.*, 1992; Lindbo and Dougherty, 1992a; Van der Vlugt *et al.*, 1992). In addition to the coat protein sequences of viruses, a number of other parts of viral genomes have been shown to induce RNA-mediated silencing (resistance). These include the 6- and 21-kDa reading frames of the VPg of TEV (Swaney *et al.*, 1995), and the NIb RNA replicase of PPV (Guo *et al.*, 1999). RNA-mediated resistance to PVX has also been demonstrated (Mueller *et al.*, 1995), and satellite virus sequences have been shown to induce RNA-mediated resistance (Taliansky *et al.*, 1998).

The first explanation of RNA-mediated resistance came when Lindbo *et al.* (1993a) showed an association between transgenic resistance to TEV and post-transcriptional suppression (gene silencing) of transgene expression. Baulcombe (1996a) pointed out that as gene silencing with non-viral transgenes has also been shown to be due to a post-transcriptional mechanism (Ingelbrecht *et al.*, 1994; de Carvalho Niebel *et al.*, 1995), this substantiates the explanation put forward by Lindbo *et al.* (1993a). The post-transcriptional mechanism operates at the RNA level and, therefore, has the potential to suppress the accumulation of viral RNA that shares sequence identity with the silenced transgene. The mechanism has been referred to as homology-dependent gene silencing (Mueller *et al.*, 1995) and post-transcriptional gene silencing (Baulcombe, 1996b). The proposed mechanisms for this phenomenon have been reviewed in detail by Baulcombe (1996a, 1996b) and Van den Boogaart *et al.* (1998). They involve the suppression of accumulation of nucleus-derived RNA (gene silencing) and virus-derived RNAs with homology to the transgene (resistance). The mechanism probably requires a high degree of sequence homology because the resistance is highly strain-specific.

To explain the sequence specificity, it has been suggested that base-pairing interactions occur involving RNA produced from the transgene. This transgenic RNA would interact with viral RNA to account for homology-dependent virus resistance and with nucleus-derived RNA to account for homology-dependent gene silencing (Mueller *et al.*, 1995). One model is the formation of a dsRNA in the cytoplasm by hybridisation of the transgene transcript to the virus strand of opposite polarity. This model requires the production of the negative strand of the viral RNA, which is produced as an intermediate in the replication cycle of most viral RNAs by the RNA-dependent RNA polymerase. The model could explain homology-dependent virus resistance but it is difficult to see how it could explain the transgene silencing. Furthermore, viral helicase activity would normally separate the dsRNAs. A more likely explanation is that the interaction is indirect, as proposed by Lindbo *et al.* (1993a), and mediated by an RNA complement of the transgene. Such antisense RNA could be produced by a host RNA-dependent RNA polymerase using the transgene as a template (Schiebel *et al.*, 1993). However, the predicted antisense RNA has not been detected in plants showing homology-dependent resistance. Schiebel *et al.* (1993) suggested that methods used to search for the predicted antisense RNA would not have detected small or dispersed RNAs, which are the likely product of the host-encoded RNA-dependent RNA polymerase.

Recently, Marano and Baulcombe (1998) showed that TMV replicase-mediated resistance in transgenic tobacco was due to a post-transcriptional gene silencing mechanism. However, unlike other examples of gene silencing associated with virus resistance, the silencing was specific for the antisense rather than the coding strand of the target RNA. Baulcombe (1996a) discussed the ways in which the formation of duplex RNA could influence accumulation of host and viral RNAs to cause virus resistance and gene silencing. The model leaves many observations unexplained, for instance why do some lines show RNA-mediated resistance whereas others carrying the same transgene do not? Lindbo *et al.* (1993a) and others suggest that there is a threshold level of transgene expression that activates homology-dependent resistance. English *et al.* (1996) suggest that there is a qualitative factor that affects production of aberrant RNA in plants exhibiting homology-dependent resistance. For the first explanation, it has been proposed that, in some instances, the RNA-mediated resistances were only induced when the combined levels of viral and transgene RNA surpassed a threshold, thus explaining the 'recovery' phenomenon (Lindbo *et al.*, 1993a; Swaney *et al.*, 1995). Baulcombe (1996a) considered that this explanation did not account for all the data from several systems. For example, in plant lines displaying RNA homology-dependent resistance to isolates of PVX, the transgene transcription level was as high as but not higher than that in susceptible lines transformed with the same construct (Mueller *et al.*, 1995; English *et al.*, 1996). The alternative explanation based on the qualitative factor affecting production of aberrant RNA was considered to accommodate these problems. There are also many reports of an association between post-transcriptional gene silencing/RNA homology-dependent resistance and transgene methylation (English *et al.*, 1996; Jones *et al.*, 1998b; Guo *et al.*, 1999; Sonoda *et al.*, 1999).

As mentioned previously, RNA-mediated resistance is very specific and the broad resistance provided by coat-protein-mediated resistance is not observed (Mueller *et al.*, 1995; Malpica *et al.*, 1998). However, the resistance can be extreme. A potential problem in deploying homology-dependent gene silencing is that some viruses have been found to counteract the silencing. Béclin *et al.* (1998) demonstrated that CMV infection counteracted post-transcriptional gene silencing of nitrate reductase (*Nia*) or glucuronidase (*uidA*) transgenes in newly developing leaves of tobacco and *Arabidopsis* plants. The helper component protein (HC-Pro) of potyviruses has also been shown to be capable of counteracting post-transcriptional gene silencing (Anandalakshmi *et al.*, 1998; Brigneti *et al.*, 1998; Kasschau and Carrington, 1998). These observations suggest that some forms of RNA-mediated resistance could break down in the field if plants become infected by viruses capable of suppressing silencing. Evidence for a post-transcriptional gene-silencing-like mechanism for natural cases of virus resistance in wild-type plants (Covey *et al.*, 1997; Ratcliff *et al.*, 1997) suggests that there could be a selective advantage for viruses to counteract this type of resistance.

9.2.7 Antisense-mediated resistance

This type of resistance involves viral genes being transgenically expressed in such a way that the negative (antisense) viral strand is transcribed. The aim being to downregulate gene expression and thereby induce virus resistance. Antisense-mediated resistance has been obtained for CMV (Cuozzo *et al.*, 1988), PVX (Hemenway *et al.*, 1988), TMV (Powell *et al.*, 1989) and potyviruses (Fang and Grumet, 1993; Lindbo *et al.*, 1993b; Hammond and Kamo, 1995). In all these examples, resistance is weak. There are examples of high levels of resistance resulting from this approach, e.g. for PLRV (Kawchuk *et al.*, 1991), tomato ringspot virus (Yepes *et al.*, 1996) and potyviruses (Lindbo and Dougherty, 1992b; Ravelonandro *et al.*, 1993). Antisense resistance has also been induced against the ssDNA geminivirus, tomato golden mosaic virus (Bejarano and Lichtenstein, 1992), and this resistance source was also reported to be effective against beet curly top virus, but not African cassava mosaic virus (Bejarano and Lichtenstein, 1994).

Baulcombe (1996a) suggested that when resistance is weak, it is probably related to direct RNA-mediated resistance, in which the transgenic RNA either serves as a decoy molecule or hybridises with and thereby neutralises *cis*-acting elements in the viral genome. The examples of high levels of resistance had many of the attributes of homology-dependent resistance, including strain specificity and, with potyviral transgenes, a recovery phenotype (Ravelonandro *et al.*, 1993). Baulcombe (1996a) therefore thought it likely that homology-dependent resistance is activated by antisense transgenes, but less efficiently than it is with sense transgenes.

Antisense-mediated resistance suffers from unpredictability, and even when it confers strong resistance this is not as strong as other forms of transgenic resistance.

9.2.8 Ribozymes

Ribozymes are small RNA molecules, derived from certain viral satellite RNAs and viroids (as well as animals, yeasts and bacteriophages). They are capable of highly specific, intramolecular catalytic cleavage of RNA. The expression of many of these genes is stimulated in tobacco by TMV infection, particularly in those tobacco genotypes possessing the *N* gene. Atkins *et al.* (1995) introduced ribozyme genes to target the negative strand of citrus exocortis viroid in tomato and obtained a moderate reduction in accumulation of viral RNA. Ribozyme targeted against TMV also conferred resistance in tobacco (de Feyter *et al.*, 1996). Ribozymes and their potential to control plant viruses have been reviewed by Edington *et al.* (1993).

Synthetic hammerhead ribozymes specifically bind and cleave other RNA molecules and typically consist of a 22-nucleotide catalytic domain flanked by sequences that are complementary to the target RNA. The flanking sequences also known as 'hybridising arms', align the catalytic domain to juxtapose with the target RNA and provide specificity in the interaction with the target. Ribozyme-mediated resistance to TMV was conferred primarily by an antisense mechanism, although involvement of gene silencing cannot be excluded (de Feyter *et al.*, 1996).

Much more needs to be known about the mechanism of ribozyme activity in plants in order to assess their potential to confer transgenic resistance. They could be useful in providing transgenic resistance against viroids, which as a group are particularly difficult to control by genetic engineering approaches.

9.2.9 Recombination of plant RNA viruses

Recombination in RNA viruses is a general phenomenon and is considered to be a major factor in virus variability and evolution (Roossinck, 1997). An ever-increasing number of plant RNA viruses have been shown to undergo RNA recombination (Aaziz and Tepfer, 1999). The detection of recombination between host transcripts and infecting viral RNA genomes has given rise to concerns about the release of pathogen-derived, virus-resistant transgenic crops. Recombination could generate viruses with properties different from the infecting strain.

Greene and Allison (1994, 1996) have demonstrated recombination between an isolate of the bromovirus, cowpea chlorotic mottle virus, and the 3′ noncoding region and two-thirds of the coat protein gene expressed in transgenic plants. The presence of different mutation markers in the virus infecting the plants demonstrated that RNA recombination took place between the transgene mRNA and the challenging movement-defective viral genome. Sequence analysis revealed numerous modifications flanking the putative crossover site. When tested on a range of host plants, four out of seven displayed novel symptoms (Allison *et al.*, 1997). This example involved conditions of high selection pressure for recombination. As the transgenic plants were susceptible to virus infection, they would not have come through the screening phase in the development of transgenic resistance, where selection for virus resistance takes place. In terms of risk-assessment, a more realistic approach is to search for recombination between wild-type viruses and transgenic plants expressing viral sequences that confer extreme levels of resistance, i.e. under very low or zero selection pressure. Although there are examples of recombination of a DNA virus (cauliflower mosaic virus) (Wintermantel and Schoelz, 1996) and a RNA virus (tomato bushy stunt virus) (Borja *et al.*, 1999) with transgenically expressed viral sequences under conditions of moderate selection pressure, so far, there appear to be no examples of recombination between viruses and transgenically expressed viral sequences under low selection pressures, despite attempts to demonstrate this (Beachy *et al.*, 1990; Fuchs *et al.*, 1998).

Falk and Bruening (1994) concluded that the frequency of recombination between transgene RNA and viral RNA in transgenic plants is unlikely to be greater than that of virus RNAs in natural plant infections. These facts, combined with the fact that recombinant viruses are usually significantly less competitive than parental viruses, suggests that the risk of creating new and harmful viruses in transgenic plants is small, particularly when compared with the potential benefits of transgenic virus resistance. Risks associated with recombination of RNA viruses in virus-resistant transgenic plants have been reviewed previously (de Zoeten, 1991; Falk and Bruening, 1994; Robinson, 1996; Aaziz and Tepfer, 1999).

9.3 Prokaryotic genes

Although there is little information on the use of prokaryotic genes to achieve virus resistance, Wilson (1993) cites the use of the PVX subgenomic RNA promoter and diphtheria toxin mRNA to obtain transgenic tobacco plants that showed a 20-fold reduction in PVX concentration in upper, uninoculated leaves, following challenge with PVX. PVX inoculated leaves turned yellow and fell off. Watanabe *et al.* (1995) expressed fusion yeast dsRNAse in plants and obtained a broad type of resistance against a range of unrelated plant RNA viruses, based on degradation of the dsRNA intermediates produced during virus replication. Sano *et al.* (1997) produced transgenic potato lines expressing the yeast-derived dsRNA-specific ribonuclease, *pac1*, which suppressed potato spindle tuber viroid infection and accumulation and prevented viroid spread to progeny potato tubers.

This approach may be another that is particularly suited to plant viroids, and could potentially provide broad-spectrum resistance.

9.4 Non-plant eukaryotic genes

9.4.1 Plantibodies

Expression of a range of recombinant antibodies (single-chain fragment and full-length antibodies) in plants (plantibodies) has become possible due to the development of hybridoma-derived monoclonal antibodies and gene cloning techniques (Plückthun, 1991). Expression and assembly of functional antibodies in plants was first demonstrated by Hiatt *et al.* (1989). Since then, there have been further reports; however, there are only two plant viruses, artichoke mottle crinkle (Tavladoraki *et al.*, 1993) and TMV (Voss *et al.*, 1995) for which antibody-mediated protection has been obtained. The production and uses of antibodies in plants has been reviewed by Hiatt and Mostov (1993).

The level of protein expression by transgenic plants is likely to be very important for inducing effective resistance. Targeting in terms of viral epitopes is also important, an epitope of a conserved coat protein motif is likely to give broad resistance, effective against many strains of the same virus group. Plantibodies directed against coat proteins may also interfere with insect transmission.

It will be interesting to see whether plantibodies to non-structural proteins or structural proteins other than the coat protein are effective in inducing resistance, and whether such resistance is broad-spectrum. If so, it may be possible to produce resistance that is effective against a broad range of different plant virus groups that share a conserved region of such proteins, e.g. possibly replication associated genes.

9.4.2 Interferon

Human interferon is transiently active against viruses in plants and has been found to protect against TMV infection (Orchansky *et al.*, 1982). However, several reports have shown a lack of protection (Huisman *et al.*, 1985; Loesch-Fries *et al.*, 1985). Inconsistencies may be due to the concentration of interferon applied; only low levels are active in plants (Kaplan *et al.*, 1988). Transgenic plants expressing mammalian interferon have been obtained (Edelbaum *et al.*, 1992; Nakamura *et al.*, 1994) and transgenic potato plants expressing rat oligoadenylate synthetase were shown to be protected from PVX infection under field conditions (Truve *et al.*, 1993).

9.5 Plant genes

9.5.1 Disease resistance genes

The growth area in terms of transgenic resistance to viruses (also fungi and bacteria) over the next few years is likely to be the deployment of plant resistance genes. The *N* gene of tobacco was the first resistance gene to a virus to be cloned (Whitham *et al.*, 1994). This confers a 'gene-for-gene' resistance to TMV and most other members of the tobamovirus family. Some TMV strains can infect more than 200 plant species, including most members of the Solanaceae, so that immediately it is possible to see the potential of deploying the *N* gene in other plant species. A TMV-susceptible tomato cultivar has already been transformed with the *N* gene and found to be resistant to TMV (Baker *et al.*, 1996). *N* is a member of the class of disease resistance genes whose predicted protein product possesses a putative nucleotide binding site and leucine-rich repeat region (Whitham *et al.*, 1994). The transformation of this gene into another crop plant species, where it conferred resistance to TMV, demonstrates that disease resistance genes of this class can be successfully moved across the species barrier. Tobacco and tomato are closely related, being members of the Solanaceae and there are a number of *N* homologues in tomato. It will be interesting to see if the *N* gene will function in plant species that are more distantly related to tobacco. The *Rx* gene from potato that confers resistance to some isolates of PVX has recently been cloned and found to confer resistance when transformed into potato, *N. benthamiana* and *N. tabacum* (Bendahmane *et al.*, 1999).

Now that resistance genes have been cloned, the mechanistic basis of their action against pathogens is starting to be revealed. There are at least four broad mechanistic classes. Some resistance genes encode components of receptor systems that detect, either directly or indirectly, the presence of potential pathogens. Activation of such receptors probably initiates a signal transduction pathway that results in the induction of the generic response genes (Godiard *et al.*, 1994). Genes involved in gene-for-gene interactions, including *N*, belong to this class. A second class of resistance gene encodes products that detoxify and inactivate compounds that the pathogen requires to cause disease. A third class of resistance gene may code for altered targets for pathogen-derived molecules that are required for pathogenesis. A resistance allele could code for a product that did not interact with the pathogen and would tend to be recessive. The absolute dependence of viruses on the host's biochemical machinery provides particular opportunities for such resistance to evolve. Some of the recessive genes for resistance to viruses may be of this type (Fraser, 1992). Another category of resistance gene includes those encoding structural or constitutive biochemical barriers to pathogens.

Problems associated with the 'conventional' deployment of resistance genes will also apply to their transgenic deployment. It will be possible to combine (pyramid) resistance genes in particular formats that will confer resistance to all known biotypes of particular pathogens. Combining resistance genes of different mechanistic classes may increase their durability compared with their use individually (Wolfe, 1993). It will be important to obtain information on virus variation, particularly in terms of the genes/proteins interacting with the different plant resistance genes, so that this can be integrated into resistance gene deployment strategies. Molecular tests will be required to distinguish virulent and avirulent isolates for the different resistance genes, so that large numbers of virus isolates can be characterised rapidly. This will indicate which resistance genes are most likely to be effective and the frequency of resistance breaking isolates.

Strategies developed for reducing the selection pressure for virulence, such as the use of multi-lines and cultivar mixtures could be adapted for transgenic crops. A transgenic approach would offer a number of advantages; resistance genes could be introgressed into desirable plant backgrounds more rapidly and without the genetic or linkage drag that is experienced with classical backcrossing (such as observed with *Tm-2* for tobamovirus resistance in tomato) (Young and Tanksley, 1989). Also, different resistance genes could be transformed into identical genetic backgrounds, so that crops with synthetic mixtures of such resistance genes would be uniform in terms of product quality and time to harvest, etc. Transformation might facilitate the deployment of non-host resistance to viruses in crop plants. Deploying the same, or similar resistance genes widely across species barriers in many crops could decrease the durability of resistance, particularly to those pathogens that cross species barriers. Molecular approaches to manipulating plant disease resistance genes have been reviewed by Michelmore (1995).

Another potential approach to utilising natural plant resistance, which is also another form of pathogen-derived resistance, would be to transgenically express viral elicitors of natural host resistance responses. There are now a number of virus-plant interactions in which these elicitors have been identified. The tobamovirus elicitor of *N′*- (Culver and Dawson, 1991), L^1-, L^2- (de la Cruz *et al.*, 1997) and L^3-mediated resistances (Berzal-Harranz *et al.*, 1995) in tobacco and *Capsicum* and the PVX elicitor of *Rx1*-, *Rx2*- and *Nx*-mediated resistances in potato (Santa Cruz and Baulcombe, 1993; Bendahmane *et al.*, 1995; Querci *et al.*, 1995) is the viral coat protein. The movement protein of a tobamovirus is an elicitor of resistance genes, *Tm-2* (Meshi *et al.*, 1989) and $Tm\text{-}2^2$ (Weber and Pfitzer, 1998), and the PVX movement protein elicits *Nb*-mediated resistance (Malcuit *et al.*, 1998). The replicase gene of TMV confers virulence to *Tm-1* (Meshi *et al.*, 1988) and *N* (Padgett and Beachey, 1993).

It would be interesting to see whether controlled expression of these elicitors could be achieved in transgenic plants and whether such expression in plants carrying the appropriate resistance gene would elicit natural resistance to other viruses and pathogens without major effects on plant growth (Köhm *et al.*, 1993). Expression of the fungal *Avr9* and oomycete *inf1* genes coding for elicitors of the hypersensitive response (HR) from engineered PVX genomes, resulted in HR lesions and inhibition of virus spread when the recombinant virus was inoculated to tobacco plants (Kamoun *et al.*, 1999). This demonstrated that expression of fungal and oomycete avirulence genes in plants can induce virus resistance. The tomato gene *Cf-9* conferring resistance to races of the fungus *Cladosporium fulvum* that express the corresponding avirulence gene *Avr9* has been expressed in potato and tobacco (Hammond-Kosack *et al.*, 1998). When tobacco plants expressing this gene were crossed to Avr9-producing tobacco, developmentally regulated seedling death occurred in the progeny. The authors suggested various methods to utilise this gene combination to control destructive pathogens. These included using weak alleles of these genes that are known and could be used to express a very weak defence response throughout the plant. Also, a pathogen-inducible promoter system has been suggested to limit defence gene induction to the precise site of pathogen (De Wit, 1992). Alternatively, by combining three transgenes into one plant line, with the *Cf-9* gene inactivated by a maize *Dissociation* transposable element, a stabilised maize *Activator* transposable source and the *35S:Avr9* transgene, genetically engineered acquired resistance to various fungal pathogens can be obtained (Hammond-Kosack *et al.*, 1998). Somatic excision of the *Dissociation* transposon restores *Cf-9* function, which results in the localised activation of plant defence responses because the apoplasmically produced Avr9 peptide is now recognised. This could possibly provide broad-spectrum resistance to viruses; this approach could also be taken utilising plant resistance genes to viruses with the viral avirulence genes.

9.5.2 Pathogenesis-related proteins

Pathogenesis-related proteins are a family of plant polypeptides that are synthesised and accumulate in high levels following the hypersensitive reaction induced by a pathogen. Tobacco plants have been transformed to express the PR1b gene. This gene codes for a pathogenesis-related protein induced by viral infections in certain tobacco cultivars. Transgenic TMV-susceptible and TMV-resistant genotypes expressing this gene were no more resistant to TMV than parental control lines (Cutt *et al.*, 1989).

9.5.3 Ribosome-inactivating proteins

A number of plants produce a class of proteins that have become known as ribosome-inactivating proteins. Many are believed to enter the cytoplasm of plant cells whenever the cell wall is damaged, and a number have been shown to protect against viral infection. One such protein has been identified in pokeweed (*Phytolacca americana*). The pokeweed antiviral protein consists of three forms that are active against a range of viruses (Barbieri *et al.*, 1982). It was shown to be a ribosomal-binding protein, deglycosylating a specific base on eukaryotic 28S rRNA, thereby preventing binding of elongation factor 2 and arresting protein synthesis. It has recently been cloned into tobacco, *N. benthamiana*, and potato where it protected against infection by PVX, CMV and PVY (including aphid inoculation of PVY) (Lodge *et al.*, 1993). Another ribosome-inactivating protein, trichosanthin, which also acts as a *N*-glycosidase has been transgenically expressed in tobacco plants and shown to induce complete resistance to mechanical inoculation of turnip mosaic virus (Lam *et al.*, 1996). A particularly elegant strategy for deploying another ribosome-inactivating protein, dianthin, has been reported by Hong *et al.* (1996). The promoter of the geminivirus, ACMV, was coupled to the coding sequence of dianthin, and the fusion construct was transformed into *N. benthamiana*. The transgenic viral promoter controlling dianthin production was activated *in trans* following infection by the virus, causing the infected cells to die and restricting the virus to the inoculated cells.

9.6 Perceived risks associated with transgenic resistance

Perceived risks associated with the deployment of transgenic resistance to viruses have been reviewed by de Zoeten (1991), Falk and Bruening (1994) and Robinson (1996). Benefits and risks, regulation and public acceptance of genetically modified foods in general in the UK have also been reviewed (Anon, 1998).

The phenomenon of transencapsidation is discussed in section 9.2.1, and recombination in section 9.2.9. Transencapsidation is mostly pertinent to coat-protein-mediated resistance, whereas recombination is relevant whenever any viral sequence is expressed in a plant. Although it has been established that both of these phenomena occur in transgenic plants when artificial selection is applied, little is known about whether they would occur and at what frequencies under low or zero selection pressure. Both are natural phenomena and not peculiar to transgenic plants. More important than whether these phenomena will occur in transgenic plant is whether there is any real risk as a consequence of their occurrence. The information available so far has not provided any evidence of risk over and above what occurs in nature. In order to verify any perceived risks, it will be important to provide information on the frequency of recombination and transencapsidation in nature and in transgenic plants under commercial conditions and to quantify any consequent risk that might arise from this mathematically. In quantifying any risks associated with transgenic virus resistance, it is very important to do this in the context of the risks associated with current practice in terms of virus control. There are risks associated with applying large amounts of soil sterilants and fungicides to control viruses transmitted by fungi and with applying large amounts of insecticides to control insect vectors of viruses, and these need to be weighed against any risks established for transgenic resistance. The reduction in energy inputs that will result from transgenic resistance also needs to be taken into account. It may be necessary to carry out feeding experiments where transgenic virus resistance is deployed in edible crops.

As we are already consuming plant viruses in infected plants and proteins associated with plant resistance genes, again it is difficult to see how transgenic approaches deploying unaltered viral sequences and plant resistance genes can pose any risk over and above any current risks. To examine this, research is needed to determine whether insertion of transgenes conferring resistance to viruses have any effects on adjacent and distant regions of the plant genomes into which they are inserted. If there are effects, then it will be important to determine what the consequences of any such changes are. The points raised above concerning transgenes also apply to the regulatory sequences inserted into plants to control the expression of the transgenes. The use of plant promoters will remove objections currently levelled at some other promoters used to express transgenes. New transformation protocols should mean that the necessity to incorporate marker genes, such as herbicide tolerance or antibiotic resistance, will be a thing of the past. Also new vectors will minimise the amounts of extra DNA inserted at the borders of the transgenes.

Concern has been expressed about the escape of transgenes in pollen from transgenic plants to relatives of crop plants. This topic has been reviewed by Hancock *et al.* (1996). Again research is needed and is being carried out to quantify the risk of transgene escape and to determine if there are any consequences of such escapes. Disease resistant crops have been bred and widely grown for many years. In assessing consequences of transgene escape, it is important to make comparisons with what is currently happening in terms of escape of plant disease resistance genes already present in widely grown cultivars. The concept of super weeds with multiple virus resistance arising from transgene escape seems unlikely. Weeds are usually tolerant to plant viruses anyway and hybrids between crop and weed species are often sterile. If any problems were established as a consequence of escape of transgenes conferring a particular type of virus resistance, strategies such as cytoplasmic and other forms of male sterility are available to ensure the transgene does not escape. Moreover, many edible leafy vegetable and root crops, such as brassicas, carrots and lettuce, are harvested before they flower. In such crops, risk of transgene escape would be confined to seed production, which can be, and usually is (in terms of F_1 hybrid production), carried out under controlled conditions. In terms of environmental impact of transgenic virus resistance, it seems more likely that they will be beneficial rather than damaging. Targeted resistance to specific important virus diseases will reduce the dependence on blanket pesticide treatments and this will be beneficial for non-target insects (some of which are predators and parasites of pest insect species) and wildlife further up the food chain.

9.7 Concluding remarks

The transgenic approaches described in this chapter offer new opportunities to control plant viruses and other pathogens. Pathogen-derived resistance will be particularly important for those viruses for which no useful natural plant resistances have been found and also for severe viral epidemics, where there is major crops loss. In the former case, it has the potential to provide novel resistance where currently there is no resistance and in the latter case it will be much quicker to develop pathogen-derived resistant plants that are agronomically acceptable than it would be to produce virus resistant cultivars by conventional breeding strategies. As such, it could offer a first line of defence against newly introduced pathogens.

With all the research and subsequent development required to produce transgenic virus-resistant cultivars, attention needs to be paid to how best to deploy these genes so that the resistance is durable. Problems relating to the transgenic deployment of plant resistance genes are the same as those for current conventional deployment of the genes and are discussed in section 9.5.1. Deployment of pathogen-derived resistance also shows some similar features. Although the vast literature on pathogen-derived resistance, since its first demonstration in 1986, reveals many details unique to each virus-plant-transgene system, recent advances have started to unravel phenomena and establish some patterns. There are, of course, as there always will be, exceptions that do not fit the proposed models. In retrospect, it is now known that many early examples of what were thought to be viral protein-mediated resistance were in fact RNA-mediated. For many examples where plants were transformed to express viral proteins, appropriate experiments were not performed to determine whether the resistance was protein or RNA-mediated. In such cases, data provided on the specificity, strength of resistance and mechanisms, is likely to confuse any proposed mechanistic models.

In general, protein-mediated resistance, particularly coat-protein-mediated resistance confers fairly broad-spectrum but not very extreme resistance. Replicase proteins seem to give higher levels of resistance that are more specific than those derived from coat proteins. Movement-protein-mediated resistance appears potentially very useful in that it can confer extreme resistance by confining infection to inoculated cells, and the movement proteins of some viruses seem capable of providing very broad-spectrum resistance. RNA-mediated resistance often protects against very high inoculum levels of virus as well as RNA, but is highly specific (homology dependent). Where transgenic resistance is dependent on sequence homology between transgene and challenging virus, cultivation of such plants is likely to lead to selection for other viral genotypes. With detailed information on viral genotypes in particular geographic locations, it would be possible to deploy a number of transgenes that would be effective against all genotypes. The different transgenes could be deployed in different ways, for example pyramided in plants or in synthetic mixtures within cultivars as described for plant resistance genes (see section 9.5.1), in order to reduce selection pressure for resistance-breaking genotypes.

Another important area of research that would inform deployment of transgenes is viral fitness. If the fittest genotypes of viruses could be identified and characterised, these would provide the most important sequences to use for pathogen-derived resistance. It would be beneficial to combine different types of pathogen-derived resistance, for instance the broad-spectrum resistance conferred by coat-protein-mediated resistance combined with a more extreme form of resistance, such as RNA-mediated resistance targeted against the fittest viral genotype, could possibly provide broad-spectrum, extreme and durable resistance. Combining conventional and transgenic resistance is another alternative and has already been shown to enhance resistance (Barker *et al.*, 1994).

Stability of transgenes and their expression in plant genomes is important in terms of utilising transgenic virus resistance. The increased production of F_1 hybrid crop cultivars may lessen potential problems in this respect, and as detailed molecular maps of plant genomes become available, it will make it possible to follow transgene insertions and hopefully predict stability. Eventually, it should be possible to target transgene insertions to particular parts of plant genomes in order to ensure they function correctly and are stable. The use of tissue-specific and inducible promoters will give rise to improved strategies for transgenic resistance to pathogens. Expression of viral proteins or pathogen avirulence determinants in combination with resistance genes in plants and subsequent pathogen infection could potentially have yield penalties. This needs to be determined and quantified on a case-by-case basis.

Further basic research is needed to provide a greater understanding of the detailed mechanisms of the different forms of pathogen-derived resistance. Strategic research on the stability and durability of the different types of pathogen-derived resistance and different combinations of these under field conditions will allow strategies for their deployment to be developed. Basic research on viral fitness and the viral determinants of virulence and avirulence for different conventional plant resistance genes is needed, together with the mapping and cloning of such genes, so that hypotheses on the durability of different resistance gene formats can be tested. For example, it will be interesting to see whether pyramiding resistance genes in one plant genotype will be less, or more durable than growing multi-lines, where different plants within the mixture possess different resistance genes. This will allow transgenic deployment of plant resistance genes in combinations that will be durable. In future it will be possible to modify plant resistance genes in a rational manner. It may be possible to engineer genes that provide broad-spectrum recognition and resistance, targeted to conserved pathogen sequences and functions. Research on transgenic resistance to viruses has already shed considerable light on how viruses and plants interact and no doubt will continue to do so.

Acknowledgements

The author would like to thank colleagues Dr Dez Barbara and Dr Ian Puddephat for helpful discussions on transgenic resistance to plant viruses, Dr Carol Jenner for useful discussions on plant-virus interactions and Jane Morris for typing this document. Work at Horticulture Research International is supported by the Biotechnology and Biological Sciences Research Council and the Ministry of Agriculture Fisheries and Food.

References

Aaziz, R. and Tepfer, M. (1999) Recombination in RNA viruses and in virus-resistant transgenic plants. *J. Gen. Virol.*, **80** 1339-46.

Allison, R.F., Greene, A.E. and Schneider, W.L. (1997) Significance of RNA recombination in capsid-protein-mediated virus-resistant transgenic plants, in *Virus-Resistant Transgenic Plants: Potential Ecological Impact* (eds. M. Tepfer and E. Balázs), INRA and Springer-Verlag, pp. 40-44.

Anandalakshmi, R., Pruss, G.J., Ge, X., Marathe, R., Mallory, A.C., Smith, T.M. and Vance, V.B. (1998) A viral suppressor of gene silencing in plants. *Proc. Natl. Acad. Sci. USA*, **95** 13079-84.

Anderson, E.J., Stark, D.M., Nelson, R.S., Powell, P.A., Tumer, N.E. and Beachy, R.N. (1989) Transgenic plants that express the coat protein genes of tobacco mosaic virus or alfalfa mosaic virus interfere with disease development of some non-related viruses. *Phytopathology*, **79** 1284-90.

Anderson, J.M., Palukaitis, P. and Zaitlin, M. (1992) A defective replicase gene induces resistance to cucumber mosaic virus in transgenic tobacco plants. *Proc. Natl. Acad. Sci. USA*, **89** 8759-63.

Anon (1998) *Genetically Modified Foods, Benefits and Risks, Regulation and Public Acceptance.* Parliamentary Office of Science and Technology, London.

Atkins, D., Hull, R., Wells, B., Roberts, K., Moore, P. and Beachy, R.N. (1991) The tobacco mosaic virus 30K movement protein in transgenic tobacco plants is localized to plasmodesmata. *J. Gen. Virol.*, **72** 209-11.

Atkins, D., Young, M., Uzzell, Z., Kelly, L., Fillatti, J. and Gerlach, W.L. (1995) The expression of the antisense and ribozyme genes targeting citrus exocortis viroid in transgenic plants. *J. Gen. Virol.*, **76** 1781-90.

Audy, P., Palukaitis, P., Slack, S.A. and Zaitlin, M. (1994) Replicase-mediated resistance to potato virus Y in transgenic tobacco plants. *Mol. Plant-Microbe Interact.*, **7** 15-22

Baker, B., Whitham, S. and McCormick, S. (1996) The *N* gene of tobacco confers resistance to tobacco mosaic virus in transgenic tomato, in *Biology of Plant-Microbe Interactions* (eds. G. Stacey, B. Mullin and P.M. Gresshoff), International Society for Molecular Plant Microbe Interactions, St. Paul, Minnesota, pp. 65-70.

Barbieri, L., Aron, G.M., Irvin, J.D. and Stirpe, F. (1982) Purification and partial characterization of another form of the antiviral protein from seeds of *Phytolacca americana* L. (pokeweed). *Biochem. J.*, **203** 55-59.

Bardonnet, N., Hans, F., Serghini, M.A. and Pinck, L. (1994) Protection against virus infection in tobacco plants expressing the coat protein of grapevine fanleaf nepovirus. *Plant Cell Rep.*, **13** 357-60.

Barker, H., Reavy, B., Kumar, A., Webster, K.D. and Mayo, M.A. (1992) Restricted virus multiplication in potatoes transformed with the coat protein gene of potato leafroll luteovirus: similarities with a type of host gene-mediated resistance. *Ann. of Appl. Biol.*, **120** 55-64.

Barker, H., Webster, K.D., Jolly, C.A., Reavy, B., Kumar, A. and Mayo, M.A. (1994) Enhancement of resistance to potato leafroll virus multiplication in potato by combining the effects of host genes and transgenes. *Mol. Plant-Microbe Interact.*, **7** 528-30.

Baulcombe, D. (1994a) Novel strategies for engineering virus resistance in plants. *Curr. Opin. Biotechnol.*, **5** 117-24.

Baulcombe, D. (1994b) Replicase-mediated resistance: a novel type of virus resistance in transgenic plants. *Trends Microbiol.*, **2** 60-63.

Baulcombe, D.C. (1996a) Mechanisms of pathogen-derived resistance to viruses in transgenic plants. *Plant Cell*, **8** 1833-44.

Baulcombe, D. (1996b) RNA as a target and an initiator of post-transcriptional gene silencing in transgenic plants. *Plant Mol. Biol.*, **32** 79-88.

Beachy, R.N. (1988) Virus cross-protection in transgenic plants, in *Temporal and Spatial Regulation of Plant Genes* (eds. D.P.S. Verma and R.B. Goldberg), Springer-Verlag, New York, pp. 313-331.

Beachy, R.N., Loesch-Fries, S. and Tumer, N.E. (1990) Coat protein-mediated resistance against virus infection. *Annu. Rev. Phytopathol.*, **28** 451-74.

Beachy, R.N., Padgett, H.S., Kahn, T., Bendahmane, M., Fitchen, J.H., Heinlein, M., Watanabe, Y. and Epel, B.L. (1996) Cellular and molecular mechanisms of coat protein and movement protein mediated resistance against TMV, in *Biology of Plant-Microbe Interactions* (eds. G. Stacey, B. Mullin and P.M. Gresshoff), International Society for Molecular Plant-Microbe Interactions, St. Paul, Minnesota, pp. 259-64.

Beck, D.L., Guilford, P.J., Voot, D.M., Andersen, M.T. and Forster, R.L.S. (1991) Triple gene block proteins of white clover mosaic potexvirus are required for transport. *Virology*, **183** 695-702.

Beck, D.L., Van Dolleweerd, C.J., Lough, T.J., Balmori, E., Voot, D.M., Andersen, M.T., O'Brien, I.E.W. and Forster, R.L.S. (1994) Disruption of virus movement confers broad-spectrum resistance against systemic infection by plant viruses with a triple gene block. *Proc. Natl. Acad. Sci. USA*, **91** 10310-14.

Béclin, C., Berthome, R., Palauqui, J.-C., Tepfer, M. and Vaucheret, H. (1998) Infection of tobacco or *Arabidopsis* plants by CMV counteracts systemic post-transcriptional silencing of non-viral (trans)genes. *Virology*, **252** 313-17.

Bejarano, E.R. and Lichtenstein, C.P. (1992) Prospects for engineering virus resistance in plants with antisense RNA. *Trends Biotechnol.*, **10** 383-87.

Bejarano, E.R. and Lichtenstein, C.P. (1994) Expression of TGMV antisense RNA in transgenic tobacco inhibits replication of BCTV but not ACMV geminiviruses. *Plant Mol. Biol.*, **24** 241-48.

Bendahmane, A., Köhm, B.A., Dedi, C. and Baulcombe, D.C. (1995) The coat protein of potato virus X is a strain-specific elicitor of *Rx1*-mediated virus resistance in potato. *Plant J.*, **8** 933-41.

Bendahmane, A., Kanyuka, K. and Baulcombe, D.C. (1999) The *Rx* gene from potato controls separate virus resistance and cell death responses. *Plant Cell*, **11** 781-91.

Bertioli, D.J., Cooper, J.I., Edwards, M.L. and Hawes, W.S. (1992) Arabis mosaic nepovirus coat protein in transgenic tobacco lessens disease severity and virus replication. *Ann. Appl. Biol.*, **120** 47-54.

Berzal-Herranz, A., de la Cruz, A., Tenllado, F., Díaz-Ruíz, J.R., Lopez, L., Sanz, A.I., Vaquero, C., Serra, M.T. and García-Luque, I. (1995) The *Capsicum* L^3 gene-mediated resistance against the tobamoviruses is elicited by the coat protein. *Virology*, **209** 498-505.

Bevan, M.W., Mason, S.E. and Goelet, P. (1985) Expression to tobacco mosaic virus coat protein by a cauliflower mosaic virus promoter in plants transformed by *Agrobacterium*. *EMBO J.*, **4** 1921-26.

Birch, R.G. (1997) Plant transformation: problems and strategies for practical application. *Annu. Rev. Plant Physiol. Plant Mol. Biol.*, **48** 297-326.

Borja, M., Rubio, T., Scholthof, H.B. and Jackson, A.O. (1999) Restoration of wild-type virus by double recombination of tombusvirus mutants with a host transgene. *Mol. Plant-Microbe Interact.*, **12** 153-62.

Brault, V., Candresse, T., Le Gall, O., Delbos, R.P., Lanneau, M. and Dunez, J. (1993) Genetically engineered resistance against grapevine chrome mosaic nepovirus. *Plant Mol. Biol.*, **21** 89-97.

Braun, C.J. and Hemenway, C.L. (1992) Expression of amino-terminal portions or full-length viral replicase genes in transgenic plants confers resistance to potato virus X infection. *Plant Cell*, **4** 735-44.

Brigneti, G., Voinnet, O., Li, W.-X., Ji, L.-H., Ding, S.-W. and Baulcombe, D. (1998) Viral pathogenicity determinants are suppressors of transgene silencing in *Nicotiana benthamiana. EMBO J.*, **17** 6739-46.

Candelier-Harvey, P. and Hull, R. (1993) Cucumber mosaic virus genome is encapsidated in alfalfa mosaic virus coat protein expressed in transgenic tobacco plants. *Transgen. Res.*, **2** 277-85.

Carr, J.P. and Zaitlin, M. (1993) Replicase mediated resistance. *Semin. Virol.*, **4** 339-47.

Carrington, J.C., Cary, S.M., Parks, T.D. and Dougherty, W.G. (1989) A second proteinase encoded by a plant potyvirus genome. *EMBO J.*, **8** 365-70.

Carrington, J.C., Kasschau, K.D., Mahajan, S.K. and Schaad, M. C. (1996) Cell-to-cell and long-distance transport of viruses in plants. *Plant Cell*, **8** 1669-81.

Cassidy, B.G. and Nelson, R.S. (1995) Differences in protection phenotypes in tobacco plants expressing coat protein genes from peanut stripe potyvirus with or without an engineered ATG. *Mol. Plant-Microbe Interact.*, **8** 357-65.

Citovsky, V. and Zambryski, P. (1993) Transport of nucleic acids through membrane channels: snaking through small holes. *Annu. Rev. Microbiol.*, **47** 167-97.

Citovsky, V., Knorr, D., Schuster, G. and Zambryski, P. (1990) The P30 movement protein of tobacco mosaic virus is a single-strand nucleic acid binding protein. *Cell*, **60** 637-47.

Citovsky, V., Wong, M.L., Shaw, A.L., Prasad, B.V.V. and Zambryski, P. (1992) Visualization and characterization of tobacco mosaic virus movement protein binding to single-stranded nucleic acids. *Plant Cell*, **4** 397-411.

Clark, W.G., Fitchen, J.H. and Beachy, R.N. (1995) Studies of coat-protein-mediated resistance to TMV using mutant CP. I; the PM2 assembly defective mutant. *Virology*, **208** 485-91.

Cooper, B., Lapidot, M., Heick, J.A., Dodds, J.A. and Beachy, R.N. (1995) A defective movement protein of TMV in transgenic plants confers resistance to multiple viruses, whereas the functional analogue increases susceptibility. *Virology*, **206** 307-13.

Covey, S.N., Al-Kaff, N.S., Langara, A. and Turner, D. S. (1997) Plants combat infection by gene silencing. *Nature*, **387** 781-82.

Culver, J.N. and Dawson, W.O. (1991) Tobacco mosaic virus elicitor coat protein genes produce a hypersensitive phenotype in transgenic *Nicotiana sylvestris* plants. *Mol. Plant-Microbe Interact.*, **4** 458-63.

Cuozzo, M., O'Connell, K.M., Kaniewski, W.K., Fang, R.-X., Chua, N.-H. and Tumer, N.E. (1988) Viral protection in transgenic tobacco plants expressing the cucumber mosaic virus coat protein or its antisense RNA. *Bio/Technol.*, **6** 549-57.

Cutt, J.R., Harpster, M.H., Dixon, D.C., Carr, J.P., Dunsmuir, P.A. and Klessig, D.F. (1989) Disease response to tobacco mosaic virus in transgenic tobacco plants that constitutively express the pathogenesis-related PR1b gene. *Virology*, **173** 89-97.

de Carvalho Niebel, F., Frendo, P., Van Montagu, M. and Cornelissen, M. (1995) Post-transcriptional co-suppression of β-1,3-glucanase genes does not affect accumulation of transgene nuclear mRNA. *Plant Cell*, **7** 347-58.

de Feyter, R., Young, M., Schroeder, K., Dennis, E.S. and Gerlach, W. (1996) A ribozyme gene and an antisense gene are equally effective in conferring resistance to tobacco mosaic virus on transgenic tobacco. *Mol. Gen. Genet.*, **250** 329-38.

De Haan, P., Gielen, J.J.L., Prins, M., Wijkamp, I.G., Van Schepen, A., Peters, D., Van Grinsven, M.Q.J.M. and Goldbach, R.W. (1992) Characterization of RNA-mediated resistance to tomato spotted wilt virus in transgenic tobacco plants. *Bio/Technol.*, **10** 1133-37.

De la Cruz, A., Lopez, L., Tenllado, F., Díaz-Ruíz, J.R., Sanz, A.I., Vaquero, C., Serra, M.T. and García-Luque, I. (1997) The coat protein is required for the elicitation of the *Capsicum* L^2 gene-mediated resistance against the tobamoviruses. *Mol. Plant-Microbe Interact.*, **10** 107-13.

Devic, M., Jaegle, M. and Baulcombe, D. (1990) Cucumber mosaic virus satellite RNA (strain Y): analysis of sequences which affect systemic necrosis on tomato. *J. Gen. Virol.*, **71** 1443-49.

De Wit, P.J.G.M. (1992) Molecular characterization of gene-for-gene systems in plant-fungus interactions and the application of avirulence genes in control of plant pathogens. *Annu. Rev. Phytopathol.*, **30** 391-418.

de Zoeten, G.A. (1991) Risk assessment: Do we let history repeat itself? *Phytopathology*, **81** 585-86.

Dinant, S., Blaise, F., Kusiak, C., Astier-Manifacier, S. and Albouy, J. (1993) Heterologous resistance to potato virus Y in transgenic tobacco plants expressing the coat protein gene of lettuce mosaic potyvirus. *Phytopathology*, **83** 818-24.

Donson, J., Kearney, C.M., Turpen, T.H., Khan, I.A., Karuth, G., Turpen, A.M., Jones, G.E., Dawson, W.O. and Lewandoski, D.J. (1993) Broad resistance to tobamoviruses is mediated by a modified tobacco mosaic virus replicase transgene. *Mol. Plant-Microbe Interact.*, **6** 635-642.

Edington, B.V., Dixon, R.A. and Nelson, R.S. (1993) Ribozymes: descriptions and uses, in *Transgenic Plants, Fundamentals and Applications* (ed. A. Hiatt), Marcel Dekker, New York, pp. 301-323.

Edelbaum, O., Stein, D., Holland, N., Gafni, Y., Livneh, O., Novick, D., Rubinstein, M. and Sela, I. (1992) Expression of active human interferon-β in transgenic plants. *J. Interferon Res.*, **12** 449-53.

English, J.J., Mueller, E. and Baulcombe, D.C. (1996) Suppression of virus accumulation in transgenic plants exhibiting silencing of nuclear genes. *Plant Cell*, **8** 179-88.

Falk, B.W. and Bruening, G. (1994) Will transgenic crops generate new viruses and new diseases? *Science*, **263** 1395-96.

Falk, B.W., Passmore, B.K., Watson, M.T. and Chin, L.S. (1995) The specificity and significance of heterologous encapsidation of virus and virus-like RNAs, in *Biotechnology and Plant Protection, Viral Pathogenesis and Disease Resistance* (eds. D.D. Bills and S.-D. Kung), World Scientific, Singapore, pp. 391-415.

Fang, G. and Grumet, R. (1993) Genetic engineering of potyvirus resistance using constructs derived from zucchini yellow mosaic virus coat protein gene. *Mol. Plant-Microbe Interact.*, **6**, 358-67.

Farinelli, L. and Malnoë, P. (1993) Coat protein gene-mediated resistance to potato virus Y in tobacco: examination of the resistance mechanisms—is the transgenic coat protein required for protection? *Mol. Plant-Microbe Interact.*, **6** 284-92.

Farinelli, L., Malnoë, P. and Collet, G.F. (1992) Heterologous encapsidation of potato virus Y strain O (PVY^{O}) with the transgenic coat protein of PVY strain N (PVY^{N}) in *Solanum tuberosum* cv. Bintje. *Bio/Technol.*, **10** 1020-25.

Fehér, A., Skryabin, K.G., Balázs, E., Preiszner, J., Shulga, O.A., Zakharyev, V.M. and Dubits, D. (1992) Expression of PVX coat protein gene under the control of extensin gene promoter confers resistance on transgenic potato plants. *Plant Cell Rep.*, **11** 48-52.

Fraser, R.S.S. (1992) The genetics of plant-virus interactions: implications for plant breeding, in *Breeding for Disease Resistance: Euphytica 63* (eds. R. Johnson and G.J. Jellis), Kluwer, Dordrecht, pp. 175-85.

Frischmuth, T. and Stanley, J. (1994) Beet curly top virus symptom amelioration in *Nicotiana benthamiana* transformed with a naturally occurring viral subgenomic DNA. *Virology*, **200** 826-30.

Fuchs, M. and Gonsalves, D. (1995) Resistance of transgenic hybrid squash ZW-20 expressing coat protein genes of zucchini yellow mosaic virus and watermelon mosaic virus 2 to mixed infections by both potyviruses. *Bio/Technol.*, **13** 1466-73.

Fuchs, M., Klas, F.E., McFerson, J.R. and Gonsalves, D. (1998) Transgenic melon and squash expressing coat protein genes of aphid-borne viruses do not assist the spread of an aphid non-transmissible strain of cucumber mosaic virus in the field. *Transgen. Res.*, **7** 449-62.

Gal-On, A., Wolf, D., Wang, Y., Faure, J.-E., Pilowsky, M. and Zelcer, A. (1998) Transgenic resistance to cucumber mosaic virus in tomato: blocking of long-distance movement of the virus in lines harbouring a defective viral replicase gene. *Phytopathology*, **88** 1101-107.

Gerlach, W.L., Llewellyn, D. and Haseloff, J. (1987) Construction of a disease resistance gene using the satellite RNA of tobacco ringspot virus. *Nature*, **328** 802-806.

Gielen, J., Ultzen, T., Bontems, S., Loots, W., Van Schepen, A., Westerboek, A., de Haan, P. and Van Grinsven, M. (1996) Coat-protein-mediated protection to cucumber mosaic virus infections in cultivated tomato. *Euphytica*, **88** 139-49.

Godiard, L., Grant, M.R., Dietrich, R.A., Kiedrowski, S. and Dangle, J.L. (1994) Perception and response in plant disease resistance. *Curr. Opin. Genet. Dev.*, **4** 662-71.

Golemboski, D.B., Lomonossoff, G.P. and Zaitlin, M. (1990) Plants transformed with a tobacco mosaic virus non-structural gene sequence are resistant to the virus. *Proc. Natl. Acad. Sci. USA*, **87** 6311-15.

Gonsalves, D. (1998) Control of papaya ringspot virus in papaya: a case study. *Annu. Rev. Phytopathol.*, **36** 415-37.

Gonsalves, D., Chee, P., Provvidenti, R., Seem, R. and Slightom, J.L. (1992) Comparison of coat-protein-mediated and genetically-derived resistance in cucumbers to infection by cucumber mosaic virus under field conditions with natural challenge inoculations by vectors. *Bio/Technol.*, **10** 1562-70.

Greene, A.E. and Allison, R.F. (1994) Recombination between viral RNA and transgenic plant transcripts. *Science*, **263** 1423-25.

Greene, A.E. and Allison, R.F. (1996) Deletions in the 3′ untranslated region of cowpea chlorotic mottle virus transgene reduce recovery of recombinant viruses in transgenic plants. *Virology*, **225** 231-34.

Grumet, R. (1994) Development of virus resistant plants via genetic engineering. *Plant Breeding Rev.*, **12** 47-79.

Guo, H.S. and García, J.A. (1997) Delayed resistance to plum pox potyvirus mediated by a mutated RNA replicase gene: involvement of a gene-silencing mechanism. *Mol. Plant-Microbe Interact.*, **10** 160-70.

Guo, H.S., López-Moya, J.J. and García, J.A. (1999) Mitotic stability of infection-induced resistance to plum pox potyvirus associated with transgene silencing and DNA methylation. *Mol. Plant-Microbe Interact.*, **12** 103-11.

Hackland, A.F., Rybicki, E.P. and Thomson, J.A. (1994) Coat-protein-mediated resistance in transgenic plants. *Arch. Virol.*, **139** 1-22.

Hamilton, R.I. (1980) Defenses triggered by previous invaders: viruses, in *Plant Disease: An Advanced Treatise* (eds. J.G. Horsfall and E.B. Cowling), Academic Press, New York, Vol. 5, pp. 279-303.

Hammond, J. (1993) Transcapsidation in transgenic plants expressing bean yellow mosaic virus or chimeric coat proteins is not correlated with resistance to various potyviruses. *Phytopathology*, **83** 1349.

Hammond, J. and Kamo, K.K. (1995) Resistance to bean yellow mosaic virus (BYMV) and other potyviruses in transgenic plants expressing BYMV antisense RNA, coat protein or chimeric coat proteins, in *Biotechnology and Plant Protection, Viral Pathogenesis and Disease Resistance* (eds. D.D. Bills and S.-D. Kung), World Scientific, Singapore, pp. 369-389.

Hammond-Kosack, K.E., Tang, S., Harrison, K. and Jones, J.D.G. (1998) The tomato *Cf-9* disease resistance gene functions in tobacco and potato to confer responsiveness to the fungal avirulence gene product, Avr9. *Plant Cell*, **10** 1251-66.

Hancock, J.F., Grumet, R. and Hokanson, S.C. (1996) The opportunity for escape of engineered genes from transgenic crops. *HortScience*, **31** 1080-85.

Harrison, B.D., Mayo, M.A. and Baulcombe, D.C. (1987) Virus resistance in transgenic plants that express cucumber mosaic virus satellite RNA. *Nature*, **328** 799-802.

Hassairi, A., Masmoudi, K., Albouy, J., Robaglia, C., Jullien, M. and Ellouz, R. (1998) Transformation of two potato cultivars 'Spunta' and 'Claustar' (*Solanum tuberosum*) with lettuce mosaic virus coat protein gene and heterologous immunity to potato virus Y. *Plant Sci.*, **136** 31-42.

Hayakawa, T., Zhu, Y.F., Itoh, K., Kimura, Y., Izawa, T., Shimamoto, K. and Toriyama, S. (1992) Genetically engineered rice resistant to rice stripe virus, an insect transmitted virus. *Proc. Natl. Acad. Sci. USA*, **89** 9865-69.

Hefferon, K.L., Khalilian, H. and AbouHaidar, M.G. (1997) Expression of the PVY^{o} coat protein (CP) under the control of the PVX CP gene leader sequence: protection under greenhouse and field conditions against PVY^{o} and PVY^{N} infection in three potato cultivars. *Theoret. Appl. Genet.*, **94** 287-92.

Hellmann, G.M., Shaw, J.G. and Rhoads, R.E. (1988) *In vitro* analysis of tobacco vein mottling virus NIa cistron: evidence for a virus-encoded protease. *Virology*, **163** 554-62.

Hemenway, C., Fang, R.-X., Kaniewski, W.K., Chua, N.-H. and Tumer, N.E. (1988) Analysis of the mechanism of protection in transgenic plants expressing the potato virus X coat protein or its antisense RNA. *EMBO J.*, **7** 1273-80.

Hiatt, A. and Mostov, K. (1993) Assembly of multimeric proteins in plant cells: characteristics and uses of plant-derived antibodies, in *Transgenic Plants, Fundamentals and Applications* (ed. A. Hiatt), Marcel Dekker, New York, pp. 221-237.

Hiatt, A., Cafferkey, R. and Bowdish, K. (1989) Production of antibodies in transgenic plants. *Nature*, **342** 76-78.

Holt, C.A. and Beachy, R.N. (1991) *In vivo* complementation of infectious transcripts from mutant tobacco mosaic virus cDNAs in transgenic plants. *Virology*, **181** 109-17.

Hong, Y. and Stanley, J. (1996) Virus resistance in *Nicotiana benthamiana* conferred by African cassava mosaic virus replication-associated protein (AC1) transgene. *Mol. Plant-Microbe Interact.*, **9** 219-25.

Hong, Y., Saunders, K., Hartley, M.R. and Stanley, J. (1996) Resistance to geminivirus infection by virus-induced expression of dianthin in transgenic plants. *Virology*, **220** 119-27.

Huisman, M.J., Broxterman, H.J.G., Schellekens, H. and Van Vloten-Doting, L. (1985) Human interferon does not protect cowpea plant cell protoplasts against infection with alfalfa mosaic virus. *Virology*, **143** 622-25.

Huntley, C.C. and Hall, T.C. (1996) Interference with brome mosaic virus replication in transgenic rice. *Mol. Plant-Microbe Interact.*, **9** 164-70.

Ingelbrecht, I., Van Houdt, H., Van Montagu, M. and Depicker, A. (1994) Post-transcriptional silencing of reporter transgenes in tobacco correlates with DNA methylation. *Proc. Natl. Acad. Sci. USA*, **91** 10502-506.

Jones, A.L., Johansen, I.E., Bean, S.J., Bach, I. and Maule, A.J. (1998a) Specificity of resistance to pea seed-borne mosaic potyvirus in transgenic peas expressing the viral replicase (NIb) gene. *J. Gen. Virol.*, **79** 3129-37.

Jones, A.L., Thomas, C.L. and Maule, A.J. (1998b) De novo methylation and co-suppression induced by a cytoplasmically replicating plant RNA virus. *EMBO J.*, **17** 6385-93.

Kaido, M., Mori, M., Mise, K., Okuno, T. and Furusawa, I. (1995) Inhibition of brome mosaic virus (BMV) amplification in protoplasts from transgenic tobacco plants expressing replicable BMV RNAs. *J. Gen. Virol.*, **76** 2827-33.

Kallerhoff, J., Perez, P., Bouzoubaa, S., Ben Tahar, S. and Peret, J. (1990) Beet necrotic yellow vein virus coat-protein-mediated protection in sugarbeet (*Beta vulgaris* L.) protoplasts. *Plant Cell Rep.*, **9** 224-28.

Kamoun, S., Honée, G., Weide, R., Laugé, R., Kooman-Gersmann, M., de Groot, K., Govers, F. and de Wit, J.G.M. (1999) The fungal gene, *Avr9*, and the oomycete gene, *inf1*, confer avirulence to potato virus X on tobacco. *Mol. Plant-Microbe Interact.*, **12** 459-62.

Kaplan, I.B., Taliansky, M.E., Malyshenko, S.I., Ogarkov, V.I. and Atabekov, J.G. (1988) Effect of human interferon on reproduction of plant and myxoviruses. *Achiv fuer Phytopathologie und Pflanzenschutz*, **24** 3-8.

Kasschau, K.D. and Carrington, J.C. (1998) A counterdefensive strategy of plant viruses: suppression of post-transcriptional gene silencing. *Cell*, **95** 461-70.

Kawchuk, L.M., Martin, R.R. and McPherson, J. (1990) Resistance in transgenic potato expressing the potato leafroll virus coat protein gene. *Mol. Plant-Microbe Interact.*, **3** 301-307.

Kawchuk, L.M., Martin, R.R. and McPherson, J. (1991) Sense and antisense RNA-mediated resistance to potato leafroll virus in Russet Burbank potato plants. *Mol. Plant-Microbe Interact.*, **4** 247-53.

Köhm, B.A., Goulden, M.G., Gilbert, J.E., Kavanagh, T.A. and Baulcombe, D.C. (1993) A potato virus X resistance gene mediates an induced non-specific resistance in protoplasts. *Plant Cell*, **5** 913-20.

Kollàr, À., Dalmay, T. and Burgyàn, J. (1993) Defective interfering RNA-mediated resistance against cymbidium ringspot tombusvirus in transgenic plants. *Virology*, **193** 313-18.

Kunik, T., Salomon, R., Zamir, D., Navor, N., Zeidan, M., Michelson, I., Gafni, Y. and Czosnek, H. (1994) Transgenic plants expressing the tomato yellow leaf curl virus capsid protein are resistant to the virus. *Bio/Technol.*, **12** 500-504.

Lam, Y.-H., Wong, Y.-S., Wang, B., Wong, R.N.-S., Yeung, H.-W. and Shaw, P.-C. (1996) Use of trichosanthin to reduce infection by turnip mosaic virus. *Plant Sci.*, **114** 111-17.

Lapidot, M., Gafny, R., Ding, B., Wolf, S., Lucas, W.J. and Beachy, R.N. (1993) A dysfunctional movement protein of tobacco mosaic virus that partially modifies the plasmodesmata and limits virus spread in transgenic plants. *Plant J.*, **4** 959-70.

Lawson, C., Kaniewski, W.K., Haley, L., Rozman, R., Newell, C., Sanders, P. and Tumer, P.E. (1990) Engineering resistance to mixed virus infection in a commercial potato cultivar: resistance to potato virus X and potato virus Y in transgenic Russet Burbank. *Bio/Technol.*, **8** 127-34.

Leclerc, D. and AbouHaidar, M.G. (1995) Transgenic tobacco plants expressing a truncated form of the PAMV capsid protein (CP) gene show CP-mediated resistance to potato aucuba mosaic virus. *Mol. Plant-Microbe Interact.*, **8** 58-65.

Lecoq, H., Ravelonandro, M., Wipf-Scheibel, C., Monsion, M., Raccah, B. and Dunez, J. (1993) Aphid transmission of a non-aphid-transmissible strain of zucchini yellow mosaic potyvirus from transgenic plants expressing the capsid protein of plum pox potyvirus. *Mol. Plant-Microbe Interact.*, **6** 403-406.

Lehmann, P., Walsh, J., Jenner, C., Kozubek, E. and Greenland, A. (1996) Genetically engineered protection against turnip mosaic virus infection in transgenic oilseed rape (*Brassica napus* var *oleifera*). *J. Appl. Genet.*, **37A** 118-21.

Lindbo, J.A. and Dougherty, W.G. (1992a) Untranslatable transcripts of tobacco etch virus coat protein gene sequence can interfere with tobacco etch virus replication in transgenic plants and protoplasts. *Virology*, **189** 725-33.

Lindbo, J.A. and Dougherty, W.G. (1992b) Pathogen-derived resistance to a potyvirus: immune and resistant phenotypes in transgenic tobacco expressing altered forms of a potyvirus coat protein nucleotide sequence. *Mol. Plant-Microbe Interact.*, **5** 144-53.

Lindbo, J.A., Silva-Rosales, L., Proebsting, W.M. and Dougherty, W.G. (1993a) Induction of a highly specific antiviral state in transgenic plants: implications for regulation of gene expression and virus resistance. *Plant Cell*, **5** 1749-59.

Lindbo, J.A., Silva-Rosales, L. and Dougherty, W.G. (1993b) Pathogen-derived resistance to potyviruses: Working, but why? *Semin. Virol.*, **4** 369-79.

Lius, S., Manshardt, R.M., Fitch, M.M.M., Slightom, J.L., Sanford, J.C. and Gonsalves, D. (1997) Pathogen-derived resistance provides papaya with effective protection against papaya ringspot virus. *Mol. Breed.*, **3** 161-68.

Lodge, J.K., Kaniewski, W.K. and Tumer, N.E. (1993) Broad-spectrum virus resistance in transgenic plants expressing pokeweed antiviral protein. *Proc. Natl. Acad. Sci. USA*, **90** 7089-93.

Loesch-Fries, L.S., Halk, E.L., Nelson, S.E. and Krahn, K.J. (1985) Human leukocyte interferon does not inhibit alfalfa mosaic virus in protoplasts or tobacco tissue. *Virology*, **143** 626-29.

Loesch-Fries, L.S., Merlo, D., Zinnen, T., Burhop, L., Hill, K., Krahn, K., Jarvis, N., Nelson, S. and Halk, E. (1987) Expression of alfalfa mosaic virus RNA 4 in transgenic plants confers virus resistance. *EMBO J.*, **6** 1845-51.

Lomonossoff, G.P. (1995) Pathogen-derived resistance to plant viruses. *Annu. Rev. Phytopathol.*, **33** 323-43.

Lomonossoff, G.P. and Davies, J.W. (1992) Replicase-mediated resistance to plant viruses. *Sci. Prog.*, **76** 537-51.

Longstaff, M., Brigneti, G., Boccard, F., Champan, S. and Baulcombe, D. (1993) Extreme resistance to potato virus X infection in plants expressing a modified component of the putative viral replicase. *EMBO J.*, **12** 379-86.

MacFarlane, S.A. and Davies, J.W. (1992) Plants transformed with a region of the 201-kDa replicase gene from pea early browning virus RNA1 are resistant to virus infection. *Proc. Natl. Acad. Sci. USA*, **89** 5829-33.

MacKenzie, D.J. and Ellis, P.J. (1992) Resistance to tomato spotted wilt virus infection in transgenic tobacco expressing the viral nucleocapsid gene. *Mol. Plant-Microbe Interact.*, **5** 34-40.

MacKenzie, D.J., Tremaine, J.H. and McPherson, J. (1991) Genetically engineered resistance to potato virus S in potato cultivar Russet Burbank. *Mol. Plant-Microbe Interact.*, **4** 95-102.

Maiti, I.B., Murphy, J.F., Shaw, J.G. and Hunt, A.G. (1993) Plants that express a potyvirus proteinase gene are resistant to virus infection. *Proc. Natl. Acad. Sci. USA*, **90** 6110-14.

Malcuit, I., Marano, M., Kavanagh, T.A. and Baulcombe, D.C. (1998) Molecular characterisation of the PVX elicitor of *Nb*-mediated hypersensitive resistance in *Solanum tuberosum*. Abstract 1.1.28. *Proceedings of 7th International Congress of Plant Pathology*, Edinburgh.

Malpica, C.A., Cervera, M.T., Simoens, C. and Van Montagu, M. (1998) Engineering resistance against viral disease in plants, in *Plant-Microbe Interactions* (eds. B.B. Biswas and H.K. Das), Subcellular Biochemistry, Vol. 29, Plenum Press, New York, pp. 287-320.

Malyshenko, S.I., Kondakova, O.A., Nazarova, J.V., Kaplan, I.B., Taliansky, M.E. and Atabekov, J.G. (1993) Reduction of tobacco mosaic virus accumulation in transgenic plants producing non-functional viral transport proteins. *J. Gen. Virol.*, **74** 1149-56.

Marano, M.R. and Baulcombe, D. (1998) Pathogen-derived resistance targeted against the negative strand RNA of tobacco mosaic virus: RNA strand-specific silencing? *Plant J.*, **13** 537-46.

Meshi, T., Motoyoshi, F., Adachi, A., Watanabe, Y., Takamatsu, N. and Okada, Y. (1988) Two concomitant base substitutions in the putative replicase genes of tobacco mosaic virus confer the ability to overcome the effects of a tomato resistance gene, *Tm-1*. *EMBO J.*, **7** 1575-81.

Meshi, T., Motoyoshi, F., Maeda, T., Yoshiwoka, S., Watanabe, H. and Okada, Y. (1989) Mutations in the tobacco mosaic virus 30-kDa protein gene overcome *Tm-2* resistance in tomato. *Plant Cell*, **1** 515-22.

Michelmore, R. (1995) Molecular approaches to manipulation of disease resistance genes. *Annu. Rev. Phytopathol.*, **15** 393-427.

Moreno, M., Bernal, J.J., Jiménez, I. and Rodriguez-Cerezo, E. (1998) Resistance in plants transformed with the P1 or P3 gene of tobacco vein mottling potyvirus. *J. Gen. Virol.*, **79** 2819-27.

Mori, M., Mise, K., Okuno, T. and Furusawa, I. (1992) Expression of brome mosaic virus-encoded replicase genes in transgenic tobacco plants. *J. Gen. Virol.*, **73** 169-72.

Mueller, E., Gilbert, J.E., Davenport, G., Brigneti, G. and Baulcombe, D.C. (1995) Homology-dependent resistance: transgenic virus resistance in plants related to homology-dependent gene silencing. *Plant J.*, **7** 1001-13.

Murray, L.E., Elliot, L.G., Capitant, S.A., West, J.A., Hanson, K.K., Scarafia, L., Johnston, S., De Luca-Flaherty, C., Nichols, S., Cunanan, D., Dietrich, P.S., Mettler, S.D., Warnick, D.A., Rhodes, C., Sinibaldi, R.M. and Brunke, K.J. (1993) Transgenic corn plants expressing MDMV strain B coat protein are resistant to mixed infections of maize dwarf mosaic virus and maize chlorotic mottle virus. *Bio/Technol.*, **11** 1559-64.

Nakajima, M., Hayakawa, T., Nakamura, I. and Suzuki, M. (1993) Protection against cucumber mosaic virus (CMV) strains O and Y and chrysanthemum mild mottle virus in transgenic tobacco plants expressing CMV-O coat protein. *J. Gen. Virol.*, **74** 319-22.

Nakamura, S., Yoshikawa, M., Taira, H. and Ehara, Y. (1994) Plants transformed with mammalian 2′–5′ oligoadenylate synthase gene show resistance to virus infections. *Ann. Phytopathol. Soc. Japan*, **60** 691-93.

Namba, S., Ling, K., Gonsalves, C., Slightom, J.L. and Gonsalves, D. (1992) Protection of transgenic plants expressing the coat protein gene of watermelon mosaic virus II or zucchini yellow mosaic virus against six potyviruses. *Phytopathology*, **82** 940-46.

Nejidat, A. and Beachy, R.N. (1989) Decreased levels of TMV coat protein in transgenic tobacco plants at elevated temperatures reduced resistance to TMV infection. *Virology*, **173** 531-38.

Nejidat, A. and Beachy, R.N. (1990) Transgenic tobacco plants expressing a coat-protein gene of tobacco mosaic virus are resistant to some other tobamoviruses. *Mol. Plant-Microbe Interact.*, **3** 247-51.

Nelson, R.S., Powell Abel, P. and Beachy, R.N. (1987) Lesions and virus accumulation in inoculated transgenic tobacco plants expressing the coat protein gene of tobacco mosaic virus. *Virology*, **158** 126-32.

Okuno, T., Nakayama, M. and Furusawa, I. (1993) Cucumber mosaic virus coat protein-mediated resistance. *Semin. Virol.*, **4** 357-61.

Orchansky, P., Rubinstein, M. and Sela, I. (1982) Human interferons protect plants from virus infection. *Proc. Natl. Acad. Sci. USA*, **79** 2278-80.

Osbourn, J.K., Watts, J.W., Beachy, R.N. and Wilson, T.M.A. (1989) Evidence that nucleocapsid disassembly and a later step in virus replication are inhibited in transgenic tobacco protoplasts expressing TMV coat protein. *Virology*, **172** 370-73.

Osbourn, J.K., Sarkar, S. and Wilson, T.M.A. (1990) Complementation of coat protein-defective TMV mutants in transgenic tobacco plants expressing TMV coat protein. *Virology*, **179** 921-25.

Padgett, H.S. and Beachy, R.N. (1993) Analysis of a tobacco mosaic virus strain capable of overcoming *N* gene-mediated resistance. *Plant Cell*, **5** 577-86.

Pang, S.-Z., Bock, J.H., Gonsalves, C., Slightom, J.L. and Gonsalves, D. (1994) Resistance of transgenic *Nicotiana benthamiana* plants to tomato spotted wilt virus and impatiens necrotic spot virus tospoviruses: evidence of involvement of the N protein and *N* gene RNA in resistance. *Phytopathology*, **84** 243-49.

Passelègue, E. and Kerlan, C. (1996) Transformation of cauliflower (*Brassica oleracea* var. *botrytis*) by transfer of cauliflower mosaic virus genes through combined co-cultivation with virulent and avirulent strains of *Agrobacterium*. *Plant Sci.*, **113** 79-89.

Pehu, T.M., Makivalkonen, T.K., Valkonen, J.P.T., Koivu, K.T., Lehto, K.M. and Pehu, E.P. (1995) Potato plants transformed with a potato virus Y P1 gene sequence are resistant to PVY. *Am. Potato J.*, **72** 523-32.

Pierpoint, W.S., Hughes, K.J.D. and Shewry, P.R. (1995) *Targets for Introduction of Pest and Disease Resistance into Crops by Genetic Engineering.* IACR-Long Ashton Research Station, UK.

Plafker, T. (1994) First biotech safety rules don't deter Chinese efforts. *Science*, **266** 966-67.

Ploeg, A.T., Mathis, A., Bol, J.F., Brown, D.J.F. and Robinson, D.J. (1993) Susceptibility of transgenic tobacco plants expressing tobacco rattle virus coat protein to nematode-transmitted and mechanically inoculated tobacco rattle virus. *J. Gen. Virol.*, **74** 2709-15.

Plückthun, A. (1991) Antibody engineering. *Curr. Opin. Biotechnol.*, **2** 238-46.

Ponz, F. and Bruening, G. (1986) Mechanisms of resistance to plant viruses. *Ann. Rev. Phytopathol.*, **24** 355-81.

Powell Abel, P., Nelson, R.S., De, B., Hoffmann, N., Rogers, S.G., Fraley, R.T. and Beachy, R.N. (1986) Delay of disease development in transgenic plants that express the tobacco mosaic virus coat protein gene. *Science*, **232** 738-43.

Powell, P.A., Stark, D.M., Sanders, P.R. and Beachy, R.N. (1989) Protection against tobacco mosaic virus in transgenic plants that express tobacco mosaic virus antisense RNA. *Proc. Natl. Acad. Sci. USA*, **86** 6949-52.

Powell, P.A., Sanders, P.R., Tumer, N., Fraley, R.T. and Beachy, R.N. (1990) Protection against tobacco mosaic virus infection in transgenic plants requires accumulation of coat protein rather than coat protein RNA sequences. *Virology*, **175** 124-30.

Querci, M., Baulcombe, D.C., Goldbach, R.W. and Salazar, L.F. (1995) Analysis of the resistance-breaking determinants of potato virus X (PVX) strain HB on different potato genotypes expressing extreme resistance to PVX. *Phytopathology*, **85** 1003-10.

Raffo, A.J. and Dawson, W.O. (1991) Construction of tobacco mosaic virus subgenomic replicons that are replicated and spread systemically in tobacco plants. *Virology*, **184** 277-89.

Ratcliff, F., Harrison, B.D. and Baulcombe, D.C. (1997) A similarity between viral defense and gene silencing in plants. *Science*, **276** 1558-60.

Ravelonandro, M., Monsion, M., Delbos, R. and Dunez, J. (1993) Variable resistance to plum pox virus and potato virus Y infection in transgenic *Nicotiana* plants expressing plum pox virus coat protein. *Plant Sci.*, **91** 157-69.

Register, J.C. and Beachy, R.N. (1988) Resistance to TMV in transgenic plants results from interference with an early event in infection. *Virology*, **166** 524-32.

Regner, F., da Câmara Machado, A., Laimer da Câmara Machado, M., Steinkellner, H., Mattanovich, D., Hanzer, V., Weiss, H. and Katinger, H. (1992) Coat-protein-mediated resistance to plum pox virus in *Nicotiana clevlandii* and *N. benthamiana. Plant Cell Rep.*, **11** 30-33.

Reimann-Philipp, U. and Beachy, R.N. (1993) Plant resistance to virus diseases through genetic engineering: Can a similar approach control plant-parasitic nematodes? *J. Nematol.*, **25** 541-47.

Robinson, D.J. (1996) Environmental risk assessment of releases of transgenic plants containing virus-derived inserts. *Transgen. Res.*, **5** 359-62.

Roossinck, M.J. (1997) Mechanisms of plant virus evolution. *Annu. Rev. Phytopathol.*, **35** 191-209.

Rubino, L. and Russo, M. (1995) Characterization of resistance to cymbidium ringspot virus in transgenic plants expressing a full-length viral replicase gene. *Virology*, **212** 240-43.

Rubino, L., Caprioti, G., Lupo, R. and Russo, M. (1993) Resistance to cymbidium ringspot tombusvirus infection in transgenic *Nicotiana benthamiana* plants expressing the virus coat protein gene. *Plant Mol. Biol.*, **21** 665-72.

Sanders, P.R., Sammons, B., Kaniewski, W., Haley, L., Layton, J., La Vallee, B.J., Delannay, X. and Tumer, N.E. (1992) Field resistance of transgenic tomatoes expressing the tobacco mosaic virus or tomato mosaic virus coat protein genes. *Phytopathology*, **82** 683-90.

Sanford, J.C. and Johnston, S.A. (1985) The concept of parasite-derived resistance-deriving resistance genes from the parasite's own genome. *J. Theor. Biol.*, **113** 395-405.

Sano, T., Nagayama, A., Ogawa, T., Ishid, I. and Okada, Y. (1997) Transgenic potato expressing a double-stranded RNA-specific ribonuclease is resistant to potato spindle tuber viroid. *Nat. Biotechnol.*, **15** 1290-94.

Santa Cruz, S. and Baulcombe, D.C. (1993) Molecular analysis of potato virus X isolates in relation to the potato hypersensitivity gene, *Nx. Mol. Plant-Microbe Interact.*, **6** 707-14.

Schiebel, W., Haas, B., Marinkovic, S., Klanner, A. and Sanger, H.L. (1993) RNA-directed RNA polymerase from tomato leaves. I. Purification and physical properties. *J. Biol. Chem.*, **268** 11851-57.

Scorza, R., Levy, L., Damsteegt, V., Yepes, L.M., Cordts, J., Hadidi, A., Slightom, J. and Gonsalves, D. (1995) Transformation of plum with the papaya ringspot virus coat protein gene and reaction of transgenic plants to plum pox virus. *J. Am. Soc. Hort. Sci.*, **120** 943-52.

Sijen, T., Wellink, J., Hendriks, J., Verver, J. and Van Kammen, A. (1995) Replication of cowpea mosaic virus RNA1 or RNA2 is specifically blocked in transgenic *Nicotiana benthamiana* plants expressing the full-length replicase or movement protein genes. *Mol. Plant-Microbe Interact.*, **8** 340-47.

Sleat, D.E. and Palukaitis, P. (1992) A single nucleotide change with a plant virus satellite RNA alters the host specificity of disease induction. *Plant J.*, **2** 43-49.

Sonoda, S., Mori, M. and Nishiguchi, M. (1999) Homology-dependent virus resistance in transgenic plants with the coat protein gene of sweet potato feathery mottle potyvirus: target specificity and transgene methylation. *Virology*, **89** 385-91.

Stanley, J., Frischmuth, T. and Ellwood, S. (1990) Defective viral DNA ameliorates symptoms of geminivirus infection in transgenic plants. *Proc. Natl. Acad. Sci. USA*, **87** 6291-95.

Stark, D.M. and Beachy, R.N. (1989) Protection against potyvirus infection in transgenic plants: evidence for broad spectrum resistance. *Bio/Technol.*, **7** 1257-62.

Suzuki, M., Masuta, C., Takanami, Y. and Kuwata, S. (1996) Resistance against cucumber mosaic virus in plants expressing the viral replicon. *FEBS Lett.*, **379** 26-30.

Swaney, S., Powers, H., Goodwin, J., Rosales, L.S. and Dougherty, W.G. (1995) RNA-mediated resistance with non-structural genes from tobacco etch virus genome. *Mol. Plant-Microbe Interact.*, **8** 1004-11.

Syller, J., Marczewski, W. and Pawlowicz, J. (1997) Transmission by aphids of potato spindle tuber viroid encapsidated by potato leafroll luteovirus particles. *Eur. J. Plant Pathol.*, **103** 285-89.

Taliansky, M.E., Ryabov, E.V. and Robinson, D.J. (1998) Two distinct mechanisms of transgenic resistance mediated by groundnut rosette virus satellite RNA sequences. *Mol. Plant-Microbe Interact.*, **11** 367-74.

Tavladoraki, P., Benvenuto, E., Trinca, S., De Martinis, D., Cattaneo, A. and Galeffi, P. (1993) Transgenic plants expressing a functional single-chain Fv antibody are specifically protected from virus attack. *Nature*, **366** 469-72.

Tenllado, F., García-Luque, I., Serra, M.T. and Díaz-Ruíz, J.R. (1995) *Nicotiana benthamiana* plants transformed with the 54-kDa region of the pepper mild mottle tobamovirus replicase gene exhibit two types of resistance responses against viral infection. *Virology*, **211** 170-83.

Tricoli, D.M., Carney, K.J., Russell, P.F., McMaster, J.R., Groff, D.W., Hadden, K.C., Himmel, P.T., Hubbard, J.P., Boeshore, M.L. and Quemada, H.D. (1995) Field evaluation of transgenic squash containing single or multiple virus coat protein gene constructs for resistance to cucumber mosaic virus, watermelon mosaic virus 2, and zucchini yellow mosaic virus. *Bio/Technol.*, **13** 1458-65.

Truve, E., Aaspollu, A., Honkanen, J., Puska, R., Mehto, M., Hassi, A., Teerie, T.H., Kelve, M., Seppänen, P. and Saarma, M. (1993) Transgenic potato plants expressing mammalian 2′–5′ oligoadenylate synthetase are protected from potato virus X infection under field conditions. *Bio/Technol.*, **11** 1048-52.

Van den Boogaart, T., Lomonossoff, G.P. and Davies, J.W. (1998) Can we explain RNA-mediated virus resistance by homology-dependent gene silencing? *Mol. Plant-Microbe Interact.*, **7** 717-23.

Van den Heuvel, J.F.J.M., Van der Vlugt, R.A.A., Verbeek, M., de Haan, P.T. and Huttinga, H. (1994) Characteristics of a resistance-breaking isolate of potato virus Y causing potato tuber necrotic ringspot disease. *Eur. J. Plant Pathol.*, **100** 347-56.

Van der Vlugt, R.A.A., Ruiter, R.K. and Goldbach, R.W. (1992) Evidence for sense RNA-mediated protection to PVY^N in tobacco plants transformed with the viral coat protein cistron. *Plant Mol. Biol.*, **20** 631-39.

Van Dun, C.M.P., Bol, J.F. and Van Vloten-Doting, L. (1987) Expression of alfalfa mosaic virus and tobacco rattle virus coat protein genes in transgenic tobacco plants. *Virology*, **159** 299-305.

Van Dun, C.M.P., Overduin, B., Van Vloten-Doting, L. and Bol, J.F. (1988a) Transgenic tobacco expressing tobacco streak virus or mutated alfalfa mosaic virus coat protein gene does not cross-protect against alfalfa mosaic virus infection. *Virology*, **164** 383-89.

Van Dun, C.M.P., Van Vloten-Doting, L. and Bol, J.F. (1988b) Expression of alfalfa mosaic virus cDNA 1 and 2 in transgenic tobacco plants. *Virology*, **163** 572-78.

Vardi, E., Sela, I., Edelbaum, O., Livneh, O., Kuznetsova, L. and Stram, Y. (1993) Plants transformed with a cistron of a potato virus Y protease (NIa) are resistant to virus infection. *Proc. Natl. Acad. Sci. USA*, **90** 7513-17.

Voss, A., Niersbach, M., Hain, R., Hirsch, J.J., Liao, Y.C., Kreuzaler, F. and Fischer, R. (1995) Reduced virus infectivity in *N. tabacum* secreting a TMV-specific full-size antibody. *Mol. Breed.*, **1** 39-50.

Watanabe, Y., Ogawa, T., Takahashi, H., Ishida, I., Takeuchi, Y., Yamamoto, M. and Okada, Y. (1995) Resistance against multiple plant viruses in plants mediated by double-stranded-RNA specific ribonuclease. *FEBS Lett.*, **372** 165-68.

Weber, H. and Pfitzner, A.J.P. (1998) *Tm-2*^{2} resistance in tomato requires recognition of the carboxy terminus of the movement protein of tomato mosaic virus. *Mol. Plant-Microbe Interact.*, **11** 498-503.

Wen, F. and Lister, R.M. (1991) Heterologous encapsidation in mixed infections among four isolates of barley yellow dwarf virus. *J. Gen. Virol.*, **72** 2217-24.

Whitham, S., Dinesh-Kumar, S.P., Choi, D., Hehl, R., Corr, C. and Baker, B. (1994) The product of tobacco mosaic virus resistance gene, *N*: similarity to Toll and the interleukin-1 receptor. *Cell*, **78** 1101-15.

Wilson, T.M.A. (1993) Strategies to protect crop plants against viruses: pathogen-derived resistance blossoms. *Proc. Natl. Acad. Sci. USA*, **90** 3134-41.

Wintermantel, W.M. and Schoelz, J.E. (1996) Isolation of recombinant viruses between cauliflower mosaic virus and a viral gene in transgenic plants under conditions of moderate selection pressure. *Virology*, **223** 156-64.

Wittner, A., Palkovics, L. and Balazs, E. (1998) *Nicotiana benthamiana* plants transformed with the plum pox virus helicase gene are resistant to virus infection. *Virus Res.*, **53** 97-103.

Wolf, S., Deom, C.M., Beachy, R.N. and Lucas, W.J. (1991) Movement protein of tobacco mosaic virus modifies plasmodesmatal size exclusion limit. *Science*, **246** 377-79.

Wolf, S. and Lucas, W.J. (1994) Virus movement proteins and other molecular probes of plasmodesmal function. *Plant Cell Environ.*, **17** 573-85.

Wolfe, M.S. (1993) Can the strategic use of disease-resistant hosts protect their inherent durability? In *Durability of Disease Resistance* (eds. T. Jacobs and J.E. Parlevliet), Kluwer, Dordrecht, pp. 83-96.

Yamaya, J., Yoshioka, M., Meshi, T., Okado, Y. and Ohno, T. (1988a) Expression of tobacco mosaic virus RNA in transgenic plants. *Mol. Gen. Genet.*, **211** 520-25.

Yamaya, J., Yoshioka, M., Meshi, T., Okada, Y. and Ohno, T. (1988b) Cross-protection in transgenic tobacco plants expressing a mild strain of tobacco mosaic virus. *Mol. Gen. Genet.*, **215** 173-75.

Yepes, L.M., Fuchs, M., Slightom, J.L. and Gonsalves, D. (1996) Sense and antisense coat protein gene constructs confer high levels of resistance to tomato ringspot nepovirus in transgenic *Nicotiana* species. *Phytopathology*, **86** 417-24.

Young, N.D. and Tanksley, S.D. (1989) RFLP analysis of the size of thea chromosomal segments retained around the *Tm-2* locus of tomato during backcross breeding. *Theor. Appl. Genet.*, **77** 95-101.

Zaccomer, B., Cellier, F., Boyer, J.-C., Haenni, A.-L. and Tepfer, M. (1993) Transgenic plants that express genes including the 3′ untranslated region of turnip yellow mosaic virus (TYMV) genome are partially protected against TYMV infection. *Gene*, **136** 87-94.

Zaitlin, M., Anderson, J.M., Perry, K.L., Zhang, L. and Palukaitis, P. (1994) Specificity of replicase-mediated resistance to cucumber mosaic virus. *Virology*, **201** 200-205.

10 Emerging technologies and their application in the study of host-pathogen interactions

Robert A. Dietrich

10.1 Introduction

Advances in the techniques of molecular biology have made significant contributions to our understanding of host pathogen interactions in the past decade. We now know the structure of many resistance (*R*) genes and the proteins they encode. In some cases, the structure of the corresponding avirulence gene product in the pathogen has been determined as well. We also know the identity of some of the downstream genes that are induced as a result of pathogen infection, such as the pathogenesis-related (PR) genes. In some cases, particularly in *Arabidopsis*, some of the genes which encode proteins that function at intermediate steps in the signal transduction pathway for response to pathogens have been identified and cloned. Some examples are: NDR1 (necessary for defence response) (Century *et al.*, 1997); EDS1 (enhanced disease symptoms) (Falk *et al.*, 1999); and NPR1/NIM1 (non-inducible PR1/non-inucible immunity) (Cao *et al.*, 1997; Ryals *et al.*, 1997). These genes are required for a functional resistance response to some, though not all, pathogens. While we now know of some of the genes that participate in defence response pathways, how they all interact to mount an effective resistance response is still very much a mystery. This chapter discusses some new techniques and tools that may be helpful in bridging the gap between *R* gene structure and functional resistance and in studying other plant defence mechanisms as well. This includes new methods that can be used to enhance traditional genetics-based approaches; techniques that will enhance our ability to identify and clone genes acting at all stages in a signal transduction pathway, from the initial regulatory steps through intermediate signalling steps to the final response genes at the end of the pathway. Strategies that are being used to facilitate genetic approaches are discussed in the first section.

The second section covers some of the new genomics approaches. Genomics-based approaches take a more global perspective in examining specific responses. For example, instead of focusing on the expression of one or a few genes, gene expression profiling analyses the expression of large numbers of genes at the same time, with the ultimate goal of assaying changes in expression of all the genes in an organism. This will not only provide information on every individual gene but also on how different pathways may interact and impact each other. This information will be essential in attempts to modify pathways to optimize desirable traits in crops, including disease resistance.

10.2 Genetic approaches

With the advent of large scale sequencing projects, vast amounts of sequence information are becoming available, both in the form of genomic sequence and as partial complementary deoxyribonucleic acid (cDNA) sequences, known as expressed sequence tags (ESTs). The complete genomic sequence of *Arabidopsis* will soon be available, with sequences for other plant genomes soon to follow as past experience and improved technology lead to increased sequencing efficiency. Sequence information alone, however, is of little value. The challenge is now to link functions to specific sequences. Functional genomics refers to the identification of genes and determination of their functions. This will require genetic analysis, both forward genetics, where a function is known and the corresponding gene sequence must be determined, and reverse genetics, where a gene sequence is known, and the function must be determined. Many new techniques and tools are facilitating both forward and reverse genetic approaches to functional genomics.

10.2.1 Forward genetics

The classical way to identify a gene with a specific function is to use forward genetics. Screens are performed on a mutagenized population to identify mutants with a specific altered phenotype. By carefully choosing the mutant phenotype and designing elegant screens, mutations can be targeted to a specific pathway, or even to specific points within a pathway. Despite the amount of DNA sequence that is becoming available, forward genetics may still be the most efficient way to identify a gene with a specific function. The two most common methods that are currently used to clone genes using forward genetics are insertion mutagenesis and map-based cloning (also known as positional cloning).

10.2.1.1 Gene tagging

Insertion mutagenesis refers to the insertion of a known piece of DNA into a gene to modify expression of that gene. At the present time, insertions cannot be targeted to a specific gene, so entire populations with insertions at random loci are generated and screened for individuals with the desired phenotype. Insertions usually result in null alleles by disrupting the open reading frame (ORF) of the gene, though insertions into regulatory elements can alter expression levels. There are currently two types of insertion elements that are being used to generate insertion mutants in plants, transposable elements and random transferred-DNA (T-DNA) insertions. The main advantage of insertion mutagenesis is that the inserted DNA serves as a convenient tag to facilitate cloning of the affected gene. Thus, the much more laborious map-based cloning approach is avoided. Insertion mutants can be used both for forward genetics and also for reverse genetics, as described in the reverse genetics section.

Endogenous transposable elements have long been used to generate mutant populations in maize, using the various endogenous transposable elements, including Mu, Ac/Ds and En/Spm. The Ac/Ds and En/Spm maize elements have been transferred to other species, such as *Arabidopsis*, flax and tomato, where they also transpose, and have been used for insertion mutagenesis. For example, two resistance genes in flax, *L6* and *M*, which encode for resistance to two different strains of flax rust, were cloned using the Ac/Ds transposon system (Anderson *et al.*, 1991; Lawrence *et al.*, 1995).

Transposable elements have features that can be either an advantage or a disadvantage, or both, depending on their desired use. Some endogenous transposable elements, such as the maize mutator element (Mu), can accumulate to high copy numbers in the genome, with a high rate of transposition. This is helpful in that a large number of mutations is present per plant, reducing the size of the mutant population that must be screened to recover an insertion in a specific gene. This high copy number, however, makes it much more difficult to genetically link a specific element with the mutant phenotype. Polymerase chain reaction (PCR) based techniques have been developed to speed the process, including transposon display (Van der Broeck *et al.*, 1998) and amplification of insertion mutagenized sites (AIMS) (Frey *et al.*, 1998).

Mutations generated by transposable elements can be unstable, since the element can transpose out of the gene as well as into it. Stable mutants can be generated by using a two element system (Smith *et al.*, 1996). One element lacks the *cis* elements required for transposition and is therefore unable to transpose, but it expresses the tranposase protein required for transposition. The second element contains the required *cis* elements but does not express transposase. Only the second element is capable of transposition but does so only in the presence of the transposase expressing element. Mutants are generated by combining the two elements in a single plant by crossing the two lines to initiate transposition. When a plant with the desired phenotype is identified, the transposase expressing element is segregated out, stabilizing the mutation. The instability of transposon mutagenized alleles can also be an advantage, since excision of transposons is imprecise and an allelic series can be generated by independent excision events.

With some transposable elements, transposition occurs primarily into closely-linked loci. This is a disadvantage if genome-wide mutagenesis is desired, though selection strategies have been developed so that only transposition events into unlinked loci are recovered (Sundaresan *et al.*, 1995). Local transposition is an advantage if the goal is to obtain mutations in a specific locus and the map position of the locus is known. By starting with a transposable element closely linked to the desired target, the chance of transpositions into that locus is greatly enhanced.

A second type of insertion mutagenesis, the random insertion of T-DNA from *Agrobacterium* into the plant genome, is also being used extensively as a tool for mutagenesis (Koncz *et al.*, 1989; Feldmann, 1991; Mathur *et al.*, 1998). The current populations have an average of 2–2.5 T-DNA inserts per line, based on genomic Southern analysis. Genetically, these represent an average of 1.5 loci per line. Thus, some of the loci represent complex insertions of tandem T-DNAs, which can complicate attempts to clone the flanking genomic DNA. One caveat with the T-DNA mutagenized populations of *Arabidopsis* is that approximately 50% of the mutations are not due to T-DNA insertions but are apparently a consequence of the transformation process. Thus, before attempting to clone any flanking DNA, genetic analysis must be performed to determine whether the mutant phenotype co-segregates with any of the T-DNAs in the genome. At 1.5 mutant loci per plant, it takes a large number of plants to saturate the genome with mutations. Even in *Arabidopsis*, where transformation is relatively simple, this is not a trivial task. There are however, T-DNA mutagenized populations publicly available through the *Arabidopsis* stock centre (http://aims.cps.msu.edu/aims/).

There are several variations on the insertion mutagenesis theme (Maes *et al.*, 1999). These include elements designed to identify regulatory elements, both promoters and enhancers, as well as insertion elements to turn on expression of endogenous genes near the site of insertion. These are summarized in Figure 10.1.

Gene trap and enhancer trap insertion elements have been developed to identify insertions into or near functional *cis* regulatory elements in the genome. A reporter gene, such as the beta-glucuronidase (GUS) gene, either without a promoter (gene trap) or with a minimal promoter (enhancer trap), is placed adjacent to the right border in the T-DNA. Thus, the reporter gene will be expressed only if it inserts adjacent to a functional promoter or near a functional enhancer. The expression pattern of the reporter will reflect the expression pattern of the gene normally controlled by the promoter. Genes expressed under specific conditions, or in specific developmental stages or tissues, can be identified by screening the mutant population for individuals expressing the reporter gene in the desired stage or tissue. Since these are identifiable in the heterozygous state, insertions into essential genes can be recovered. Gene trap elements are sometimes made with a partial intron and a splice acceptor site at the 5′ end of the coding sequence of the reporter gene. Because splicing is required for its expression, the reporter gene is expressed only if it inserts into a functional gene. While this increases the likelihood that reporter gene expression results from insertion within a functional expressed gene, insertions into genes lacking introns will not be detected.

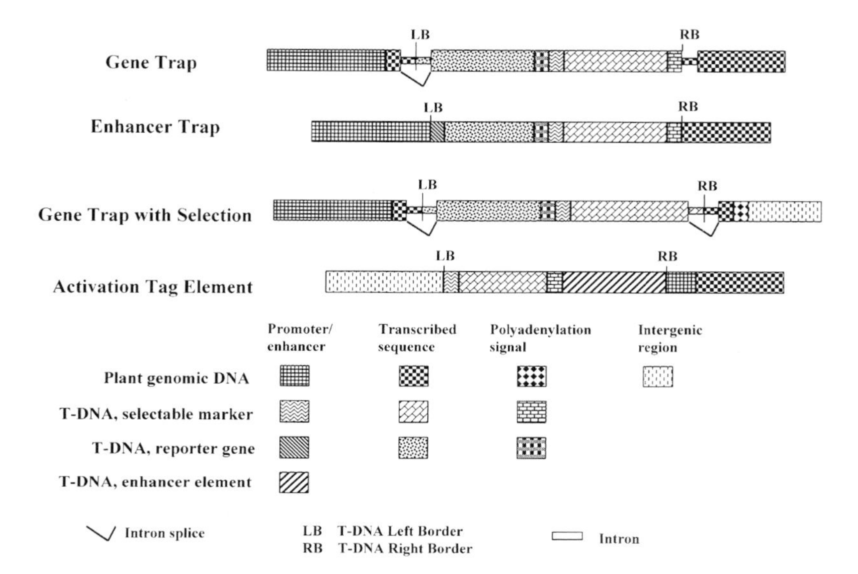

Figure 10.1 Specialized insertion elements. A number of specialized elements have been designed for different types of insertion mutagenesis to do more than simply insert into a gene and inactivate it. The gene trap element is used to identify promoters with specific expression patterns. It contains a promoterless reporter gene, such that expression of the reporter will occur only if the element inserts adjacent to a functional endogenous promoter. The element shown here has a partial intron and a splice acceptor site at the 5′ end, so that expression of the reporter protein requires not only insertion into a gene but splicing as well. This feature is optional; the element can be made without the intron and splice site, so that splicing is not required for expression. The enhancer trap element is used to identify endogenous enhancers in the plant genome. It is similar to the gene trap element but with a minimal promoter at the 5′ end of the reporter gene. Thus, it will be expressed if the transferred deoxyribonucleic acid (T-DNA) inserts near an enhancer element. The third element combines the gene trap function with selection for insertion into an expressed gene. The selectable marker lacks a polyadenylation signal but ends in a splice donor site with an intron. For the transformation event to survive selection, the element must insert such that the selection gene is spliced to a polyadenylation sequence in the genomic DNA. This type of element has been used in mouse cells (Zambrowicz *et al.*, 1998) but has not been reported in plants. The activation tag element is used to generate gain of function mutations. An enhancer element is included on the insertion element so that is will turn on expression of plant genes near the site of insertion.

Activation tagging is a variation of T-DNA insertion mutagenesis, in which an enhancer element is incorporated adjacent to the right border of the T-DNA. The enhancer element can activate expression of a gene by insertion within a few kb either 5′ or 3′ of a gene. Thus, while most insertions into coding sequence will result in loss of expression, the activation tag can also result in gain of function phenotype by inserting near a gene. With some genes, gain of function alleles may be the only type that result in a detectably altered phenotype. This is true for genes encoding proteins with essential functions, where knockouts will be lethal and never recovered, and also genes for which there is redundancy, where the elimination of expression of one gene is compensated for by the expression of a second gene of similar function.

These gene trap elements result in reporter gene expression if they insert in or near a gene that is expressed in the tissue that is screened. However, the vast majority of insertions are not in regulatory elements and, therefore, only a small percentage of the transgenic lines generated will show any detectable reporter gene expression. A very elegant complex tagging element was recently described and used in mouse embryonic stem cells, which provides for selection of insertion into any functional gene, whether it is expressed in embryonic stem cells or not (Zambrowicz *et al.*, 1998). The element contains a constitutive promoter expressing a selectable marker gene that lacks a polyadenylation sequence. Thus, the selectable marker gene will only be expressed if it inserts upstream of a functional polyadenylation site in the genome, and therefore only insertions into expressed genes will survive selection. In addition, the complex element contains a promoterless reporter gene, similar to the standard gene trap element, so that expression of the reporter will be under the control of the promoter of the gene into which the element has inserted. The advantage of this complex element over standard gene trap elements is the positive selection for insertion into expressed genes, resulting in a reduction in the number of individuals with insertions into unexpressed regions, reducing the number that must be analysed. Such a system has not yet been reported in plants, but it may be very useful.

A number of techniques have been devised to clone the genomic DNA flanking the insertion element. These include: plasmid rescue (Koncz *et al.*, 1989); inverse PCR (Ochman *et al.*, 1988); and thermal asymmetric interlaced (TAIL) PCR (Liu and Whittier, 1995). No one technique has emerged as clearly better than the others in all situations. The appropriate technique depends, to some extent, on the structure of the insertion element used for the mutagenesis. For plasmid rescue, for example, the T-DNA must contain a plasmid origin of replication as well as a bacterial selectable marker gene.

10.2.1.2 Map-based cloning

While insertion mutagenesis is generally the fastest way to connect a mutant phenotype to a gene sequence, it does have limitations. The majority of tagged alleles generated by insertion mutagenesis are nulls. Thus, it is not useful for studying any pathways that are essential during the life cycle of a plant, since null alleles would result in a lethal phenotype. Chemical mutagens, such as ethyl methanesulfonate (EMS), primarily induce point mutations, which can have a more subtle effect on gene function and may result in a wider range of alleles, including gain of function, partial loss of function, or conditional mutants. For these mutants, map-based cloning is the current method of choice for cloning the affected gene.

Also, as mentioned previously, approximately 50% of the mutations found in T-DNA insertion populations are not genetically linked to an insertion element, so that no tag is available to clone the affected genes. If the mutation selected is not genetically linked to an insertion element, there are two options: to continue screening until an insertion allele is found; or to proceed with the allele in hand and use a map-based cloning strategy. In species like *Arabidopsis*, for which many mapping tools are available, positional cloning is a viable alternative. Techniques are being developed which are also applicable in species with relatively uncharacterized genomes.

Map-based cloning is based on the co-segregation of the mutant phenotype and known, mapped markers. The resolution of mapping depends on the size of the segregating population that is scored in the mapping cross, and the number of mapping markers that are available. The larger the segregating F_2 population, and the more markers used, the higher the map resolution. Mapping is usually the rate-limiting step in a positional cloning strategy because the number of plants in the segregating population that must be scored to obtain a map position of suitable resolution is high, hundreds and sometimes into the thousands. Moreover, once the gene has been placed in a rough map position, it is likely that mapping markers will have to be generated for fine mapping around the locus. Recent developments should make it easier to find mapping markers, and also reduce the need for such high resolution mapping.

Mapping was originally performed based on co-segregation with previously mapped morphological markers, but these have been for the most part replaced by DNA-based markers for a number of reasons. Firstly, the number of morphological markers is very limited. Also, DNA-based markers are not subject to variable penetrance caused by environmental conditions as are some morphological markers. Finally, DNA-based markers can be scored when the plant is still at the seedling stage. There is no need to grow the plant to maturity as for some morphological markers, such as those affecting flower development.

Restriction fragment length polymorphisms (RFLPs) were among the first DNA-based markers. Allelic variation arises from a sequence difference between the two parents that affects a restriction enzyme recognition site, so that it is present in one parent but not the other. Detection is performed with genomic Southern blots. Because genomic Southern blots are used, microgram quantities of genomic DNA are required for each plant scored. As a result, RFLPs are being replaced by PCR-based markers, for which only nanograms of genomic DNA are required. The amount of tissue required is small; a leaf disc 0.5 cm in diameter supplies enough DNA for 50 or more PCR reactions. In addition, very simple protocols for preparation of PCR quality genomic DNA have been developed so that 100 or more preparations can be carried out in a day (Lukowitz *et al.*, 1996).

Random amplified polymorphic DNA sequences (RAPDs) are PCR-based markers generated with random PCR primers. Short primers (usually 10 base-pairs) are used and generally produce multiple (5–10) PCR products. Thus, multiple loci are assayed in each reaction. RAPD polymorphisms can result from insertions or deletions that change the size of the PCR product, but are usually the result of base differences in the priming sites that result in the presence of the PCR product in one parent and absence in the other. Identification of RAPDs can be targeted to the region around the locus of interest by performing bulked segregant analysis (Michelmore *et al.*, 1991). Because RAPDs often result in presence/absence polymorphisms, they are less informative, since heterozygous individuals cannot be distinguished from one of the homozygotes without testing in the next generation. As a result, RAPDs are being replaced in favour of co-dominant markers.

One commonly used type of co-dominant PCR-based marker is cleaved amplified polymorphic sequences (CAPS) (Konieczny and Ausubel, 1993). Like RFLPs, the basis of allelic variation is the presence of a restriction site in one parent but not in the other. Rather than scoring these on a Southern blot however, PCR primers are used to generate a genomic fragment spanning the polymorphic restriction site. The PCR products are digested with the diagnostic restriction enzyme and the digests are run on a gel for scoring. Normally, both parental genotypes and the heterozygote are all clearly distinguishable.

Given a limited amount of sequence information, CAPS markers can often be generated to refine a genetic map in a localized region of the genome. Genomic fragments of one to two kb can be cloned from the region of interest and the ends sequenced to design PCR primers. Alternatively, if genomic sequence data for the region is available, appropriately spaced PCR primers can be designed from the database sequence. PCR products from the two parents are then digested with a range of restriction enzymes to find a polymorphism. Another approach is simply to sequence the entire PCR product from both parental genotypes and subject the sequences to 'electronic' restriction digests, using sequence analysis programs to scan for restriction sites. A large number of CAPS markers have been generated in various positional cloning projects in *Arabidopsis*. Information for many of these is available via the internet (http://genome-www.stanford.edu/Arabidopsis/aboutcaps.html).

An extension of the basic CAPS procedure, sometimes referred to as dCAPS, has been developed to extend the potential for generating useful CAPS markers (Michaels and Amasino, 1998; Neff *et al.*, 1998). Many single nucleotide polymorphisms do not result in a change in a restriction enzyme recognition site and are, therefore, not directly useful for CAPS markers. To generate dCAPS markers, single nucleotide polymorphisms are identified and mismatch primers are designed that generate a restriction site in one allele. A computer program has been written to design dCAPS primers for two given sequences containing a polymorphism (Neff *et al.*, 1998).

Simple sequence length polymorphisms (SSLPs) have also been developed as mapping markers in *Arabidopsis* (Bell and Ecker, 1994). The polymorphisms are based on differences in lengths of short dinucleotide repeat sequences. They have an advantage over CAPS markers in that there is no restriction digestion required; the PCR products are simply run on a gel to check for differences in length. They can be somewhat difficult to work with because they are often small fragments, (~100 to 200 base-pairs) and the size difference between two ecotypes may be only a few base-pairs, so that it can be hard to distinguish sizes without high resolution polyacrylamide gels. A list of *Arabidopsis* SSLPs can be found at http://genome.bio.upenn.edu/SSLP_info/SSLP.html. If genomic sequence is available for a region of interest and simple sequence repeats are found there, it is worth testing to see whether there is any polymorphism in the repeat lengths in the parental genotypes, as the lengths of these repeats often vary among ecotypes.

Markers based on single nucleotide polymorphisms (SNPs) are being developed in other systems, and also have great potential for use in mapping plant genes. Since single nucleotide polymorphisms are the most common type of polymorphism, it should be possible to generate very high resolution genetic maps. For example, it has been estimated that the *Arabidopsis* ecotypes, Landsberg and Columbia, have an average of one polymorphism every 260 base-pairs (Konieczny and Ausubel, 1993), though, as with all DNA-based polymorphisms, the frequency of polymorphism is very dependent on the two parental ecotypes being compared. Also, the rate of polymorphisms may vary greatly across the genome, with some regions having more and some having less. Currently, SNPs can be detected based on differences in mobility of heteroduplex and homoduplex DNA in denaturing high performance liquid chormotography (DHPLC) (Underhill *et al.*, 1996), or high resolution polyacryamide gels (Hauser *et al.*, 1998).

In the future, SNPs may form the basis for microarray-based mapping and genotyping (Wang *et al.*, 1998). Hybridization to oligonucleotide microarrays (see section 10.3.1.1) can be sensitive to single base-pair differences. Thus, two oligonucleotides, each containing one variant of the polymorphism, could be synthesized on a microarray and used for genotyping of segregating progeny of a mapping cross. By including hundreds of pairs of oligonucleotides for known SNPs and using bulked segregants as probes, it should be possible to rapidly and accurately map any mutation. A recent example of microarray-based mapping in yeast demonstrates the power of the system (Winzeler *et al.*, 1998). Over 3700 DNA-based markers were identified as polymorphic between the parental strains. These polymorphic markers could be placed on the physical map of the yeast genome, since the entire genomic sequence is known. DNA from individual segregating progeny was hybridized to the mapping chip and co-segregation of the mutant phenotypes with specific markers was used to pinpoint the locations of the mutant genes.

While the rapid increase in the number of mapping markers will facilitate fast and accurate mapping of a gene, improvements in transformation vector technology should also contribute to increased efficiency of map-based cloning by reducing the mapping resolution needed. Two types of large insert, transformation competent cloning vectors have recently been described: binary bacterial artificial chromosome (BIBAC) vectors (Hamilton *et al.*, 1996) and transformation-competent artificial chromosome (TAC) vectors (Liu *et al.*, 1999) can both be used to clone genomic DNA fragments of 80–150 kb. The clones are stable in *Escherichia coli* and *Agrobacterium tumefacians*, and both can be used directly for *Agrobacterium* mediated transformation of the insert DNA into plants. The direct cloning into binary vectors eliminates subcloning steps, and the large insert size means that larger fragments can be used to test for complementation of the mutant phenotype. Thus, mapping resolution does not have to be so high, meaning that not as many individuals from the segregating population need to be scored.

Microarray-based mapping and the large insert transformation vectors may also facilitate mapping and cloning of quantitative trait loci (QTL). Because the microarrays will have a high density of markers throughout the genome, it will be possible to map multiple loci simultaneously. High resolution mapping of QTLs can be difficult because their effects are often subtle, but any reduced resolution of mapping may be compensated for by the increased size of the insert fragments in BIBAC and TAC vectors used for complementation.

The internet is another important resource for map-based cloning tools. In addition to those already mentioned, there are many sequencing and mapping sites, especially for model systems, such as *Arabidopsis*, maize or rice (http://www.Arabidopsis.org/maps.html; http://www.agron.missouri.edu/; http://genome.cornell.edu/cgi-bin/WebAce/webace?db = ricegenes; http://www.staff.or.jp/). Among the more useful websites for map-based cloning are those that provide information on physical maps based on overlapping inserts from YAC and BAC libraries. A chromosome walk can be rather tedious. It involves rounds of isolating the appropriate clone, subcloning the ends and orienting the clone insert with respect to the map and the mutant, then repeating the procedure until the locus is crossed. Each step takes weeks, and, if many steps have to be made, it can be a slow process. Now, for many regions of the *Arabidopsis* genome, an entire walk can be performed at a computer terminal in an afternoon. BAC contigs (contiguous sets of overlapping BACs) are being generated by 'fingerprinting' the inserts, i.e. determining the characteristic restriction digest pattern for each insert. Using computer analysis, regions of overlap between multiple BACs are detected by common restriction fragment patterns. While the entire *Arabidopsis* genome has not yet been contiged, extensive regions have been completed. 'Virtual' chromosome walks can now be performed via the internet (http://genome.wustl.edu/gsc/Projects/thaliana.shtml; http://www.mpimp-golm.mpg.de/101/mpi_mp_map/access.html). Ultimately, when all the contigs have been assembled, a minimum tiling path will be determined. This refers to the minimum set of overlapping BACs that will span the entire genome.

Map-based cloning can be performed in any species for which there are sufficiently polymorphic parents that can be crossed to generate a segregating F_2 population. While extensive mapping resources already exist for *Arabidopsis*, many of the techniques described for *Arabidopsis* are applicable to any species. The difference is in the types of tools that are already available, and how much work it takes to set them up in the plant of interest if they are not available. There are plant species in which chromosome walking is more difficult, particularly those with large amounts of repetitive DNA. The maize genome contains large amounts of dispersed repetitive DNA, mostly in the form of retroposons, that makes map-based cloning more complicated. One observation that may facilitate cloning genes in maize and other monocotyledonous species is that there is a high degree of synteny between their genomes. That is, the genes appear to be in the same order and in the same relative position in the genome. The difference between maize and rice, for example, is that in maize the genes are much farther apart due to the dispersed repetitive DNA. Thus, it may be possible to perform the mapping and cloning in a monocotyledonous species with a smaller genome, and use that information to map and clone the gene in maize or wheat. Rice is emerging as the model monocotyledon: it has a relatively small genome (around 430 Mb), about 3–4 times the size of *Arabidopsis*, but much smaller than wheat (approximately 16,000 Mb) or maize (approximately 2400 Mb) (Bevan and Murphy, 1999). Many tools are being developed for rice, including an EST database, a genomic sequencing project, and BAC and YAC libraries.

One current method of choice for cloning a gene in a species with an uncharacterized genome is amplified fragment length polymorphism (AFLP) (Vos *et al.*, 1996). AFLP is a DNA fingerprinting technique that can be applied to mapping. It is a PCR-based technique, in which genomic DNA is cut with two restriction enzymes, and adapters specific to each type of sticky end are ligated on to the fragments. Primers complementary to the two adapters are used to amplify the restriction fragments. Two to three extra bases are added at the 3′ end of the primers, meaning that only a subset of the restriction fragments serve as templates for the PCR reaction. This reduces the complexity of the mixture of PCR products (to around 50–100 products per reaction), so that when the products are run on a polyacrylamide gel, the majority of bands represent a single PCR product, and differences in band presence or absence can be detected. By using various combinations of primers with different extensions, vast numbers of genomic fragments can be screened for polymorphism. For example, in identifying markers closely linked to the barley *Mlo* locus, over 250,000 AFLP loci were screened. By pooling individuals in a segregating population based on homozygosity for one phenotype or the other, AFLP markers can be identified closely linked to the region of interest, and these can be used to screen YAC or BAC libraries and generate a contig in the region of interest. AFLP is applicable to any species, and is independent of any sequence information. Among the genes cloned using AFLP are the barley *mlo* gene, which encodes for non-race-specific resistance to *Erysiphe* (Buschges *et al.*, 1997; Simons *et al.*, 1997).

10.2.2 Reverse genetics

Reverse genetics is a relatively new approach to connecting sequence to function. Initially, DNA sequencing was focused on genes that were identified based on function. With the advent of large-scale sequencing, however, we now know the sequences of many genes for which we have no hint as to function. In addition, many genes have been identified with a suspected function, for example based on homology to known genes, or genes obtained in a yeast two-hybrid screen. Reverse genetics is one way to confirm the functions of such genes.

Increasing or decreasing expression of a gene may result in an altered phenotype that suggests function. For example, using a strong constitutive promoter to ectopically express a regulatory gene can result in constitutive expression of the downstream genes. The identity of these upregulated genes may, in turn, provide clues to the function of the overexpressed gene. Attempts to overexpress a gene often lead not only to overexpression in some of the transgenic lines but also underexpression in others, a phenomenon known as co-suppression or gene silencing. Expression of both the transgene and the endogenous copy of the gene is reduced or eliminated. Expression of a gene in the antisense orientation can also result in reduced expression of the endogenous gene. Both co-suppression and antisense suppression can be used as a complement to overexpression to assess gene function by generating plants with reduced expression of the gene.

The exact mechanisms of antisense and sense suppression are unknown, but some recent reports suggest that the mechanisms may be related. Evidence is accumulating that double-stranded (ds) RNA intermediates may be involved in at least some instances of gene silencing. Recent results from experiments in the nematode, *Caenorhabditis elegans*, have shown that injection of dsRNAs can result in complete silencing of endogenous genes containing sequences present in the dsRNA (Fire *et al.*, 1998). Experiments in plants are producing similar results. Plants expressing a gene in either the sense or antisense orientation that did not exhibit silencing were crossed to each other and a proportion of the progeny from these crosses resulted in silencing. In contrast, selfing of the parents or crosses of sense lines with other sense lines, or antisense lines with antisense lines, did not result in silencing. Thus, expression of the gene in both orientations increased the rate of silencing (Waterhouse *et al.*, 1998). In a second set of experiments, expression of transgenes containing inverted repeats capable of forming hairpin structures with a region of dsRNA upon expression have resulted in a significantly higher percentage of lines exhibiting co-suppression than either sense or antisense lines (Hamilton *et al.*, 1998). Thus, incorporating inverted repeats in transgene constructs may be an effective way of suppressing expression of endogenous genes for testing function of gene sequences.

The effectiveness of dsRNA in silencing gene expression suggests that an RNA-dependent RNA polymerase may normally be involved in the silencing process by generating a dsRNA from an overexpressed gene. Evidence of a role for RNA-dependent RNA polymerase in post-transcriptional gene silencing comes from a *Neurospora* mutant defective in silencing. The defective gene was found to encode an RNA-dependent RNA polymerase (Cogoni and Macino, 1999). The *Neurospora* gene has homologues in both tomato and *Arabidopsis*.

The mechanism of sense and antisense suppression in plants may be a defence mechanism against viral infection. Transgenic plants constitutively expressing a viral gene can result in resistance to the virus that is the source of the transgene. In at least some cases, resistance does not require translation of the viral messenger ribonucleic acid (mRNA), rather the RNA itself mediates the resistance. Resistance was shown to be associated with post-transcriptional gene silencing, which also accounts for some co-suppression silencing (Baulcombe, 1996). This observation has led to the development of viral vectors for silencing of endogenous plant genes. Such a system has been demonstrated in *Nicotiana benthamiana*, using potato virus X (PVX) vectors (Ruiz *et al.*, 1998). A plant gene of interest is cloned into the virus vector and the recombinant virus used to infect plants. The virus infection becomes systemic and results in systemic silencing of the endogenous plant gene. Because this virus-induced gene silencing (VIGS) does not require the generation of transgenic plants, it is much more efficient for screening a large number of genes. VIGS using the PVX vector is limited to solanaceous species, though it may be possible to develop similar systems for other species using similar viruses with different host ranges.

As with any strategy to generate or phenocopy null alleles, it may be impossible to recover any plants homozygous for the null mutations if the gene being tested is essential for survival. In this case, the transgene can be placed under the regulation of an inducible promoter. Systems for regulated induction have been developed based on the yeast Gal4 transcription regulator. Gal4 is a transcription factor that binds to upstream activating sequences (UASs) found in promoters of genes regulated by Gal4. There are no plant transcription factors that activate expression from the UAS elements, so a transgene containing a minimal promoter and multiple Gal4 UASs is not normally expressed in plants. Two different systems have been developed to regulate expression from transgenes under the control of Gal4. In one, the Gal4 DNA binding domain was fused to the transcription activating domain from the maize *C1* gene and constitutively expressed. A second transgenic line is generated containing the gene to be overexpressed under the control of the Gal4 UAS elements. The two lines are then crossed. The progeny express the Gal4/C1 gene, which then induces transcription of the second transgene (Guyer *et al.*, 1998). In the second system, the Gal4 DNA binding domain is fused to a transcription activation domain and to the receptor domain of a glucocorticoid receptor (Aoyama and Chau, 1997). In the absence of the glucocorticoid ligand, the glucocorticoid receptor domain represses the transcriptional activity of the fusion protein. In the presence of the ligand, however, transcriptional activation occurs, and the genes regulated by the UAS elements are expressed. Both the gene for the Gal4-DNA binding domain-glucocorticoid receptor domain fusion and the gene regulated by the Gal4 UAS elements are present in the same plant, but the second gene is not expressed until the plant is treated with the glucocorticoid ligand. This system has been shown to work in both tobacco and *Arabidopsis*. The glucocorticoid-regulated system has the advantage of being inducible at any stage of development.

Insertion mutations can be used for reverse genetics as well as for forward genetics. Instead of screening for a mutant phenotype, the population is screened at the DNA level, for insertions into a specific sequence. Screening is often performed using PCR on genomic DNA from the mutagenized plants. Primers specific for the insertion element are used in combination with primers specific for the gene of interest. A combination of an insertion element primer and gene specific primer will give a PCR product only if the element inserts in, or very near, the gene of interest. While large numbers of mutant plants must be screened, the sensitivity of PCR allows the pooling of genomic DNAs. Thus, tens or hundreds of thousand of individual plants can be screened by performing a few hundred PCR reactions.

For reverse genetics, insertion mutagenized populations are currently being screened, primarily one gene at a time. With advances in robotics, it is becoming feasible to isolate the flanking genomic DNA from a large number of insertion mutant plants on a high-throughput basis. The individual flanking DNAs can all be spotted on filters to make arrays, so searching for an insertion mutation in ones gene of interest would simply involve probing the array with ones gene and obtaining the seed corresponding to any spots that hybridized. Alternatively, the flanking genomic fragments can all be sequenced, providing a database of insertion sites. Some of these 'sequenced insertion sites' (SINs) from an *Arabidopsis* population mutagenized with the maize En/Spm transposable element are being made publicly available on the *Arabidopsis* database on line (http://www.jic.bbsrc.ac.uk/Sainsbury-lab/jonathan-jones/SINS-database/sins.htm). By searching the database with ones gene sequence, it may be possible to identify a mutant allele of a specific gene. As the size of these databases increases, the chances of success increase. One complication with the sequencing approach is that many insertion lines contain multiple insertion elements, and therefore multiple flanking sequences, so the individual flanking fragments must be separated before sequencing. When spotting the flanking fragments on a filter, it does not matter if a spot contains more than one flanking genomic fragment.

Deletion mutants can also be used for reverse genetics. The mutant population can be screened with PCR, as described in a recent example in *C. elegans* (Westlund *et al.*, 1999). PCR primers are designed to span the gene of interest. Genomic DNA from pools of mutant lines are screened to select lines in which the PCR product is shorter than the wild-type band. If a pool gives a PCR product shorter than the expected size, the individual within the pool can be identified to confirm that the gene, or a portion of it, is in fact deleted.

The ultimate system for reverse genetics involves targeted gene replacement, so specific mutations could be made in a gene *in vitro* and the mutated version of the gene then used to directly replace the endogenous copy in the genome. While homologous recombination can be performed in many organisms, including fungi and mammals, it is currently not a viable option in plants. Homologous recombination has been reported in *Arabidopsis*, but the efficiency is still too low to make it practical (Kempin *et al.*, 1997).

10.2.3 Identification of other steps in a pathway once one gene is known

Once a gene in a desired pathway has been cloned, there are a number of ways to use this gene to identify other genes whose products function in the pathway. These include the yeast-based one-hybrid and two-hybrid screens, as well as further genetic screens for extragenic suppressor mutations.

The yeast two-hybrid screen is a method to identify proteins that interact with a known protein (Finley and Brent, 1996). The interaction is detected in yeast cells, so it is an *in vivo* assay. The gene of interest is cloned into the 'bait' vector, such that it is expressed as a fusion protein with a DNA binding domain that binds to a promoter element in a yeast reporter gene or selectable marker. A cDNA library made from tissue in which interactors would be expected to be expressed is made in the 'prey' vector, resulting in fusion proteins containing the plant protein fused to a transcriptional activation domain. The prey vector containing the cDNA library is transformed into a yeast strain containing the bait vector and reporter gene. If the prey fusion contains a protein or domain that interacts with the protein of interest in the bait fusion, the transcription activation domain and the DNA binding domain are brought together on the promoter and the reporter or selectable marker gene is expressed.

While the yeast two-hybrid system has been very useful in identifying interacting proteins, evidence of interaction in yeast is not proof of biologically relevant interaction *in planta*. The interaction must be confirmed using other methods. If antibodies for the two interacting proteins are available, co-immunoprecipitation experiments can be carried out. Reverse genetic approaches, such as screening for insertion mutants or antisense suppression, can be performed to see if a knockout mutant of the protein identified in the two-hybrid screen has a phenoptype consistent with its predicted function. It is not unusual for a large number of proteins to be selected in a two-hybrid screen, so it is essential to have other independent evidence for any interactions found.

Conversely, protein-protein interactions that occur in plants may not be detected in yeast. There may be some modifications that take place *in planta* that do not occur in yeast. In addition, proteins may interact as part of a complex. Since in the two-hybrid system, only one cDNA is expressed in the prey vector in each yeast cell, mutli-component complexes cannot form.

Despite its limitations, the yeast two-hybrid system is a very useful technique to identify interacting proteins. One example in the host-pathogen field is the *AHBP-1b* gene, which was identified using the *NPR1/NIM1* gene as bait. *AHBP1* encodes for a transcription factor that was shown to bind to the promoter of *PR1*, a gene regulated by *NPR1/NIM1* (Zhang *et al*., 1999).

A genetic approach can also be taken to identify other genes acting in a pathway with a known mutation. Extrageneic suppressors of the original mutation can be selected. These can result from mutations either upstream or downstream of the original mutant, or from mutations in independent pathways that can compensate for the original mutation. This approach has been used to identify genes required for the cell death phenotype of some lesion mimic mutants (Morel and Dangl, 1999), and genes required in two *Erysiphe* resistance pathways in barley (Freialdenhoven *et al*., 1994, 1996).

If the downstream genes in a pathway are known, it may be possible to identify the regulatory genes that control their expression. Promoters of co-regulated genes can be examined for common regulatory elements. These can be used to isolate the transcription factors that bind to the regulatory elements. Among the methods available to isolate the transcription factors is the yeast one-hybrid screen (Li and Herskowitz, 1993). The one hybrid system is similar to the two-hybrid, but in the one-hybrid the yeast reporter gene is placed under the control of the promoter element of interest. The promoter element becomes the bait to trap the DNA binding protein. A cDNA library is expressed as a protein fusion with a transcriptional activation domain in yeast cells with the reporter construct. Any fusion with the activation domain that results in binding to the element will lead to expression of the reporter. This system was used to identify the *CBF1* gene, which regulates cold adaptation in *Arabidopsis* (Stockinger *et al*., 1997).

10.3 Genomics-based approaches

Among the new technologies that can be applied to study plant defence responses are the so-called genomics approaches. The ultimate goal is to be able to examine changes in the levels of accumulation of all the mRNAs, proteins or metabolites in an organism rather than focusing on limited subsets. Results from genomics experiments are not only more complete, they are less biased by previous results. If you can look at the expression levels of all genes in a plant, you are not limited to the genes you have on hand or your ability to identify new genes. In addition, by comparing changes in mRNAs, proteins and metabolites, the relative contributions of regulation at the different levels can be assessed, permitting more logical approaches for modifying specific pathways.

10.3.1 Gene expression profiling

Gene expression profiling refers to the parallel analysis of the expression of large numbers of genes at the level of mRNA accumulation. By comparing the expression profiles from different treatments or conditions, changes in expression patterns can be seen, indicating which genes may play a role in the particular response under study. For example, one could identify all genes that are up- or downregulated in a plant following pathogen infection.

10.3.1.1 Microarrays for expression profiling

Microarrays, also known as chips, are emerging as the method of choice for large scale analysis of gene expression. The small size of microarrays means that the expression of thousands of genes can be examined in parallel in a single experiment. The ultimate expression profiling experiment is possible when the entire genomic sequence is known. This is not yet true for any plant species, but arrays are being made based on extensive EST collections, supplemented with predicted ORFs from genomic sequencing projects.

Microarrays for gene expression profiling are a refinement of reverse Northern dot blot procedures. In Northern blots, RNA is bound to a filter and hybridized with a DNA probe to determine how much of the specific RNA homologous to the DNA probe is present in the RNA sample. In a reverse Northern dot blot, the DNA is applied to the filter (or glass substrate in the case of microarrays) and probed with the RNA, in the form of labelled cDNA made from polyA mRNA. Because the DNA on the substrate is in excess, the hybridization signal is proportional to the amount of the specific RNA in the total RNA sample that is homologous to the DNA in that particular DNA spot. By putting a number of different DNA spots on the filter, the expression levels of many genes can be determined in a single experiment. The power of microarrays is in their miniaturization, and the vast number of genes that can be analysed simultaneously.

There are two types of microarrays currently in use, those in which oligonucleotides of specific sequences are synthesized *in situ* on a glass substrate, and those in which DNA fragments are spotted on a glass substrate (Table 10.1). The first type is sometimes referred to as a synthesis array. The fabrication is based on a light-directed combinatorial chemical synthesis of the oligonucleotides *in situ* on the substrate (Lipshutz *et al.*, 1999). Adapting photolithography

Table 10.1 Comparison of microarray technologies

	Synthesis (oligonucleotide) array	Deposition (cDNA) array
Size of DNA on chip	< 25 bp due to technical limits	Up to kbs long
Density of elements	$150,000/cm^2$	$3,000/cm^2$
Discrimination	Sensitive to single bp changes	May get cross-hybridization between gene family members
Sequencing requirements	Need complete sequence information for any DNA on chip	No sequence information required
Logistics	Do not have to handle tens of thousands of DNAs	Lots of DNA manipulations, organization required
Cost	Prototype expensive, mass production cheaper	Cheaper to make, though cost of production does not drop significantly
Sensitivity	1:100,000–1:1,000,000	1:100,000–500,000
Amount of mRNA for probe	3–5 μg	600 ng

Abbreviations: mRNA, messenger ribonucleic acid; bp, base-pair; cDNA, complementary deoxyribonucleic acid.

technology developed for the semiconductor industry, this can be carried out on a very small scale. With existing commercial technology, around 300,000 different oligonucleotides, representing around 8000 different genes, can be synthesized in an area of approximately $1.6\,cm^2$, with the number likely to increase as the technology improves. Due to limitations in the *in situ* synthesis process, the length of the oligonucleotides is currently limited to around 25 bases, but as with the density of the oligonucleotides, with improved technology, the length will probably increase as well. The normal procedure is to synthesize a small number of oligonucleotides (up to 20) spanning different regions of each cDNA. Any region that is repetitive or homologous to other known genes can be avoided. The multiple oligonucleotides for each gene provide an internal positive control. The amount of hybridization should be constant across all the oligonucleotides for a single gene, provided there is no alternate processing of the transcript. A second set of oligonucleotides is often included as a negative control. For each oligonucleotide, a second is synthesized that is identical to the first except for a single mismatch in the central base of the oligonucleotide. Since hybridization is sensitive to single base-pair mismatches, if the signal intensity is not stronger on the wild-type sequence than on the mismatch sequence, the signal probably does not reflect specific hybridization.

The second type of microarray is sometimes referred to as a deposition or cDNA microarray. The deposition type is directly analogous to a dot blot, in that DNA fragments are deposited directly onto the substrate, but since a non-porous glass substrate is used for the microarray instead of a membrane, the density of the spots can be much greater. The direct spotting of the DNA fragments means that the fragment length is not limited to short oligonucleotides as with the synthesis arrays. Full length cDNAs or genomic fragments can be used. With current technology, deposition arrays can be generated at a density of approximately 10,000 elements (representing up to 10,000 genes) per 3.24 cm^2 (Kehoe *et al.*, 1999). It is estimated that all *Arabidopsis* genes (approximately 25,000) could fit onto a single microscope slide. It should be noted that the total number of genes that are represented on the two types of microarrays do not differ as dramatically as the differences in element density might suggest. While the element density of deposition arrays is well below the density for synthesis arrays, on the deposition arrays, each element can represent an entire gene, while in a synthesis array, each element is a short oligonucleotide, and multiple elements are used for each gene.

Hybridization to the arrays is performed using fluorescence-labelled cDNA probes generated from mRNA of the samples to be tested. Because the DNA on the array is in excess, the degree of hybridization of each spot as measured by fluorescence is proportional to the amount of mRNA corresponding to that gene in the total sample. With the deposition arrays, two separate RNA samples are labelled with different fluorescent labels and combined in a hybridization mix on a single array. After hybridization and washing, the microarray is scanned sequentially at the two emission wavelengths of the probes. Thus, the relative expression of each gene under two different conditions can be determined in a single hybridization experiment. With synthesis arrays, a single probe is hybridized to each array. For comparisons between arrays, internal standards are used to compensate for array-to-array variation.

Expression profiling reaches its fullest potential when sequence of the entire genome is known and the expression levels of all the genes in an organism can be examined. While there are so far no plant genomes that have been completely sequenced, expression profiling can still be a useful technique in species for which there are extensive EST collections. Expression profiling can be performed in a more directed form as well. Rather than looking at expression of all genes, expression of a subset of genes known or suspected to be involved in a specific response can be examined. Expression profiling is being used to look at changes in gene expression following a wide range of inductions, both endogenous, such as during developmental changes, or for exogenous inducers, including response to pathogens.

Expression profiling data can be used in a number of ways. It can provide useful information suggesting gene function. Where and under what conditions a gene is expressed may suggest possible functions or eliminate others. By identifying genes that are induced in response to certain stimuli, it may be possible to select genes whose products play a role in that response. The technique can also be used to compare the response profiles for different inducers.

For example, the set of genes that is induced by a fungal pathogen could be compared to the set of genes induced by a bacterial pathogen. Genes that are induced specifically in response to each type of pathogen could be identified, as well as genes that are induced by both and that may be part of a general response to pathogens.

In addition to information on specific genes, expression profiling can provide more global information on patterns of gene expression during physiological and developmental processes. By comparing expression patterns of individual genes through developmental stages, or during specific responses, sets of genes that are induced and repressed coordinately may be found. By identifying these sets of co-regulated genes (regulons), it may be possible to determine the molecular basis of their co-regulation. Promoters of genes in a regulon can be searched to identify shared regulatory elements. The regulatory elements can be used, in turn, to isolate the transcription factors that bind to the elements, using procedures such as yeast one-hybrid screens (Li and Herskowitz, 1993). This has been demonstrated in yeast, where experiments were performed to examine changes in gene expression profiles during progression through the cell cycle. Sets of co-regulated genes were identified and common regulatory elements were found in the promoters of some of the genes, including a recognition site for a transcription factor that had previously been shown to regulate gene expression in a cell-cycle-dependent manner (Cho *et al.*, 1998). Thus, expression profiling data may be helpful in identifying components of signal transduction pathways.

While expression profiling is a relatively new tool in plant research, impressive results from other systems hint at its potential. Because the entire yeast genomic sequence is known, some of the most complete results have been obtained in experiments carried out using yeast microarrays. Examples include the studies on changes in gene expression during the mitotic cell cycle mentioned previously, and during the shift from anaerobic to aerobic growth (DeRisi *et al.*, 1997). The expression patterns of many genes involved in these responses had previously been determined using other methods, and the results from the microarrays were completely consistent with the previous results. Thus, these experiments were very useful in demonstrating the validity of results obtained using microarrays. In addition, the expression patterns of many previously uncharacterized genes suggested a possible function in the shift from fermentation to respiration.

Many interesting experiments have also been carried out in mammalian systems using microarrays based on extensive EST collections. Comparisons have been made between healthy and diseased tissue (DeRisi *et al.*, 1996; Gerhold *et al.*, 1999). Knowing the genes that are expressed specifically in diseased or in healthy tissue may provide clues to the molecular basis of the disease, and suggest methods of treatment. Microarrays can also be used in drug discovery. If altering flux through a pathway by up- or downregulation is shown to be an effective disease treatment, microarrays can be used to assess the ability of test drugs to have the desired effect on the pathway of interest. In addition, the specificity of the test drug can be determined by checking for altered expression of other pathways that might lead to undesirable side-effects.

Compared to yeast and mammalian systems, relatively few expression profiling experiments using plant systems have been reported. There are a number of reasons for this, including the high cost of either making or buying arrays and the relatively low amount of sequence data available compared to humans and yeast. One of the first microarrays was in fact performed with *Arabidopsis* cDNAs (Schena *et al.*, 1995), and a recent publication describes results from microarray gene expression experiments performed using a set of over 1400 *Arabidopsis* cDNAs (Ruan *et al.*, 1998). The situation is likely to change very soon, as the *Arabidopsis* genomic sequence nears completion and more EST sequences become available in *Arabidopsis* as well as in crop species.

10.3.1.2 Expression profiling for reverse genetics

Microarrays can also be used in reverse genetic studies to determine the function of regulatory genes. This can be done by comparing expression profiles of mutants with reduced expression of the gene to wild-type, and expression profiles of individuals overexpressing the gene to wild-type. Target genes can be identified, providing hints to function of the gene (DeRisi *et al.*, 1997).

10.3.1.3 Limitations in microarray experiments

There are a number of considerations that must be kept in mind when analysing microarray data. Foremost among these is not a limitation of the technique but rather a limitation of the approach. Expression profiling examines only one aspect of the regulation of gene expression. It provides information on changes in the steady-state levels of individual mRNAs. This is useful for genes that are regulated at the level of mRNA accumulation, but there are many other levels of gene regulation as well. These include rate of protein synthesis and turnover, and also post-translational modifications, such as phosphorylation. Also, genes involved in early recognition events in response pathways tend to be relatively low abundant transcripts, and often do not show significant induction of expression. In a yeast expression profiling experiment studying changes in gene expression of all yeast genes during the shift from fermentation to respiration, of 149 genes encoding known or putative transcription factors, only two were induced by a factor of more than threefold following the shift (DeRisi *et al.*, 1997). This may mean that all the responses are controlled by two transcription factors, but it is also possible that other transcription factors involved are regulated by mechanisms other than increased rate of transcription.

Expression profiling may not be appropriate for identifying genes involved in early recognition events in signal transduction pathways. For example, most plant resistance genes are constitutively expressed at low levels, with little or no increase in expression following inoculation. Conversely, the genes at the downstream end of the defence response pathway are often strongly induced and are easily detected in expression profiling experiments. Thus, the types of genes identified in any differential expression screen, including expression profiling, may be more representative of the downstream end of the signal transduction pathway than the upstream regulatory events.

Cross-hybridization of transcripts from closely-related members of gene families can be a problem, primarily with cDNA arrays. A member of a gene family that is constitutively expressed at a high level could cross-hybridize and mask the signal from family members that are differentially regulated. This is less of a problem with the synthesis arrays, where it may be possible to design gene-specific oligonucleotides for members of gene families. A similar situation exists for genes that are regulated by differential splicing. This would be difficult to detect if it was not known *a priori*, though the chances of detection are greater on oligonucleotide arrays.

Cost is an important consideration in expression profiling. The arraying robots and chip scanners required to make deposition arrays are available commercially, so the entire process can be set up in any laboratory but the equipment is not cheap. Depending on the size of the arrays to be made, maintaining thousands of cDNA clones and generating the PCR products to put on the arrays is an organizational challenge. One advantage of deposition arrays is that if new ESTs are discovered, they can easily be incorporated on new arrays.

Synthesis arrays are currently only available commercially due to the technical complexities of making the arrays. They are very expensive to set up, though once a prototype has been made, the cost of production drops and they can be produced in large numbers, with no maintenance manipulations of bacteria or DNA fragments. When an entire genomic sequence is known, synthetic oligonucleotide arrays will probably become the standard. They are cheaper to mass produce, and since they will be based on the entire genome, there will be no need to allow for the addition of more genes later.

There are a couple of limitations, which are likely to be only short-term. One of these is data management. A single expression profiling experiment can result in tens or hundreds of thousands of datapoints. The ability of software to handle and analyse all this data has yet to catch up with our ability to generate the data. The challenge is no longer in obtaining data but in mining the data to get the most useful information out of it. The field of bioinformatics is developing rapidly to fill this void. Another limitation in many plant species is a lack of sequence information. This will become less of a problem as large-scale sequencing continues, but for some species, it will be a while before complete genomic or EST sequence data are available. Some methods for isolating sets of cDNAs are discussed in the next section.

In spite of these limitations, gene expression profiling offers tremendous potential in enhancing our understanding of plant responses. It will become an invaluable tool in studying many processes in plants, including defence responses.

10.3.1.4 Microarray websites

A large amount of information concerning various aspects of microarray technology is available on the internet, including sites from public and private laboratories, as well as a very extensive journal site. A few are listed below, and most of these have links to other relevant sites as well.

Nature Genetics recently (January, 1999) published a supplementary issue, entitled 'The Chipping Forecast', devoted entirely to various aspects of microarray technology. This provides a wealth of useful information on various aspects of microarrays, the different types and applications. The entire issue is available on line at http://genetics.nature.com/chips_interstitial.html.

A couple of sites have information on plans and protocols for making cDNA microarrays, the first is from the Pat Brown laboratory at Stanford and the second is maintained by the National Human Genome Research Institute. The third site has a program to analyse expression profiling data and group co-regulated genes into clusters:

- http://cmgm.stanford.edu/pbrown/mguide/index.html
- http://www.nhgri.nih.gov/DIR/LCG/15K/HTML/
- http://rana.stanford.edu

Two commercial sites have information on the two different types of microarrays. Synteni makes deposition arrays and Affymetrix makes synthesis arrays:

- Synteni/Incyte Inc http://gem.incyte.com/gem/
- Affymetrix Inc. http://www.affymetrix.com/

10.3.1.5 Alternatives to microarrays for gene expression profiling

Arrays can also be performed on nylon filters. Though they do not quite qualify as microarrays, by using robots a high density of spots can be obtained, up to 120 spots per cm^2, either by spotting DNA on the filters or allowing bacterial colonies to grow directly on the filters. This method is much cheaper than microarrays on glass, though it does have some drawbacks. Because nylon is a porous support, the DNA cannot be arrayed so densely. Because larger filters are used, more probe is required. Samples are usually compared by hybridizing two separate filters with the two RNA samples, necessitating normalization of signal to correct for filter-to-filter variation. A recent paper described the use of colorimetric-based detection for use on nylon-based arrays. A dual colour detection system was described, so that two probes could be hybridized simultaneously to a single filter, as done with the dual fluorescenece labelling system for deposition microarrays (Chen *et al.*, 1998). While filter-based arrays are cheaper to make and use than microarrays, filters are not likely to offer the sensitivity and dynamic range of microarrays. Filter-based arrays are most appropriate when only a subset of genes is to be analysed.

Serial analysis of gene expression (SAGE) is a PCR-based technique that can be used for expression profiling (Velculescu *et al.*, 1995). SAGE is a method of generating concatemers of small cDNA fragments (10–14 base-pairs long) alternating with 4 base-pair tags that mark the limit of the cDNA sequence. The concatemers are sequenced and the individual short cDNA fragment sequences are compared to a sequence database to identify the gene that was the source of the fragment. The 10–14 base-pair fragments have been shown to be sufficient to identify a single cDNA. Because the frequency of obtaining a fragment from a specific cDNA is proportional to the relative expression level of the corresponding mRNA, SAGE can be considered a form of expression profiling. SAGE fragment sequences can also be used to identify and confirm predicted ORFs in genomic sequence if a corresponding EST has not been found. As with oligonucleotide microarrays, the effectiveness of SAGE analysis is dependent on how complete the corresponding sequence databases are.

10.3.1.6 Targeted cloning of cDNAs for use in expression profiling

Data from expression profiling experiments using arrays are only as complete as the collection of genes on the chip. This varies from species to species, depending on how much sequence is available. Even in species with extensive EST collections, genes that are expressed at low levels or genes which are induced only under specific conditions may be underrepresented. There are number of methods to expand an EST collection, including some recently developed PCR techniques. With some of these, it is possible to target the cloning to cDNAs representing subsets of genes induced in a specific response.

Differential display (Liang and Pardee, 1992, 1997) and representational difference analysis (RDA) (Hubank and Schatz, 1999) were two of the first PCR-based methods for identification of differentially expressed transcripts. Differential display is based on amplification of PCR products using a short ($\sim$10 base-pairs) random primer and an oligo dT primer with a two base-pair extension. The primers are used in PCR reactions, with cDNA generated from mRNA from the two tissues to be compared. The two base extension means that the oligo dT primer will prime from only a subset of the templates in the total cDNA sample, while the random primer will likewise prime from only a subset of the templates. The products from the PCR reaction using the primers on the two cDNA samples to be compared are run on a polyacrylamide gel to find bands that are present in one sample and not the other. These are eluted from the gel and cloned, then tested as probes on RNA blots to confirm differential expression. By using different combinations of $5'$ and $3'$ primers, much of the cDNA population can be sampled. RDA is based on successive rounds of subtractive hybridization, followed by PCR to selectively amplify sequences unique to one cDNA population. While differential display and RDA are useful techniques, both result in a high rate of false positives.

cDNA-AFLP is an RNA fingerprinting technique that is based on the AFLP DNA fingerprinting technique used in mapping (Bachem *et al.*, 1996). In cDNA-AFLP, total cDNA is made by reverse transcription of mRNA, and the cDNA is digested with two restriction enzymes. Linkers are ligated onto the ends of the restriction fragments and PCR is performed using primers based on the linker sequences plus two base-pair extensions. As with differential display, the two base-pair extensions at the 3′ ends of the primers reduce the number of cDNA fragments that can serve as templates for PCR, reducing the complexity of the products. The PCR products from the two samples to be compared are run on a polyacrylamide gel and differential bands are eluted and cloned. Since longer primers based on known sequence are used for cDNA-AFLP, it is more reproducible and generates fewer false positives than differential display. By using various combinations of two base-pair extenstion primers, it is possible to systematically screen the cDNA populations from two mRNA samples for any differences.

Suppression subtractive hybridization (SSH) (Diatchenko *et al.*, 1996, 1999) is another PCR-based technique for isolating differentially expressed genes. Unlike cDNA-AFLP and differential display, polyacrylamide gels are not required to select differentially expressed products. Rather, SSH generates a library of cDNA clones that is enriched in differentially expressed genes. While not all cDNAs obtained with SSH represent differentially expressed genes, the method is fast and the cDNA population produced is enriched for differentially expressed genes. Actual differentially expressed genes can be identified by using each clone individually to probe RNA blots, or by making arrays of the library on filters and probing these arrays with the labelled cDNA from the two RNA samples used to generate the subtracted library. One advantage of SSH is that a normalization step occurs during hybridization, increasing the chances of obtaining cDNAs for low abundance messages.

These methods can be used to generate collections of cDNAs representing differentially expressed genes. The collections can be used for smaller expression profiling experiments targeted to a specific response, or they can be incorporated into a total EST collection for large-scale expression profiling experiments.

10.3.2 Proteomics

Proteomics refers to characterization of all the proteins expressed by the genome of an organism. While gene expression profiling provides information only at the level of transcript accumulation, proteomics can provide a range of information on the proteins. Depending on how the experiment is set up, information such as where and when each protein accumulates, to what level it accumulates, and post-translational modifications to the proteins can be obtained. It provides a good complement to gene expression profiling, since many genes are not regulated at the transcription level. Ultimately, proteomics should provide more precise information on gene expression, since the functional product of most genes is not an RNA but a protein.

The basis of most separations for protein analysis is two-dimensional (2-D) gel electrophoresis. This is not a new technique but recent refinements have made it useful on a genome-wide scale (for examples of 2-D gels, see http://www.expasy.ch/ch2d/). These include improvements in the reproducibility, along with better computer imaging to allow different gels to be dependably aligned. A number of new techniques have been developed to identify proteins from the small amount of sample that can be obtained from a 2-D protein gel. In addition, identification of proteins has been facilitated by the vast amounts of sequence available in databases. A protein spot is eluted from the gel and the protein is subjected to proteolytic cleavage into peptide fragments. These fragments are analysed by matrix-assisted laser desorption/ionization-time of flight (MALDI-TOF) mass spectrometric analysis. This generates a list of peptide masses for the peptide fragments. The combination of fragment sizes is characteristic of a specific protein and can be predicted from a gene sequence, so the results for a protein can be compared to a database of calculated peptide masses for each ORF in the genome. Using this method in yeast, 90% of the proteins analysed were identified by screening the sequence database (Shevchenko *et al.*, 1996). If no match can be found in the sequence database, proteins can be analysed by peptide sequencing using nanoelectrospray ionization. This technique allows identification of peptides at femtamole levels. These sorts of experiments can also be performed in plants with extensive sequence databases, either genomic or in the form of ESTs, though the efficiency will increase as the complexity of the sequence databases increase.

In the future, array technology may be adapted for use in proteomics. For example, it may be possible to array antibodies for a large number of proteins on a chip to analyse for changes in protein levels the way changes in RNA levels can now be measured. Alternatively, protein arrays could be used to probe for interacting proteins.

10.3.3 Metabolite profiling

The last of the global approaches to studying plant responses is at the level of metabolites, using metabolic profiling analogous to mRNA or protein profiling (Trethewey *et al.*, 1999). The changes in the chemical constituents that make up the plant are clearly a part of the plants response system. For example, metabolites such as phytoalexins are thought to be an important component of a plants defence response, and there are probably many other equally important metabolites that are as yet unknown. One of the difficulties in metabolite analysis is their diversity. Metabolites encompass a wide range of small molecules with highly diverse structures. As a result, global analysis of all metabolites is a much greater technical challenge than analysis of mRNAs or proteins.

As techniques for extraction, purification and especially detection improve, metabolic profiling will no doubt become an increasingly useful tool. Initial efforts may be directed at the analysis of a specific subset of metabolites, much in the way that the initial efforts in gene expression profiling were targeted to known cDNAs. With both techniques, the ultimate goal is global analysis of all genes, or all metabolites. An example of a targeted metabolic profiling experiment was performed in barley to identify differences in composition of phenolics in plants containing the *mlo* resistance allele. *mlo* Mediated resistance is not accompanied by hypersensitive cell death, but is characterized by the accumulation of autofluorescent material, suggesting the involvement of phenolics in resistance. By comparing high-performance liquid chromatography (HPLC) profiles of soluble phenolics from epidermal cells of resistant and susceptible plants following infection, one phenolic compound, p-coumaroyl-hydrozyagmatine (p-CHA), was shown to accumulate differentially in the resistant versus susceptible leaves. In addition, p-CHA was shown to have antifungal activity (von Ropenack *et al.*, 1998). Metabolic profiling is the least developed of the three levels of genomic analysis, but given the extensive secondary metabolism of plants, it should be a very interesting area in the future.

10.4 Bioinformatics

The new genomics-based approaches will generate vast amounts of data. Processing all these data also requires new techniques, and to meet these needs, the new field of bioinformatics is emerging. Bioinformatics is something of an imprecise term, and the definition is still evolving, but generally it refers to the intersection of computers and biology. In various contexts, it can refer to topics as diverse as electronic publishing to complicated database analysis to making large sets of data publicly available via the internet. Bioinformatics is all this, and more, and is clearly having a major impact on the types of experiments that can be performed, and the types of analyses that can be carried out on the data. Examples are genomic sequencing and gene expression profiling. With the huge amounts of data that can be generated, the more important challenge is to mine the data for useful information. Computer technology has, for the most part, been keeping pace with the ability to generate the data, but the biologist then faces the challenge of keeping pace with the computer technology. Elsevier Trends Journals published a very useful guide to bioinformatics in 1998 (Brenner and Lewitter, 1998). It contains tutorials on topics such as database searching, sequence analysis, protein analysis and databases of biological information.

Another important component of the information explosion is the internet. Relevant internet addresses have been included throughout the rest of this chapter. In addition to those sites on specific topics, there are many sites that catalogue other related sites. As such, they are good places to start looking for a specific program or database. As with many aspects of the computer age, the amount of information that is accessible is overwhelming. The challenge is to find the specific information that one is looking for in the shortest amount of time. These sites may be a good place to start looking. Note that what is available on the internet is changing almost daily, and web sites can have a short half-life, so some of the sites listed in this section, and throughout the chapter, may soon be obsolete. Moreover, it is important to keep in mind the dynamic nature of the internet, and the best sites today may not be the best sites next month.

The Agricultural Genome Information service of the United States Department of Agriculture has a site with links to websites for 15 different plant species from *Arabidopsis* to maize to rice to beans and cassava (http://ars-genome.cornell.edu/).

The following three sites are very useful sources for finding publicly accessible analysis programs and databases:

- DATABANKS (http://srs.ebi.ac.uk/) is a database of databanks on public servers around the world. It provides direct access to around 350 different databases relating to DNA sequence, protein sequence and structure, and many others (Kreil and Etzold, 1999). DATABANKS is searchable, so it is possible to quickly access the specific links of interest.
- A similar resource, DBCAT, is a database catalogue of approximately 400 public databases on topics such as DNA, RNA, genomics, mapping protein and literature (http://www.infobiogen.fr/services/dbcat/).
- Nucleaic Acids Research (Burks, 1999) maintains a website listing links to many useful database and analysis programs (http://www3.oup.co.uk/nar/Volume_27/Issue_01/summary/gkc105_gml.html).

Obviously there are many other ways to access the information that is collected in these sites, but these are good places to start in a search for a particular database or analysis program.

10.5 Conclusions

There are many newly emerging technologies that should be very useful in helping us understand plant defence response pathways and to use this information to enhance disease resistance in agricultural settings. Two recent papers describe results that demonstrate how these approaches can be used to modify pathways to make plants more able to withstand specific stresses.

One example of how this information may be useful in enhancing disease resistance involved a forward genetic approach leading to the cloning of the NPR1/NIM1 gene. This gene is an essential component of the signal transduction pathway resulting in systemic acquired resistance (SAR). SAR is an induced immunity that is triggered by a localized infection that causes necrosis. This results in increased resistance to secondary infection throughout the plant. Screens were performed to select *Arabidopsis* mutants that were unable to develop SAR. The NPR1/NIM1 gene was identified and subsequently cloned. Because the gene is required for the induction of SAR, experiments were carried out to see if overexpression of the gene would result in enhanced resistance; which was in fact the result. Plants overexpressing the gene have increased resistance to both fungal and bacterial pathogens (Cao *et al.*, 1998), and show enhanced responsiveness to inducers of SAR (L. Friedrich, K. Lawton, J. Ryals, and R. Dietrich, unpublished results). Thus, by identifying a key regulatory gene in the SAR pathway and modifying its expression, resistance could be enhanced.

The second example deals with the response to cold stress in *Arabidopsis*. The strategy to increase cold tolerance was similar to the strategy for the SAR pathway, i.e. identify a key regulatory gene and modify its expression, but a different approach was used to get the desired regulatory gene. *Arabidopsis* plants gradually shifted to cold temperatures adapt to cold and can withstand colder temperatures than if they were shifted directly to the low temperature. The expression of a set of cold-response (COR) genes is correlated with the acclimatisation process. A specific *cis*-acting element was found to be present in the promoters of many COR genes. Experiments fusing this element to a minimal promoter showed that it directed expression in response to low temperature stress. This cold response element was then used in a yeast one-hybrid screen to identify the cold-response element binding factor gene, CBF1 (Stockinger *et al.*, 1997). Transgenic plants expressing CBF1 from a strong constitutive promoter constitutively expressed the COR genes, and showed increased cold tolerance, with no acclimation period required (Jaglo-Ottosen *et al.*, 1998). This example demonstrates how identification of genes at the downstream end of a pathway can be used to work back up the signal transduction pathway to identify key regulatory genes.

These two examples demonstrate the potential for modifying response pathways in plants to enhance their ability to survive stresses, including pathogens. In the case of NPR1/NIM1, a forward genetic approach was used to identify a mutant, and the mutant phenotype was used to clone the affected gene. Genetics approaches will continue to be important in identifying genes with specific functions. In the second example, genes at the end of the signal transduction pathway, the COR genes, were used to isolate the gene CBF1 that regulated their expression. This approach will be facilitated by gene expression profiling techniques, which can be used to rapidly identify sets of co-regulated genes.

The ultimate goal is to understand plant defence responses and to use this information to enhance plant resistance to pests and pathogens. The more completely we understand the systems and pathways involved in the plant defence responses, the better we will be able to modify them to make them more effective. Molecular techniques and genomics approaches will play a very important role in expanding our knowledge of how plant defence responses are regulated, and in using this information more effectively to enhance disease resistance in crop plants.

References

Anderson, P.A., Lawrence, G.J., Morrish, B.C., Ayliffe, M.A., Finnegan, E.J. and Ellis, J.G. (1991) Inactivation of the flax rust resistance gene, *M*, associated with loss of a repeated unit within the leucine-rich repeat coding region. *Plant J.*, **9** 641-51.

Aoyama, T. and Chau, N.H. (1997) A glucocorticoid-mediated transcriptional induction system in transgenic plants. *Plant J.*, **11** 605-12.

Bachem, C.W., van der Hoeven, R.S., de Bruijn, S.M., Vreugdenhil, D., Zabeau, M. and Visser, R.G. (1996) Visualization of differential gene expression using a novel method of RNA fingerprinting based on AFLP: analysis of gene expression during potato tuber development. *Plant J.*, **9** 745-53.

Baulcombe, D. (1996) Mechanisms of pathogen-derived resistance to viruses in transgenic plants. *Plant Cell*, **8** 1833-44.

Bell, C.J. and Ecker, J.R. (1994) Assignment of 30 microsatellite loci to the linkage map of *Arabidopsis*. *Genomics*, **19** 137-44.

Bevan, M. and Murphy, G. (1999) The small, the large and the wild: the value of comparison in plant genomics. *Trends Genet.*, **15** 211-14.

Brenner, S. and Lewitter, F. (Eds.) (1998) *Trends Guide to Bioinformatics*, Elsevier Trends Journals, West Sussex, UK.

Burks, C. (1999) Molecular biology database list. *Nucleic Acids Res.*, **27** 1-9.

Buschges, R., Hollrichter, K., Panstruga, R., Simons, G., Wolter, M., Frijters, A., van Daelen, R., van der Lee, T., Diergaarde, P., Groenendijk, J., Topsch, S., Vos, P., Salamini, F. and Schulze-Lefert, P. (1997) The barley *MLO* gene: a novel control element of plant pathogen resistance. *Cell*, **88** 695-705.

Cao, H., Glazebrook, J., Clarke, J.D., Volko, S. and Dong, X. (1997) The *Arabidopsis* NPR1 gene that controls systemic acquired resistance encodes a novel protein containing ankyrin repeats. *Cell*, **88** 57-63.

Cao, H., Li, L. and Dong, X. (1998) Generation of broad-spectrum disease resistance by overexpression of an essential regulatory gene in systemic acquired resistance. *Proc. Natl. Acad. Sci. USA*, **95** 6531-36.

Century, K.S., Shapiro, A.D., Repetti, P.P., Dahlbeck, D., Holub, E. and Staskawicz, B.J. (1997) *NDR1*, a pathogen-induced component required for *Arabidopsis* disease resistance. *Science*, **278** 1963-65.

Chen, J.J., Wu, R., Yang, P.C., Huang, J.Y., Sher, Y.P., Han, M.H., Kao, W.C., Lee, P.J., Chiu, T.F., Chang, F., Chu, Y.W., Wu, C.W. and Peck, K. (1998) Profiling expression patterns and isolating differentially expressed genes by cDNA microarray system with colorimetry detection. *Genomics*, **51** 313-24.

Cho, R.J., Campbell, M.J., Winzeler, E.A., Steinmetz, L., Conway, A., Wodicka, L., Wolfsvrg, T.G., Gabrienlian, A.E., Landsman, D., Lockhart, D.J. and Davis, R.W. (1998) A genome-wide transcriptional analysis of the mitotic cell cycle. *Mol. Cell*, **2** 65-73.

Cogoni, C. and Macino, G. (1999) Gene silencing in *Neurospora crassa* requires a protein homologous to RNA-dependent RNA polymerase. *Nature*, **399** 166-69.

DeRisi, J., Penland, L., Brown, P.O., Bittner, M.L., Meltzer, P.S., Ray, M., Chen, Y., Su, Y.A. and Trent, J.M. (1996) Use of a cDNA microarray to analyse gene expression patterns in human cancer. *Nat. Genet.*, **14** 457-60.

DeRisi, J.L., Iyer, V.R. and Brown, P.O. (1997) Exploring the metabolic and genetic control of gene expression on a genomic scale. *Science*, **278** 680-86.

Diatchenko, L., Lau, Y.F., Campbell, A.P., Chenchik, A., Moqadam, F., Huang, B., Lukyanov, S., Lukyanov, K., Gurskaya, N., Sverdlov, E.D. and Siebert, P.D. (1996) Suppression subtractive hybridization: a method for generating differentially-regulated or tissue-specific cDNA probes and libraries. *Proc. Natl. Acad. Sci. USA*, **93** 6025-30.

Diatchenko, L., Lukyanov, S., Lau, Y.F. and Siebert, P.D. (1999) Suppression subtractive hybridization: a versatile method for identifying differentially expressed genes. *Methods Enzymol.*, **303** 349-80.

Falk, A., Feys, B.J., Frost, L.N., Jones, J.D., Daniels, M.J. and Parker, J.E. (1999) EDS1, an essential component of *R* gene-mediated disease resistance in *Arabidopsis* has homology to eukaryotic lipases. *Proc. Natl. Acad. Sci. USA*, **96** 3292-97.

Feldmann, K.A. (1991) T-DNA insertion mutagenesis in *Arabidopsis*: mutational spectrum. *Plant J.*, **1** 71-82.

Finley, R.L.J. and Brent, R. (1996) Interaction trap cloning with yeast, in *DNA Cloning-Expression Systems: A Practical Approach* (eds. D. Glover and B.D. Hames) Oxford University Press, Oxford, UK, pp. 169-203.

Fire, A., Xu, S., Montgomery, M.K., Kostas, S.A., Driver, S.E. and Mello, C.C. (1998) Potent and specific genetic interference by double-stranded RNA in *Caenorhabditis elegans*. *Nature*, **391** 806-11.

Freialdenhoven, A., Scherag, B., Hollricher, K., Collinge, D.B., Thordal-Christensen, H. and Schulze-Lefert, P. (1994) *Nar-1* and *Nar-2*, two loci required for *Mla12*-specified race-specified resistance to powdery mildew in barley. *Plant Cell*, **6** 983-94.

Freialdenhoven, A., Peterhansel, C., Kurth, J., Kruezaler, F. and Schulze-Lefert, P. (1996) Identification of genes required for the function of non-race-specific *mlo* resistance to powdery mildew in barley. *Plant Cell*, **8** 5-14.

Frey, M., Stettner, C. and Gierl, A. (1998) A general method for gene isolation in tagging approaches: amplification if insertion mutagenised sites (AIMS). *Plant J.*, **13** 717-21.

Gerhold, D., Rushmore, T. and Caskey, C.T. (1999) DNA chips: promising toys have become powerful tools. *Trends Biochem. Sci.*, **24** 168-73.

Guyer, D., Tuttle, A., Rouse, S., Volrath, S., Johnson, M., Potter, S., Gorlach, J., Goff, S., Crossland, L. and Ward, E. (1998) Activation of latent transgenes in *Arabidopsis* using a hybrid transcription factor. *Genetics*, **149** 633-39.

Hamilton, A.J., Brown, S., Yuanhai, H., Ishizuka, M., Lowe, A., Solis, A.-G.A. and Grierson, D. (1998) A transgene with repeated DNA causes high frequency, post-transcriptional suppression of ACC-oxidase gene expression in tomato. *Plant J.*, **15** 737-46.

Hamilton, C.M., Frary, A., Lewis, C. and Tanksley, S. D. (1996) Stable transfer of intact high molecular weight DNA into plant chromosomes. *Proc. Natl. Acad. Sci. USA*, **93** 9975-79.

Hauser, M.-T., Adhami, F., Dorner, M., Fuchs, E. and Glossl, J. (1998) Generation of co-dominant PCR-based markers by homo- and heterduplex analysis on high resolution gels. *Plant J.*, **16** 117-25.

Hubank, M. and Schatz, D. G. (1999) cDNA representational difference analysis: a sensitive and flexible method for identification of differentially-expressed genes. *Methods Enzymol.*, **303** 325-49.

Jaglo-Ottosen, K.R., Gilmour, S.J., Zarka, D.G., Schabenberger, O. and Thomashow, M.F. (1998) *Arabidopsis CBR1* overexpression induces *COR* genes and enhances freezing tolerance. *Science*, **280** 104-106.

Kehoe, D.M., Villand, P. and Somerville, S. (1999) DNA microarrays for studies of higher plants and other photosynthetic organisms. *Trends Plant Sci.*, **4** 38-41.

Kempin, S.A., Liljegren, S.J., Block, L.M., Rounsley, S.D., Yanofsky, M.F. and Lam, E. (1997) Targeted disruption in *Arabidopsis. Nature*, **389** 802-803.

Koncz, C., Martini, N., Mayerhofer, R., Koncz-Kalman, Z., Korber, H., Redei, G.P. and Schell, J. (1989) High-frequency T-DNA-mediated gene tagging in plants. *Proc. Natl. Acad. Sci. USA*, **86** 8467-71.

Konieczny, A. and Ausubel, F.M. (1993) A procedure for mapping *Arabidopsis* mutations using co-dominant ecotype-specific PCR-based markers. *Plant J.*, **4** 403-10.

Kreil, D.P. and Etzold, T. (1999) DATABANKS: a catalogue database of molecular biology databases. *Trends Biochem. Sci.*, **24** 155-57.

Lawrence, G.J., Finnegan, E.J., Ayliffe, M.A. and Ellis, J.G. (1995) The *L6* gene for flax rust resistance is related to the *Arabidopsis* bacteria resistance gene, *RPS2*, and the tobacco viral resistance gene, *N. Plant Cell*, **7** 1195-206.

Li, J.J. and Herskowitz, I. (1993) Isolation of ORC6, a component of the yeast origin recognition complex by a one-hybrid system. *Science*, **262** 1870-74.

Liang, P. and Pardee, A. (1992) Differential display of eukaryotic messenger RNA by means of the polymerase chain reaction. *Science*, **257** 967-71.

Liang, P. and Pardee, A.B. (1997) Differential display: a general protocol. *Methods Mol. Biol.*, **85** 3-11.

Lipshutz, R.J., Fodor, S.P.A., Gingeras, T.R. and Lockhart, D.J. (1999) High density synthetic oligonucleotide arrays. *Nat. Genet.*, **21** (Suppl. 1), 20-24.

Liu, Y.G. and Whittier, R.F. (1995) Thermal asymmetric interlaced PCR: automatable amplification and sequencing of insert end fragments from P1 and YAC clones for chromosome walking. *Genomics*, **25** 674-81.

Liu, Y.-G., Shirano, Y., Fukaki, H., Yanai, Y., Tasaka, M., Tabata, S. and Shibata, D. (1999) Complementation of plant mutants with large genomic DNA fragments by a transformation-competent artificial chromosome vector accelerates positional cloning. *Proc. Natl. Acad. Sci. USA*, **96** 6535-40.

Lukowitz, W., Mayer, U. and Jurgens, G. (1996) Cytokinesis in the *Arabidopsis* embryo involves the syntaxin-related KNOLLE gene product. *Cell*, **84** 61-71.

Maes, T., De Keukeleire, P. and Gerats, T. (1999) Plant tagnology. *Trends Plant Sci.*, **4** 90-96.

Mathur, J., Szabados, L., Schaefer, S., Grunenberg, B., Lossow, A., Jonas-Straube, E., Schell, J., Koncz, C. and Koncz-Kalman, Z. (1998) Gene identification with sequenced T-DNA tags generated by transformation of *Arabidopsis* cell suspension. *Plant J.*, **13** 707-16.

Michaels, S.D. and Amasino, R.M. (1998) A robust method for detecting single-nucleotide changes as polymorphic markers by PCR. *Plant J.*, **14** 381-85.

Michelmore, R.W., Paran, I. and Kesseli, R.V. (1991) Identification of markers linked to disease-resistance genes by bulked segregant analysis: a rapid method to detect markers in specific genomic regions by using segregating populations. *Proc. Natl. Acad. Sci. USA*, **88** 9828-32.

Morel, J.B. and Dangl, J.L. (1999) Suppressors of the *Arabidopsis* lsd5 cell death mutation identify genes involved in regulating disease resistance responses. *Genetics*, **151** 305-19.

Neff, M.M., Neff, J.D., Chory, J. and Pepper, A.E. (1998) dCAPS, a simple technique for the genetic analysis of single nucleotide polymorphisms: experimental applications in *Arabidopsis thaliana* genetics. *Plant J.*, **14** 387-92.

Ochman, H., Gerber, A.S. and Hartl, D.L. (1988) Genetic applications of an inverse polymerase chain reaction. *Genetics*, **120** 621-23.

Ruan, Y., Gilmore, J. and Conner, T. (1998) Towards *Arabidopsis* genome analysis: monitoring expression profiles of 1400 genes using cDNA microarrays. *Plant J.*, **15** 821-33.

Ruiz, M.T., Voinnet, O. and Baulcombe, D.C. (1998) Initiation and maintenance of virus-induced gene silencing. *Plant Cell*, **10** 937-46.

Ryals, J., Weymann, K., Lawton, K., Friedrich, L., Ellis, D., Steiner, H.-Y., Johnson, J., Delaney, T.P., Jesse, T., Vos, P. and Uknes, S. (1997) The *Arabidopsis NIM1* protein shows homology to the mammalian transcription factor inhibitor, IkB. *Plant Cell*, **9** 425-39.

Schena, M., Shalon, D., Davis, R.W. and Brown, P.O. (1995) Quantitative monitoring of gene expression patterns with a complementary DNA microarray. *Science*, **270** 467-70.

Shevchenko, A., Jensen, O.N., Podtelejnikov, A.V., Sagliocco, F., Wilm, M., Vorm, O., Mortensen, P., Shevchenko, A., Boucherie, H. and Mann, M. (1996) Linking genome and proteome by mass spectrometry: large-scale identification of yeast proteins from two dimensional gels. *Proc. Natl. Acad. Sci. USA*, **93** 14440-45.

Simons, G., van der Lee, T., Diergaarde, P., van Daelen, R., Groenendijk, J., Frijters, A., Buschges, R., Hollricher, K., Topsch, S., Schulze-Lefert, P., Salamini, F., Zabeau, M. and Vos, P. (1997) AFLP-based fine mapping of the *Mlo* gene to a 30-kb DNA segment of the barley genome. *Genomics*, **44** 61-70.

Smith, D., Yanai, Y., Liu, T.-G., Ishiguro, S., Okada, K., Shibata, D., Whittier, R.F. and Federoff, N.V. (1996) Characterization and mapping of Ds-GUS T-DNA lines for targeted insertional mutagensis. *Plant J.*, **10** 721-32.

Stockinger, E.J., Gilmour, S.J. and Thomashow, M.F. (1997) *Arabidopsis thaliana* CBF1 encodes an AP2 domain-containing transcriptional activator that binds to the C-repeat/DRE, a *cis*-acting DNA regulatory element that stimulates transcription in response to low temperature and water deficit. *Proc. Natl. Acad. Sci. USA*, **94** 1035-40.

Sundaresan, V., Springer, P., Volpe, T., Haward, S., Jones, J.D., Dean, C., Ma, H. and Martienssen, R. (1995) Patterns of gene action in plant development revealed by enhancer trap and gene trap transposable elements. *Genes Dev.*, **9** 1797-810.

Trethewey, R.N., Krotzky, A.J. and Willmitzer, L. (1999) Metabolic profiling: a Rosetta Stone for genomics? *Curr. Opin. Plant Biol.*, **2** 83-85.

Underhill, P.A., Jin, L., Zemans, R., Oefner, P.J. and Cavalli-Sforza, L.L. (1996) A pre-Columbian Y chromosome-specific transition and its implications for human evolutionary history. *Proc. Natl. Acad. Sci. USA*, **93** 196-200.

Van der Broeck, D., Maes, T., Sauer, M., Zethof, J., De Keukeleire, P., D'Hauw, M., Van Montagu, M. and Gerats, T. (1998) Transposon display identifies individual transposable elements in high copy number lines. *Plant J.*, **13** 121-29.

Velculescu, V.E., Zhang, L., Vogelstein, B. and Kinzler, K.W. (1995) Serial analysis of gene expression. *Science*, **270** 484-87.

von Ropenack, E., Parr, A. and Schulze-Lefert, P. (1998) Structural analyses and dynamics of soluble and cell-wall-bound phenolics in a broad-spectrum resistance to the powdery mildew fungus in barley. *J. Biol. Chem.*, **273** 9013-22.

Vos, P., Hogers, R., Bleeker, M., Reijans, M., van de Lee, T., Hornes, M., Frijters, A., Pot, J., Peleman, J., Kuiper, M. and Zabeau, M. (1996) AFLP: a new technique for DNA fingerprinting. *Nucl. Acid Res.*, **23** 4407-14.

Wang, D.G., Fan, J.B., Siao, C.J., Berno, A., Young, P., Sapolsky, R., Ghandour, G., Perkins, N., Winchester, E., Spencer, J., Kruglyak, L., Stein, L., Hsie, L., Topaloglou, T., Hubbell, E., Robinson, E., Mittmann, M., Morris, M.S., Shen, N., Kilburn, D., Rioux, J., Nusbaum, C., Rozen, S., Hudson, T.J. and Lander, E.S. (1998) Large-scale identification, mapping and genotyping of single-nucleotide polymorphisms in the human genome. *Science*, **280** 1077-82.

Waterhouse, P.M., Graham, M.W. and Wang, M.-B. (1998) Virus resistance and gene silencing in plants can be induced by simultaneous expression of sense and antisense RNA. *Proc. Natl. Acad. Sci. USA*, **95** 13959-64.

Westlund, B., Parry, D., Clover, R., Basson, M. and Johnson, C. D. (1999) Reverse genetic analysis of *Caenorhabditis elegans* presenilins reveals redundant but unequal roles for sel-12 and hop-1 in Notch-pathway signaling. *Proc. Natl. Acad. Sci. USA*, **96** 2497-502.

Winzeler, E.A., Richards, D.R., Conway, A.R., Godlstein, A.L., Kalman, S., McCullough, M.J., McCusker, J.H., Stevens, D.A., Wodicka, L., Lockhart, D.J. and Davis, R.W. (1998) Direct allelic variation scanning of the yeast genome. *Science*, **281** 1194-97.

Zambrowicz, B.P., Friedrich, G.A., Buxton, E.C., Lilleberg, S.L., Person, C. and Sands, A.T. (1998) Disruption and sequence identification of 2000 genes in mouse embryonic stem cells. *Nature*, **392** 608-11.

Zhang, Y., Fan, W., Kinkema, M. and Dong, X. (1999) Interaction of NPR1 with basic leucine zipper protein transcription factors that bind sequences required for salicylic acid induction of the *PR-1* gene. *Proc. Natl. Acad. Sci. USA*, **96** 6523-28.

Index